AF182428

Stability and Characterization of Protein and Peptide Drugs

Case Histories

Pharmaceutical Biotechnology

Series Editor: Ronald T. Borchardt
The University of Kansas
Lawrence, Kansas

Volume 1 PROTEIN PHARMACOKINETICS AND METABOLISM
Edited by Bobbe L. Ferraiolo, Marjorie A. Mohler, and Carol A. Gloff

Volume 2 STABILITY OF PROTEIN PHARMACEUTICALS, Part A: Chemical and Physical Pathways of Protein Degradation
Edited by Tim J. Ahern and Mark C. Manning

Volume 3 STABILITY OF PROTEIN PHARMACEUTICALS, Part B: *In Vivo* Pathways of Degradation and Strategies for Protein Stabilization
Edited by Tim J. Ahern and Mark C. Manning

Volume 4 BIOLOGICAL BARRIERS TO PROTEIN DELIVERY
Edited by Kenneth L. Audus and Thomas J. Raub

Volume 5 STABILITY AND CHARACTERIZATION OF PROTEIN AND PEPTIDE DRUGS: Case Histories
Edited by Y. John Wang and Rodney Pearlman

Stability and Characterization of Protein and Peptide Drugs

Case Histories

Edited by

Y. John Wang

Scios Nova, Inc.
Mountain View, California

and

Rodney Pearlman

Genentech, Inc.
South San Francisco, California

Springer Science+Business Media, LLC

Library of Congress Cataloging-in-Publication Data

Stability and characterization of protein and peptide drugs : case histories / edited by Y. Wang and Rodney Pearlman.
p. cm. -- (Pharmaceutical biotechnology ; v.5)
Includes bibliographical references and index.

1. Protein drugs. 2. Drug stability. 3. Drug Stability.
I. Wang, Y. John. II. Pearlman, Rodney. III. Series.
[DNLM: 1. Proteins--pharmacokinetics. 2. Proteins--chemistry.
3. Peptides--pharmacokinetics. 4. Peptides--chemistry. QU 55 S775 1993]
RS431.P75S69 1993
615'.3--dc20
DNLM/DLC
for Library of Congress 93-7490
CIP

10 9 8 7 6 5 4 3 2

ISBN 978-1-4899-1238-1 ISBN 978-1-4899-1236-7 (eBook)
DOI 10.1007/978-1-4899-1236-7

© 1993 Springer Science+Business Media New York
Originally published by Plenum Press, New York in 1993.
Softcover reprint of the hardcover 1st edition 1993

All rights reserved

No part of this book may be reproduced, stored in a retrieval system, or transmitted in any form or by any means, electronic, mechanical, photocopying, microfilming, recording, or otherwise, without written permission from the Publisher

To Doris and Anne

Contributors

Akwete Lex Adjei • Product Development, Pharmaceutical Products Division, Abbott Laboratories, North Chicago, Illinois 60064

Thomas A. Bewley • Department of Pharmaceutical Research and Development, Genentech, Inc., South San Francisco, California 94080

Jens Brange • Novo Research Institute, Novo Nordisk A/S, DK-2880 Bagsvaerd, Denmark

Stephen R. Davio • Drug Delivery R&D—Pharmaceutics, Upjohn Laboratories, The Upjohn Company, Kalamazoo, Michigan 49007

Jodi Fausnaugh • Institute of Analytical Research, Syntex Research, Palo Alto, California 94304

John Geigert • Quality Control Department, Chiron Corporation, Emeryville, California 94608. *Present address:* Immunex Corporation, Seattle, Washington 98101

Leo Gu • Institute of Pharmaceutical Sciences, Syntex Research, Palo Alto, California 94304

Michael J. Hageman • Drug Delivery R&D—Pharmaceutics, Upjohn Laboratories, The Upjohn Company, Kalamazoo, Michigan 49007

L. Hsu • Product Development Pharmaceutical Products Division, Abbott Laboratories, North Chicago, Illinois 60064

Daniel J. Kroon • The R. W. Johnson Pharmaceutical Research Institute, Raritan, New Jersey 08869

Lotte Langkjær • Novo Research Institute, Novo Nordisk A/S, DK-2880 Bagsvaerd, Denmark

Mark C. Manning • School of Pharmacy, University of Colorado Health Sciences Center, Denver, Colorado 80262

James W. Mitchell • Oread Laboratories, Lawrence, Kansas 66044

Tue H. Nguyen • Department of Pharmaceutical Research and Development, Genentech, Inc., South San Francisco, California 94080

Kamlesh Patel • SmithKline Beecham Pharmaceuticals, King of Prussia, Pennsylvania 19406

Rodney Pearlman • Department of Pharmaceutical Research and Development, Genentech, Inc., South San Francisco, California 94080

Denise Pretzer • Merck Research Laboratories, West Point, Pennsylvania 19486

Patricia E. Rao • The R. W. Johnson Pharmaceutical Research Institute, Raritan, New Jersey 08869. *Present address:* Ortho Diagnostic Systems, Inc., Raritan, New Jersey 08869

Niek Roosdorp • Chiron Corporation, Emeryville, California 94608

Brenda S. Schulteis • Marion Merrell Dow Research Institute, Kansas City, Missouri 64134

Christopher D. Smith • Department of Pharmaceutical Chemistry, University of Kansas, Lawrence, Kansas 66045

Nicholas Solli • Quality Control Department, Chiron Corporation, Emeryville, California 94608

David G. Vander Velde • NMR Laboratory, University of Kansas, Lawrence, Kansas 66045

Sriram Vemuri • Scios Nova Inc., Mountain View, California 94043

Carole Ward • Department of Pharmaceutical Research and Development, Genentech, Inc., South San Francisco, California 94080

Peggy Woehleke • Quality Control Department, Chiron Corporation, Emeryville, California 94608

C. Tony Yu • Bristol–Myers Squibb, Pharmaceutical Research Institute, Buffalo, New York 14213

Preface to the Series

A major challenge confronting pharmaceutical scientists in the future will be to design successful dosage forms for the next generation of drugs. Many of these drugs will be complex polymers of amino acids (e.g., peptides, proteins), nucleosides (e.g., antisense molecules), carbohydrates (e.g., polysaccharides), or complex lipids.

Through rational drug design, synthetic medicinal chemists are preparing very potent and very specific peptides and antisense drug candidates. These molecules are being developed with molecular characteristics that permit optimal interaction with the specific macromolecules (e.g., receptors, enzymes, RNA, DNA) that mediate their therapeutic effects. However, rational drug design does not necessarily mean rational drug delivery, which strives to incorporate into a molecule the molecular properties necessary for optimal transfer between the point of administration and the pharmacological target site in the body.

Like rational drug design, molecular biology is having a significant impact on the pharmaceutical industry. For the first time, it is possible to produce large quantities of highly pure proteins, polysaccharides, and lipids for possible pharmaceutical applications. The design of successful dosage forms for these complex biotechnology products represents a major challenge to pharmaceutical scientists.

Development of an acceptable drug dosage form is a complex process requiring strong interactions between scientists from many different divisions in a pharmaceutical company, including discovery, development, and manufacturing. The series editor, the editors of the individual volumes, and the publisher hope that this new series will be particularly helpful to scientists in the development areas of a pharmaceutical company (e.g., drug metabolism, toxicology, pharmacokinetics and pharmacodynamics, drug delivery, preformulation, formulation, and physical and analytical chemistry). In addition, we hope this series will help to build bridges between the development scientists and scientists in discovery (e.g., medicinal chemistry, pharmacology, immunology, cell biology, molecular biology) and in manufacturing (e.g., process chemistry, engineering). The design of successful dosage forms for the next generation of drugs will require not only a high level of expertise by individual scientists, but also a high degree of interaction between scientists in these different divisions of a pharmaceutical company.

Finally, everyone involved with this series hopes that these volumes will also be useful to the educators who are training the next generation of pharmaceutical scientists. In addition to having a high level of expertise in their respective disciplines, these young scientists will need to have the scientific skills necessary to communicate with their peers in other scientific disciplines.

RONALD T. BORCHARDT
Series Editor

Preface

This volume attempts to provide the formulation scientist with case histories involving the use of therapeutic proteins and peptides that have been marketed or are under clinical testing. In previous volumes of this series, fundamental theories and principles of protein characterization and stability were presented in depth by researchers in their fields of expertise. The way from theory to practice is not always obvious and straightforward. There is a need for practical examples of how the principles and theories are put into use, specifically in the development of a pharmaceutical product. It is our hope that this volume will fulfill such a need.

It is not a simple task to choose a panel of proteins and peptides from the over 200 agents in human clinical trials. We have tried to collect a wide representation of molecules of different sizes—from 10 amino acids (Leuprolide) to 1020 amino acids (Muromonab CD3). The examples include agents derived from various sources including monoclonal antibodies (Muromonab CD3), recombinant DNA (human and bovine growth hormones), natural source (fibrolase), and chemical synthesis (Leuprolide). Clearly this list is not intended to be encyclopedic. It is the first time a collection of this sort has been made accessible to the formulation scientists involved in developing protein and peptide products.

Although each chapter in this volume focuses primarily on the characterization and stability of a specific molecule, each has unique aspects. Chapter 1 on human growth hormone provides detailed examples of characterization and stability of the protein from both biochemical and pharmaceutical aspects. Mechanisms on protein aggregation and covalent bond modifications are described. Chapter 2 presents spectral analysis on bovine growth hormone (somatotropin), a protein homologous to human growth hormone. Solubility, solution stability, and absorption of moisture by the freeze-dried solid are described, which will aid scientists facing similar challenges.

Alteplase, tissue type plasminogen activator, is covered in Chapter 3. This monograph illustrates approaches to handling a heterogeneous product. Thermal analysis and stability–pH profile of multiple degradation pathways are valuable information contained in this chapter. Literature on the stability of monoclonal antibodies is very sparse. Chapter 4 addresses antibody stability of Muromonab CD3 (Orthoclone OKT3). The sites and mechanism of degradation were identified through isoelectric focusing and supported by peptide mapping.

In Chapter 5, the specifications required of a peptide product are described. Also of interest in this chapter on Leuprolide is the discussion of the bioavailability profile of the peptide when given by different routes of administration. Through the discussion of ACTH stability and its active hexapeptide segment, Chapter 6 presents a detailed mechanistic analysis of deamidation, which is a major route of degradation for all proteins and peptides.

Development of a reliable, quantitative bioassay is always a concern to formulation scientists. In Chapter 7, results from four different *in vitro* bioassays for human interleukin-1β are delineated. Extensive formulation studies are also included. Chapter 8, on IL-2, presents a description of manufacturing procedures and the process of determining the shelf life of a protein product.

The kinetics of aggregation, an obvious physical change which is the bane of many protein products, is discussed with regard to α_1-antitrypsin in Chapter 9. In Chapter 10, an extensive spectral analysis on the natural product fibrolase, by CD and NMR, is presented. Although insulin has been available since the 1920s, the degradation pathways have not been completely elucidated and presented in a systematic fashion. Chapter 11 consolidates all existing information in a concise monograph.

The editors wish to thank all of the contributors for their willingness to share updated information that made this volume possible. It is hoped that as more biotechnology products are approved, information related to characterization and stability can be made more accessible to the scientists in this field. Before that happens, this volume can at least provide a guide and serve as a handy reference to those in need of this kind of information.

Y. John Wang
Rodney Pearlman

Contents

Chapter 2

Characterization and Formulation Considerations for Recombinantly Derived Bovine Somatotropin

Stephen R. Davio and Michael J. Hageman

Chapter 3

Stability Characterization and Formulation Development of Alteplase, a Recombinant Tissue Plasminogen Activator

Tue H. Nguyen and Carole Ward

Chapter 4

Orthoclone OKT3: Chemical Mechanisms and Functional Effects of Degradation of a Therapeutic Monoclonal Antibody

Patricia E. Rao and Daniel J. Kroon

Chapter 5

Leuprolide and Other LH-RH Analogues

Akwete Lex Adjei and L. Hsu

Chapter 6

Stability of Adrenocorticotropic Hormone (ACTH) and Pathways of Deamidation of Asparaginyl Residue in Hexapeptide Segments

Kamlesh Patel

Chapter 7

Stability and Characterization of Human Interleukin-1β

Leo Gu and Jodi Fausnaugh

Chapter 8

Development and Shelf-Life Determination of Recombinant Interleukin-2 (Proleukin)

John Geigert, Nicholas Solli, Peggy Woehleke, and Sriram Vemuri

Chapter 9

Formulation and Stability of Recombinant α_1-Antitrypsin

Sriram Vemuri, C. Tony Yu, and Niek Roosdorp

Chapter 10

Fibrolase: A Fibrinolytic Protein from Snake Venom

Denise Pretzer, Brenda S. Schulteis, Christopher D. Smith, David G. Vander Velde, James W. Mitchell, and Mark C. Manning

Chapter 11

Insulin Structure and Stability

Jens Brange and Lotte Langkjær

Chapter 1

Stability and Characterization of Human Growth Hormone

Rodney Pearlman and Thomas A. Bewley

1. INTRODUCTION

Human growth hormone (hGH) is a protein responsible for a wide range of growth-promoting effects in the body, either by direct action on receptors in various tissues or via stimulation of the release of other growth factors (Wilhelmi, 1982). Under hypothalamic control, the hormone is secreted from anterior pituitary granules, into the bloodstream in a pulsatile fashion (Evans *et al.,* 1987). The average human pituitary weighs about 1 g, and is estimated to contain from 5 to 10 mg hGH (Daughaday, 1985). This chapter describes the history and clinical uses of the hormone, methods used for its characterization and analysis, and includes a discussion of hGH degradation reactions and its stability profile as well as a description of various hGH formulations.

1.1. History of hGH and hGH Preparations

The role of the pituitary in promoting growth was first suggested by Evans and Long in 1921, but it was not until 1944 that Li and Evans purified

Rodney Pearlman and Thomas A. Bewley • Department of Pharmaceutical Research and Development, Genentech, Inc., South San Francisco, California 94080.

Stability and Characterization of Protein and Peptide Drugs: Case Histories, edited by Y. John Wang and Rodney Pearlman. Plenum Press, New York, 1993.

the long-sought "factor" from bovine pituitaries (bGH) and proved it to be a protein molecule with distinct physicochemical properties. In 1950, the species specificity of growth hormone was established when Bennett *et al.* demonstrated that bGH is ineffective in humans. The first highly purified human and monkey GHs were reported by Li and Papkoff in 1956, and by Raben in 1957. Both primate hormones were found to exhibit the same specific activity as bGH in the rat (Li and Papkoff, 1956), but only the primate hormones were effective in man (Beck *et al.,* 1957; Raben, 1957). To date, GHs have been isolated from a large number of species, and all display a high degree of sequence homology (Bewley and Li, 1975; Nicoll *et al.,* 1986; Li *et al.,* 1986, 1987).

Until the mid-1980s, the only source of hGH was from human cadaver tissue, and is denoted herein as pituitary hGH, or pit-hGH. In the early days, material distributed for clinical use was prepared in several laboratories by various methods (Raben, 1957; Wilhelmi, 1961; Li *et al.,* 1962; Roos *et al.,* 1963; Saxena and Henneman, 1966). These preparations involved extraction of fresh-frozen or acetone-preserved tissue using either mild alkaline conditions (Li and Papkoff, 1956; Wilhelmi, 1961; Li *et al.,* 1962) or hot glacial acetic acid (Raben, 1957). Extraction was followed by various serial fractionations using pH, ammonium sulfate, or ethanol precipitations. Final purification was effected by one or more chromatographic steps, followed by lyophilization. Since purified GH was always in extremely short supply, many of these procedures were specifically designed to provide a clinically active product in the highest possible yield. As a result, their chemical and physical properties exhibited some degree of variability. For example, the method of Li *et al.* (1962) was designed to produce a more highly purified material for physicochemical and structural studies rather than for clinical use. This procedure typically yielded 1.4 mg of highly purified pit-hGH per fresh pituitary gland, but yields sometimes exceeding 5 mg per gland of so-called "clinical-grade" material were obtained by other methods (Wallace and Ferguson, 1961). It must be stated that "purity" is a relative term, linked to the purification methods and analytical techniques available at the time. The works of Skyler *et al.* (1977) and Lewis *et al.* (1980) describe a number of different forms of hGH found in what were once thought to be highly purified preparations of pit-hGH. Some of these variants include deamidated species, oxidized forms, dimers and higher-order aggregates as well as various enzymatically clipped forms. Moreover, degradation during storage of the final lyophilized preparations was almost never discussed, and perhaps rarely tested. Thus, the issue of purity must be considered in light of these findings.

Historically, because hGH is a protein hormone, its potency has been expressed in terms of international units (IU). The definition of the hGH international unit is given in the proposed USP monographs for growth hormone (USP Pharmacopeial Forum, 1990), and is based on a weight gain assay in hypophysectomized rats (Wilhelmi, 1982). Some degree of confusion still exists regarding how many international units of hGH are equivalent to a milligram of protein, with several different numbers having been used in the past. Riggin *et al.* (1988) discuss the issue of biopotency, and show that depending on the assay chosen, hGH may display a potency between 2.0 and 3.0 IU/mg. In a recently published collaborative study, the potency of a reference sample of recombinant hGH was determined in ten different laboratories worldwide (Bristow and Jeffcoate, 1992). In this study, the potency of hGH was measured by a number of different assays including two animal-based bioassays. The average potency of the hGH sample was determined to be 3.4 IU/mg, and it is possible that the World Health Organization will adopt this figure as the new reference standard. This value is considerably higher than the previously recognized activity of approximately 2.6 IU/mg (*Physicians Desk Reference,* 1992, pp. 1049 and 1266). Such differences in potency values have little effect on dosing the hormone in the United States, since dosing is performed on a weight of protein per kilogram of child's weight, but may affect levels of hGH administered in European countries and Japan, where doses are calculated in units per kilogram.

pit-hGH was removed from the U.S. market when its use was linked to deaths from a slow virus infection known as Creutzfeldt–Jakob syndrome (Brown *et al.,* 1985). The lack of an adequate supply of pit-hGH and the advent of genetic engineering had already resulted in production of a biosynthetic methionyl hGH (Met-hGH). This protein is produced in a strain of *E. coli,* modified to include the gene for hGH (Goeddel *et al.,* 1979). Met-hGH was approved for use in the United States in 1985, and has the United States Approved Name (USAN) of Somatrem.

A second form of the hormone, the methionyl-free version, was introduced into the United States in 1987, and this molecule has the USAN Somatropin. Since both of the biosynthetic forms of hGH are produced using recombinant DNA technology in bacteria (and do not involve the use of human tissues in their production), the risk of contamination by Creutzfeldt–Jakob and other pathogens is eliminated.

1.2. Clinical Use of hGH

The therapeutic use of hGH is in replacement therapy for treatment of pediatric hypopituitary dwarfism and children suffering from low levels of hGH (Preece and Tanner, 1976; Hindmarsh and Brook, 1987). These conditions occur when the pituitary is unable to secrete sufficient quantities of hGH due to genetic defects, following injury, or after hypophysectomy. Also, because of its effects on nitrogen retention, and its general anabolic effect, hGH may show significant potential toward improving the recovery of burn victims, as well as certain other hospitalized patients. It is believed that hGH functions by direct action on bone and soft tissue to cause uniform growth as well as by indirect stimulation of insulinlike growth factor-1 (IGF-1).

Diagnosis of hGH deficiency in children is not trivial, and is usually performed by pediatric endocrinologists who specialize in this field. Since the hormone is released in a pulsatile fashion over the course of a day, measurement of hGH deficiency is quite complex (Evans *et al.,* 1987; Sieradzki *et al.,* 1987).

Because hGH is a protein, it is not absorbed orally to any significant extent (Moore *et al.,* 1986) and thus must be administered by injection. Typically a dose of 0.1 mg/kg is given either intramuscularly or subcutaneously three times a week. An investigational regimen of 0.05 mg/kg daily is also being used. Currently, Met-hGH and hGH produced by recombinant DNA technology are available worldwide. Both forms of recombinant hGH have been shown to have equivalent therapeutic activity compared with the pituitary-derived material (Hintz *et al.,* 1982; Jorgensen, 1987; Milner *et al.,* 1987; Rappaport, 1987).

A recent paper by Jorgensen (1991) provides a comprehensive review of hGH therapy including a discussion of efficacy, pharmacokinetics, dosing, and the effect of route of administration (intramuscular versus subcutaneous). This review references a number of key papers in the field of GH therapy, and provides an up-to-date synopsis of the area. Interestingly, hGH is degraded significantly upon subcutaneous administration (Jorgensen, *et al.,* 1988).

Recombinant GHs are safe and effective forms of therapy with relatively few side effects (Milner *et al.,* 1987). Probably the most discussed side effect is that regarding hGH antibody formation. Antibodies to GH have been observed in some patients treated with pit-hGH and recombinant hGH preparations (Milner and Flodh, 1985). The latter reference is a collection of the proceedings of a workshop sponsored by KabiVitrum, discussing the immunological aspects of hGH. One of the topics described is the wide variability in assays for hGH antibodies, thus making the true incidence of a clinically important immune response difficult to ascertain. Reports of growth attenuation occurring in patients in response to circulating levels of antibodies to hGH are quite rare. However, one study has implicated aggregated pit-hGH in reducing the growth rate of a single patient (Moore and Leppert, 1980). In another study in Japan, it was reported that a 10-year-old patient treated with Met-hGH suffered growth attenuation due to a high plasma titer of antibodies with high binding capacity (Okada, *et al.,* 1987).

Recently, the levels of antibody occurring in patients following administration of various European preparations of hGH have been compared (Rougeot *et al.,* 1991). In this study, 260 children treated with one of six different biosynthetic GH preparations were monitored for antibody formation over 24 months of therapy. Several of the preparations did not elicit any detectable antibody response, while with other preparations, antibodies were detected in 7 to 22% of the patients. The authors concluded that the overall antibody titers were low and were only detected because of the sensitive methodology employed, rather than from a documented clinical problem such as growth attenuation. In addition to its growth-promoting effects, hGH acts to mobilize fat and possesses insulinlike activity, which in high doses can induce hypoglycemia and glucose intolerance as undesirable side effects. The disease acromegaly occurs in people who express high levels of hGH. This condition may be characterized by gigantism, and excessive growth of soft tissue and cartilage.

1.3. Structure and Properties of Pituitary and Recombinant hGH

There are numerous reviews regarding the structure (Li, 1969; Bewley and Li, 1975; Aubert *et al.,* 1986) and functional properties (Li, 1969; Wilhelmi, 1982; Daughaday, 1985) of pit-hGH, and the reader is encouraged to consult these more comprehensive references.

The most prevalent form of pit-hGH consists of a single polypeptide chain of 191 amino acids, internally cross-linked by two disulfide bonds (Fig. 1). The molecular mass is approximately 22 kDa, with a pI near 5.3. Met-hGH has an extra methionine residue at the NH_2-terminus (resulting in a 192-amino-acid protein). For most assays, and for most of the properties and behavior of the protein, the presence of the NH_2-terminal methionine is unimportant (Larhammar *et al.,* 1985). Certainly the two forms of GH possess the same efficacy and safety profiles. Thus, for simplicity, the remainder of this chapter will refer to human growth hormones as simply hGH, and will only make note of isotype differences where relevant.

The protein is not glycosylated, nor are there any prosthetic groups or cofactors known. Studies of hGH in solution have shown the hormone to resemble a typical globular protein (see, e.g., Bewley *et al.,* 1969; Aloj and Edelhoch, 1972; Holladay *et al.,* 1974; Bewley and Li, 1975; Aubert *et al.,* 1986). Thus, it behaves as a compactly folded, roughly spherical protein with

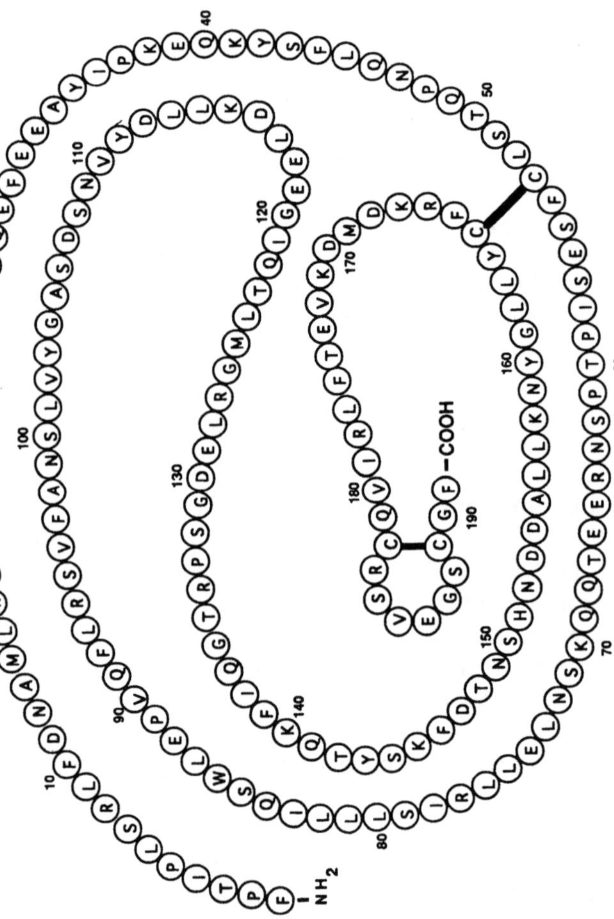

Figure 1. The primary structure of human growth hormone (Goeddel *et al.*, 1979).

a hydrophobic interior. Most of the charged side chains appear to reside on the surface, exposed to the solvent. Approximately 55 ± 5% of the polypeptide backbone exists in the right-hand α-helical conformation based on circular dichroism (CD) analysis. Roongta *et al.* (1989) used CD and two-dimensional NMR to show that a large portion of the first 28 residues of hGH exist as an α-helix, which was verified by synthesis of this peptide fragment.

These gross structural properties of hGH, deduced from numerous physicochemical investigations, compare very well with the three-dimensional structure of the porcine hormone (pGH) obtained by x-ray crystallographic analysis (Abdel-Meguid *et al.*, 1987), and shown in cartoon form in Fig. 2. Unfortunately, crystals of hGH suitable for x-ray study have not yet been reported, although considerable effort has been made toward this goal. Nev-

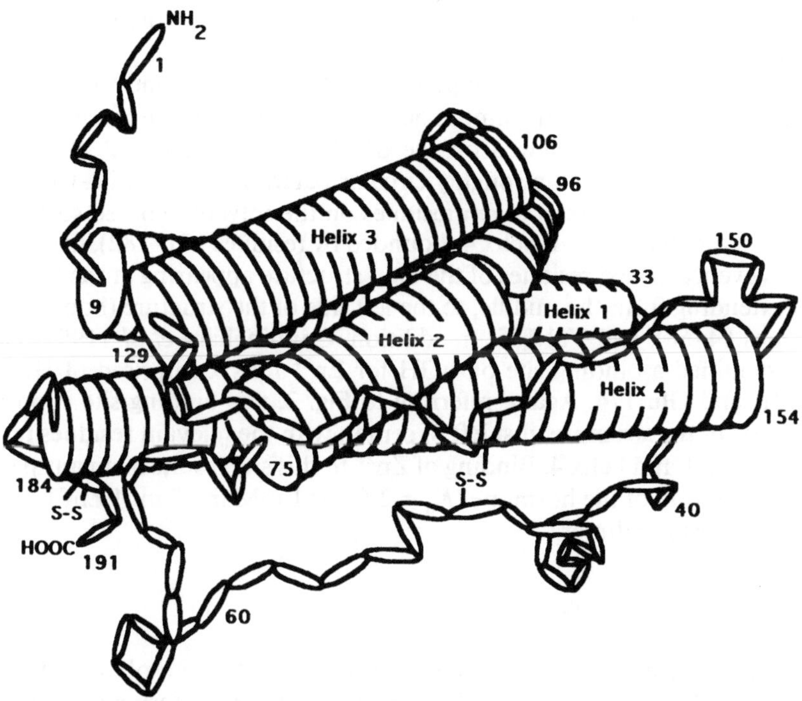

Figure 2. The three-dimensional structure of pGH modified from Abdel-Meguid *et al.* (1987). Each residue is represented by an ellipse. The four, α-helical columns are shown as larger, overlapping cylinders. Residue numbers refer to the hGH primary structure. The residue numbers of the beginning and ending amino acid in each helical column are also indicated, although there may be some ambiguity (±1 residue) in these assignments due to small sequence differences between pGH and hGH.

ertheless, the similarities between the solid-state structure of pGH and the solution properties of hGH are gratifying and undoubtedly reflect the extensive sequence homology seen in GHs isolated from a wide variety of species (Nicoll *et al.,* 1986).

The conservation of primary structures among the GHs strongly suggests that the α-helix content of hGH is distributed into four α-helical columns. These relatively rigid helices form the architectural "core" of the GH structure, around which various extended chain and complex loop structures are specifically arranged in space. The loop structures surrounding the helical core are somewhat less conserved in primary structure, and presumably are largely responsible for certain species-specific properties such as solubility, tertiary structure, quaternary structure, and differences in interactions with cellular receptors. A similar four-helix bundle is also seen in a computer-assisted prediction of the three-dimensional structure of hGH (Cohen and Kuntz, 1987), although the individual helices in this predicted version are all shorter and the loop structures are quite different from the pGH structure.

In a remarkable series of experiments, Cunningham and Wells (1989) have identified many of the amino acid residues involved in binding to the hGH receptor. When viewed on the pGH structure of Fig. 2, these side chains are distributed in a roughly circular patch including: Phe-10 on the back side of helix 1, six or seven residues on the solvent-exposed surface of the left-hand, back side of helix 4 (bounded by residues 171–182), five more-or-less evenly spaced residues of the loop structure appearing at the bottom, and extending from the middle, to the left side of the structure (bounded by residues 54–68), and Val-185. In addition, Cunningham *et al.* (1991) have also identified a binding site on hGH for Zn^{2+} ions, comprised of His-18, His-21, and Glu-174. Again, referring to Fig. 2, this binding site is on the back side of the structure, just left of center, with the relevant residues residing on helix 1 and helix 4. Binding of Zn^{2+} to this site results in an antiparallel dimerization of the hormone ($K_d = 2.6 \pm 0.1$ mM in 20 mM $ZnCl_2$) with two zinc ions per dimer.

Solubility as a Function of pH

The pI of hGH is approximately 5.3, and this is the pH of minimal solubility. The pH versus solubility profiles of the three forms of hGH described in this chapter are shown in Fig. 3: namely, pituitary derived, recombinant Met-hGH, and recombinant hGH. These studies were performed by dissolving lyophilized forms of each protein into buffer at pH 8, and then dialyzing this solution against a variety of low-ionic-strength (10 mM) buffers. After several buffer changes, the samples were filtered and protein concentrations were determined by ultraviolet (UV) spectroscopy and re-

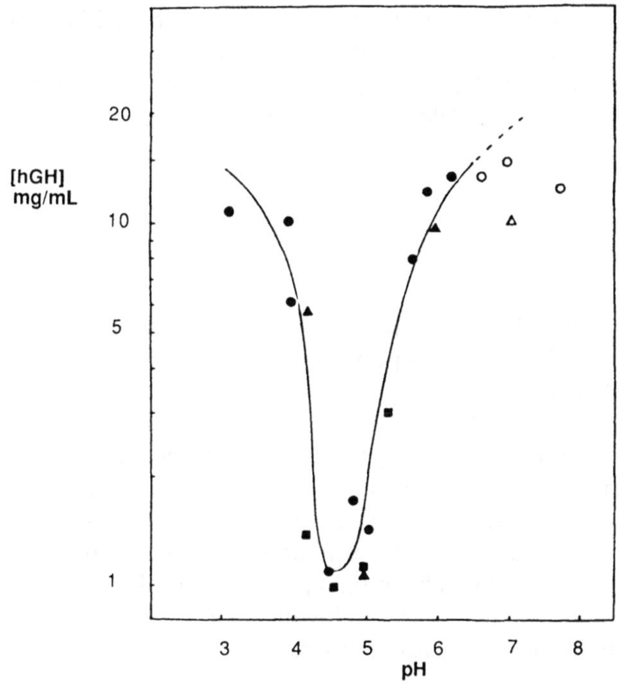

Figure 3. A plot of the solubility of various forms of hGH as a function of pH. Samples of hGH were either recombinant hGH (circles), Met-hGH (triangles), or pit-hGH (squares). Solubility was determined by dialyzing an approximately 11 mg/ml solution of each protein into an appropriate buffer for each pH. Buffers were citrate, pH 3–7, and borate, pH 8–9, all at 10 mM buffer concentrations. Concentrations of hGH were measured by UV absorbance as well as by RP-HPLC, relative to an external standard. The closed symbols indicate that precipitate was present in the dialysis tube after equilibration, whereas open symbols mean that no solid was present, and thus the solubility is at least this amount.

versed-phase high-performance liquid chromatography (RP-HPLC). At pH values above 6.5, no precipitate was observed in the dialysis bags; thus, an upper limit of solubility at pH values above 6.5 was not obtained in this study. However, it has been noted by James Oeswein (personal communication) that solubilities in excess of 25 mg/ml are readily attainable at pH 7.4. Note that each of the hGH preparations has a similar solubility profile, with a solubility minimum of about 1 mg/ml in the pH region of 4.5–5.0.

2. ANALYTICAL CHARACTERIZATION

A draft set of monographs from the USP Open Conference on Biotechnology-derived Products (Corpus Christi, Texas, September 1989) have recently been published for comment (USP Pharmacopeial Forum, 1990) and provide an excellent starting point for the analysis of Somatrem and Somatropin. In addition, an FDA Symposium on Hormone Drugs has been published (Hormone Drugs, 1982), with excellent references to analytical methods for GH.

2.1. Spectroscopy

Highly detailed reviews and specialized reports concerning the optical properties of hGH have been published (Holladay *et al.*, 1974; Bewley and Li, 1975, 1984; Bewley, 1977, 1979; Larhammar *et al.*, 1985). Perhaps no other protein hormone has received such detailed spectroscopic study in the solution state as hGH. In part, this effort has been driven by a desire to obtain information about the conformation of the protein in the frustrating absence of crystals suitable for x-ray diffraction study. Moreover, the GH molecule from a number of mammalian and nonmammalian species, as well as other closely related hormonal proteins, have had their spectral properties compared with hGH. The following is only a brief compilation of the major results from these investigations.

2.1.1. Ultraviolet Absorption

UV absorption spectroscopy is one of the most important techniques used in determining molecular properties for protein characterization. In addition to providing information regarding the immediate environments of chromophoric groups, the absorption maximum (A_{max}), in the form of a molar extinction coefficient (E_m), or as the specific absorptivity of a 1.0 mg/ml solution ($A^{0.1\%}$), is a highly sensitive and convenient parameter for determining protein concentration. However, it is rarely sufficient to simply calculate an E_m value from the chromophore composition and molecular weight of the protein (Wetlaufer, 1962). Even in cases where such data are known, the actual spectral response for folded, globular proteins like hGH may differ by as much as 5–10% from the calculated value, in addition to being redshifted 1–5 nm on the wavelength axis. These differences are due to conformational or solvent effects on the individual chromophores that result from protein folding with inclusion of many of the aromatic side chains, and disulfide bonds, within the hydrophobic, high-refractive-index interior of the native protein. Extinction coefficients are most often determined from spectra of solutions whose protein concentration is accurately known from amino acid analysis.

The UV absorption spectrum of lyophilized, recombinant hGH after reconstitution is shown in Fig. 4. Fresh, carefully prepared hGH solutions normally show low levels of light scattering which, if present, appear in the absorption spectrum at wavelengths extending well above 320 nm. However, agitation, or other conditions leading to denaturation or aggregation of the protein may produce particulate material with considerable scattering (A_{320} > 5% of A_{277}). This must be taken into account in order to accurately use the E_m or $A^{0.1\%}$ for protein concentration determination. Figure 4 demonstrates the typical level of light scattering encountered with most hGH solutions. Precise concentration measurements require that even this small amount of scattering be removed by extrapolation of the absorbance between 360 and 340 nm in a log–log form, and subtraction of the scattering component as

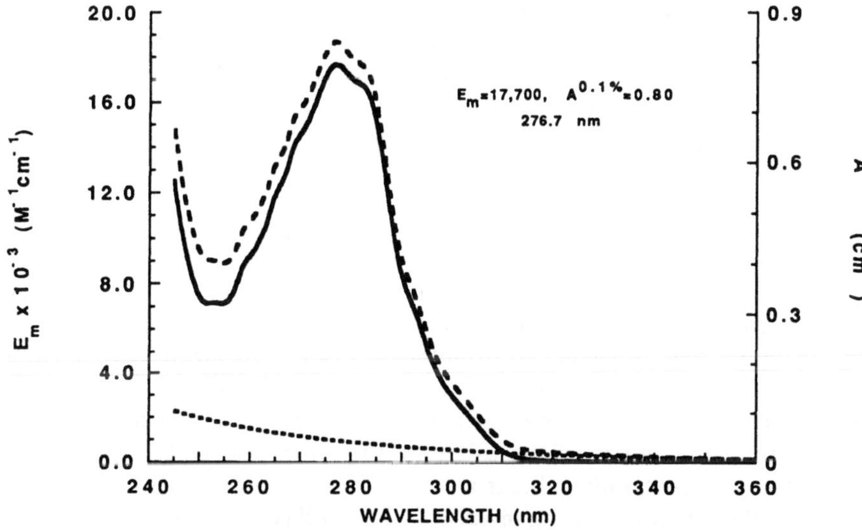

Figure 4. Ultraviolet absorption spectra of a lyophilized sample of recombinant hGH after reconstitution to a protein concentration of 1.0 mg/ml, in a formulation buffer composed of: 2.5 mM sodium phosphate, 0.34 mg/ml glycine, and 9 mg/ml mannitol (pH 7.4). Spectra were taken on an AVIV, Model 14DS dual-beam spectrophotometer in 1.0-cm-path quartz cuvettes. Data were recorded on floppy disks against a buffer blank at 25°C, in 0.5-nm intervals from 360 to 240 nm, with a spectral bandwidth of 0.5 nm. All spectra were scanned in triplicate and averaged. Each data point represents the mean of 90 readings (averaging time = 1.0 sec) for the sample at that wavelength. The baseline corrected spectrum (----) was further corrected for light scattering using a macro program written by one of the authors (T.A.B.) and subsequently incorporated into the current AVIV software package. This program extrapolates the data from 360 to 340 nm as log absorbance versus log wavelength down to 240 nm, and then subtracts this light scattering component (· · · · ·) from the baseline corrected data in log–log form to produce the final corrected spectrum (——).

described by Winder and Gent (1971). This extrapolated correction is also shown in Fig. 4. After correction for light scattering, the absorption maximum for recombinant hGH is reported to be near 277 nm with $E_m = 17{,}700$ M^{-1} cm^{-1}; this corresponds to $A^{0.1\%} = 0.8$ cm^{-1} at the same wavelength (USP Pharmacopeial Forum, 1990).

In addition to the absorption maximum, there are several other significant features visible in the spectrum. The series of weak shoulders appearing between 255 and 275 nm have been assigned to the phenylalanine chromophores (Bewley and Li, 1970; Holladay *et al.*, 1974). Also present are two weak shoulders appearing just above 290 and 300 nm. A shoulder near 290 nm is usually seen in tryptophan-containing proteins (Wetlaufer, 1962), and has been assigned to the tryptophan ^1L_b transition (Bewley and Li, 1972; Holladay *et al.*, 1974). However, an absorption band above 300 nm is rarely noted in the protein literature and this band in hGH is a highly characteristic feature of its absorption spectrum. Historically, the first interpretation of such a band in any protein was in the case of hGH. It has been assigned to the tryptophan ^1L_a transition, expressly involving a hydrogen bond between the indole ring of Trp-86 and a carboxylate group contributed by an unknown Glu or Asp side chain (Bewley and Li, 1984).

2.1.2. Circular Dichroism

CD is a specialized form of absorption spectroscopy which is extremely sensitive to the conformation of the protein. The presence of ordered secondary structures may be evaluated from the far-UV CD spectrum between 250 and 190 nm, which is dominated by amide bond absorption (Fig. 5A). This spectrum exhibits the two intense, negative maxima near 221 and 209 nm which are characteristic of the α-helix (Holzwarth and Doty, 1965). The intensities of these bands indicate an excess right-handed α-helix content of $55 \pm 5\%$, the remainder being mostly random coil (Bewley *et al.*, 1969; Aloj and Edelhoch, 1972; Holladay *et al.*, 1974). This amount of helix is virtually identical to the α-helical content indicated in the recent x-ray crystallographic study of porcine GH (Abdel-Meguid *et al.*, 1987). The secondary structure of hGH is essentially unchanged on titration of the protein from pH 2.1 to pH 12.5, and is also completely reversible on refolding of the protein after denaturation by 6 M guanidine hydrochloride or 50% acetic acid (Bewley and Li, 1972; Aloj and Edelhoch, 1972; Kawauchi *et al.*, 1976; Brems *et al.*, 1990).

All species of GH studied to date, regardless of their phylogenetic origin, display very similar amide bond CD spectra (Bewley, 1979; Bewley and Li, 1986; Li *et al.*, 1987). This portion of the CD spectrum, and the secondary structure associated with it, appear to be highly conserved in all GH species. As mentioned above, this α-helix content represents the four-helix bundle seen in the pGH x-ray structure (Fig. 2), and appears to be the architectural core of all of the GHs and related proteins.

Certain aspects of the tertiary structure of a protein may be inferred from the near-UV CD spectra, at wavelengths greater than 240 nm, where aromatic side-chain and disulfide bond absorption occurs. Dichroism in this spectral region is extremely sensitive to local environment, and conformational changes in the vicinity of the tryptophan, tyrosine, and phenylalanine side chains, in addition to the disulfide bonds. Intense dichroism is usually associated with these side-chain structures being held rigidly in a highly folded, three-dimensional structure. Denaturation of the protein will generally release this steric hindrance, resulting in the CD spectrum becoming weaker and weaker as the degree of denaturation increases.

The CD spectrum of native hGH in the region of side-chain absorption is presented in Fig. 5B. Interpretation of this spectrum by comparison with CD spectra of model compounds indicates that the average aromatic residue is indeed buried within the interior of the molecule as would be expected for a globular protein. The two negative minima near 261 and 268 nm arise from the phenylalanine side chains, while the two negative minima at 283 and 277 nm are predominantly due to tyrosine residues, although both tryptophan and the disulfide bonds may also contribute in this region (Bewley and Li, 1972; Aloj and Edelhoch, 1972; Holladay *et al.,* 1974; Bewley, 1979; Larhammar *et al.,* 1985). The positive, asymmetric band between 320 and 288 nm is caused by overlapping of both the ^1L_a and ^1L_b transitions of the single Trp-86 residue. It has also been shown that the disulfide bonds in hGH produce two weak, very broad, negative CD bands extending from beyond 300 nm to below 250 nm (Bewley, 1977).

The side-chain CD spectrum of hGH is quite sensitive to partial denaturation by acidic titration, or addition of denaturants such as guanidine hydrochloride, although all such effects have been found to be completely reversible (Bewley and Li, 1972; Aloj and Edelhoch, 1972). Certain reversible chemical alterations of the molecule such as reduction of the disulfide bonds, or alkaline titration producing ionization of the tyrosine residues ($pH > 8.5$), will also change the side-chain CD spectrum. For hGH, these spectral differences have been shown to occur almost entirely through either removal of a chromophore, or by effecting changes in a particular chromophore's CD response, and not through gross denaturation or conformational changes brought about by these same chemical actions (Bewley *et al.,* 1969; Aloj and Edelhoch, 1972; Kawauchi *et al.,* 1976). These purely spectral effects have also been shown to be fully reversible. It is worth noting that the studies of hGH clearly indicate that not all changes in CD spectra are necessarily ac-

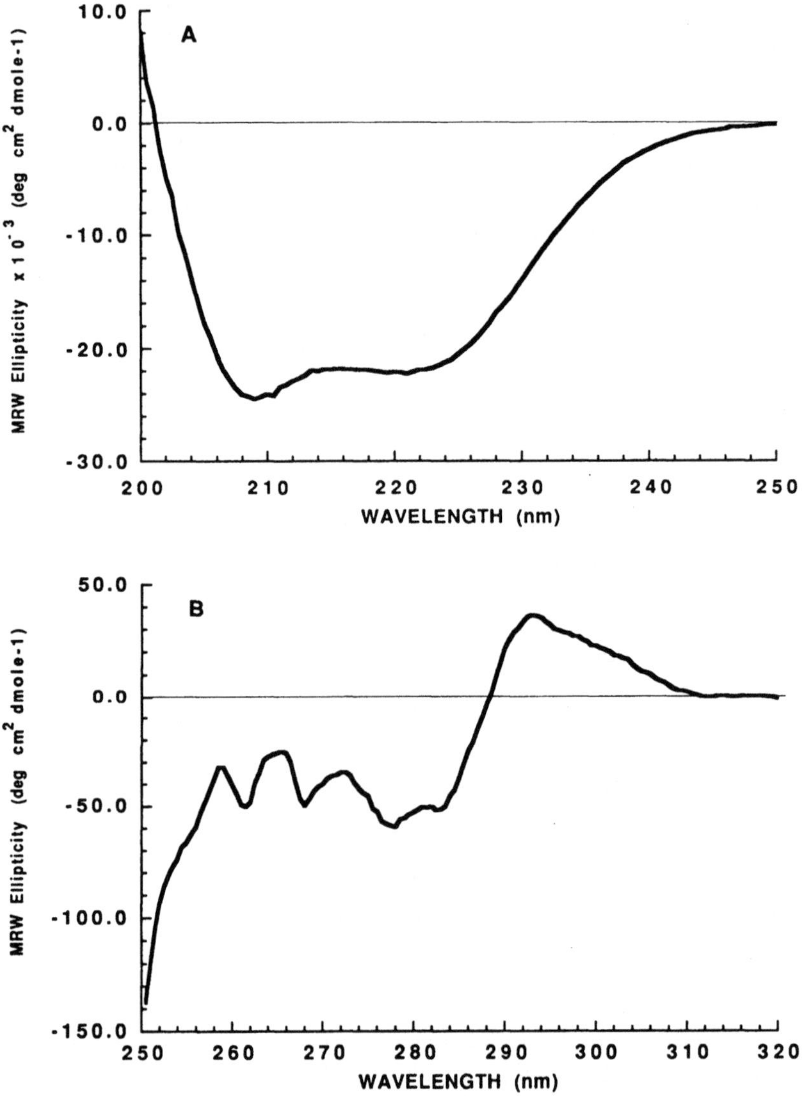

Figure 5. Ultraviolet CD spectra of a lyophilized sample of recombinant hGH after reconstitution to a protein concentration of 1.0 mg/ml, in a formulation buffer composed of: 2.5 mM sodium phosphate, 0.34 mg/ml glycine, and 9 mg/ml mannitol (pH 7.4). Spectra were taken at 25°C on an AVIV, Model 60DS spectropolarimeter in water-jacketed quartz cuvettes, and were recorded on floppy disks. The far-UV CD spectrum (A) was recorded in a 0.5-mm-path cell, at 0.5-nm intervals, and with a spectral bandwidth of 1.5 nm. Each data point (triplicate scans) represents the mean of 270 readings (averaging time = 3.0 sec) for the sample at that wavelength. The near-UV CD spectrum (B) was taken in a 2.0-cm-path cell, at 0.5-nm intervals, and with a spectral bandwidth of 0.5 nm. Each data point (5 scans) represents the mean of 450 readings

companied by significant conformational changes in the protein, a point that is often misunderstood.

In sharp contrast to the far-UV CD spectrum, the near-UV CD spectrum of hGH appears to be unique, and characteristic of the specific tertiary structure of the native molecule. Other species of mammalian GHs such as bovine, porcine, elephant, or monkey all display distinctive CD spectra in this region (Bewley, 1979; Bewley and Li, 1986; Li *et al.*, 1987).

2.1.3. Fluorescence

The fluorescence properties of hGH have been reported under a number of conditions by Aloj and Edelhoch (1972). With excitation near 295 nm, the emission maximum at pH 8.0 is near 335 nm. This is attributed primarily to tryptophan, and indicates a fairly hydrophobic environment for that residue, with relatively little intensity due to tyrosine emission near 305 nm. A 30% decrease in the intensity of the tryptophan emission peak occurs between pH 6.0 and 4.0, but little change is found at 305 nm, suggesting that only the indole emission is being affected. The authors tentatively attribute this quenching to the protonation of a side-chain carboxyl in the vicinity of the indole ring. This may be the same tryptophan–carboxylate proximity noted above from the UV absorption spectrum. At pH values below 4.0, the reverse situation prevails, and only the tyrosine emission is altered, in this case increasing by 50% at 305 nm. The tryptophan emission is also strongly quenched between pH 8.0 and 12, but without shifting of the wavelength maximum (Aloj and Edelhoch, 1972; Kawauchi *et al.*, 1976). This quenching arises both from energy transfer between the indole and ionizing tyrosine residues, as well as from direct quenching by hydroxyl ion. The lack of any "normalization" of the emission maximum to wavelengths nearer 350 nm indicates that even at pH 12, there is little unfolding of the molecule, in good agreement with results from CD spectra.

Fluorescence spectra also suggest that at concentrations below 8 M, urea has little effect on the tertiary structure of hGH between pH 5 and 8 (Aloj and Edelhoch, 1972). In contrast, Maddaiah *et al.* (1972) found the emission maximum of pit-hGH to be at 348 nm in the presence of 6 M guanidine hydrochloride, indicating significant denaturation of the tertiary structure in

---

(averaging time = 3.0 sec) for the sample at that wavelength. Buffer blanks were subtracted from the protein spectra. The baseline corrected spectra were then converted to mean residue ellipticity using a value of 115.8 for the mean residue weight of hGH. Protein concentration was taken from the absorption spectra of Fig. 4.

this solvent. Brems *et al.* (1990) have confirmed and extended this last observation with recombinant hGH.

2.2. Electrophoresis

Electrophoretic methods are widely employed as a means of characterization and identification of proteins, as well as assessing the presence of impurities (Garnick *et al.*, 1988; USP Pharmacopeial Forum, 1990; Pearlman and Nguyen, 1991). Both size and charge information may be obtained, and, while quantitation is possible from gels, chromatographic methods are the preferred means of acquiring quantitative data.

2.2.1. SDS-PAGE

SDS-PAGE is one of the most widely used gel electrophoresis methods. It is employed to obtain size information about a protein sample. Two methods for running SDS-PAGE for hGH are described in the USP Pharmacopeial Forum (1990), one for Somatrem (Met-hGH) and the other for Somatropin (hGH). These methods are very similar, and differ primarily in the preparation of gels and buffers, as well as variations in the concentrations of the ingredients. It is important to choose one method and retain that system, since even under the best circumstances comparisons between gels are often very unreliable.

SDS-PAGE is a denaturing electrophoretic system because samples are heated with SDS, which unfolds the protein and binds to it. Separation of proteins is achieved when a fixed current draws the negatively charged species toward the anode end of the gel. The speed of migration of the proteins is proportional to the logarithm of their size (approximating molecular weight), hence effecting separation.

For the SDS-PAGE analysis of hGH, samples are loaded at 2 and 20 μg, with or without a reducing agent such as β-mercaptoethanol or dithiothreitol (DTT). In the nonreduced gel, hGH runs as a monomeric species of ~22 kDa (Jones and O'Connor, 1982; Garnick *et al.*, 1988). Addition of a reducing agent causes reduction of the disulfide bonds, and yields information about disulfide-linked dimers or clips occurring in the protein that might be held together by disulfide bonds.

Visualization of the proteins is achieved by staining with Coomassie brilliant blue R-250, as described in the USP monograph, or one of the highly sensitive silver staining methods described by Jones and O'Connor (1982) or Garnick *et al.*, (1988). In addition, gels may be analyzed using the

Western blot technique, which makes proteins visible that are immunoreactive to various antibodies. This method involves transferring the proteins from the acrylamide gel onto a nitrocellulose membrane and then incubating the membrane with an antibody raised against the protein or with antibodies raised to possible contaminants, such as antibodies to *E. coli* proteins. The pattern of immunoreactive proteins is made visible with a conjugated reporter antibody, and by comparison with other staining methods, the identity of the bands on a gel may be made.

SDS-PAGE has also been used to study variants of hGH, including covalently linked dimers, higher-order aggregates, and also clipped forms of the protein. Skyler *et al.* (1977) reported the presence of a clipped form of hGH which elutes with a larger apparent molecular weight when run under nonreducing conditions, and was referred to as a two-chain form of hGH.

2.2.2. Isoelectric Focusing (IEF) Gel Electrophoresis

IEF gel electrophoresis is typically run under native conditions for the protein. This technique is carried out by setting up a potential gradient across a slab gel, into which are dispersed polyvalent ampholytes. These ampholytes migrate, and form a corresponding pH gradient, where the pH range is determined by the particular set of ampholytes in the gel. The applied protein samples migrate until their net charge is zero, i.e., at the position of their isoelectric points. Thus, a mixture of proteins of different pI values can readily be resolved and their pI's determined by comparison with standards. The gel is usually stained with Coomassie brilliant blue or silver stain.

One of the most valuable pieces of information obtained from an IEF gel is the pI of the protein. The USP Pharmacopeial Forum (1990) describes an IEF method for Somatrem, using a pH range of 3.5 to 9.0 pH units. The pI for hGH is reported to fall between pH 5.8 and 6.2. Jones and O'Connor (1982) estimate a pI of 5.2 for Met-hGH, and in our laboratories, we have determined the pI of Met-hGH and hGH to be equivalent at approximately 5.3 pH units (Pearlman *et al.*, 1987). As noted above, the pH of least solubility corresponds to the pI of hGH, and it also occurs at about pH 5.3. The pI of hGH is significantly less than that of recombinant bovine GH, which has a pI of 8.5 (Langley *et al.*, 1987). It is worth noting that considerable differences in pI values of proteins from different species may occur, and a high level of homology does not necessarily mean that the other properties of such proteins will be similar.

IEF has been used in a number of studies to characterize a variety of hGH species, which are referred to as isohormones. For example, Skyler *et al.* (1977) used IEF to catalog a variety of different isohormones of hGH extracted from the pituitary. Lewis *et al.* (1970, 1980, 1981), and Skottner *et al.* (1988) used IEF to study the deamidation of hGH.

2.2.3. Other Electrophoretic Methods

Native PAGE and two-dimensional electrophoresis have been reported for the characterization of hGH. Native PAGE, as the name implies, is not run under denaturing conditions, but under nondenaturing conditions where proteins are separated on the basis of both size and charge. Skyler *et al.* (1977) described a number of so-called isohormones of hGH, by determining the immunoreactivity of these species after native PAGE and autoradiographic techniques. This study examined eight hGH preparations, and the authors noted that all preparations contained multiple isohormone species.

The technique of two-dimensional electrophoresis separates proteins using two different electrophoretic modes. Jones and O'Connor (1982) have described the two-dimensional resolution of hGH by IEF gel separation, followed by SDS-PAGE. The hGH preparations were resolved into several bands by IEF, which were found to be either monomer or dimer when run in the second dimension. Western blot analysis showed that all of the bands were immunoreactive to anti-hGH, and that none were derived from *E. coli* proteins. Thus, the advantage of combining several electrophoretic methods can easily be seen.

2.3. Immunoassays

Immunoassays have been widely used to determine the concentration of protein in samples from various sources, including biological samples such as serum. They are also widely used for determining product purity. In general, antibodies to hGH are raised in animals (such as goats) and then are used in several types of immunoassays, including radioimmunoassays (RIA) (Nederman and Sjodin, 1987) as well as in the enzyme-linked immunosorbent assay (ELISA) format (Hashida *et al.*, 1987). The double antibody method of Schalch and Parker (1964) and its modifications have been used by a number of workers, particularly to assay hGH in plasma. Antibodies to hGH may also be produced using monoclonal antibody techniques as described by Basuyaux *et al.* (1987).

A sandwich-type RIA is available from Hybritech (Tandem-R-HGH) which uses two monoclonal antibodies directed against different antigenic sites on hGH. One of the antibodies is immobilized on plastic beads, the other containing the ^{125}I label. In this assay, hGH samples are added to the mixture of antibodies, and the amount of bound ^{125}I antibody is measured after the excess antibody is washed away.

Immunoassays are also employed to determine levels of antibodies to hGH appearing in the serum of patients (Rougeot *et al.*, 1991). Typical assays are able to detect the hGH levels normally occurring in blood. Sensitivities of immunoassays easily permit quantitation of hGH in plasma down to the ng/ml range.

Some studies have used the term "immunoreactivity" interchangeably with "bioactivity." This is certainly not recommended, since immunoreactivity may not necessarily be measuring the same attribute of a molecule which is responsible for its potency (WHO Technical Report, 1975). That is, the epitopes responsible for antibody binding are not necessarily the same as those involved in receptor binding and, hence, biological activity. This is eloquently demonstrated in the work of Cunningham *et al.* (1989, 1991).

2.4. Bioassays

Bioassays are used to measure the biological activity (or potency) of hGH. They involve dosing animals with hGH, and observing various growth effects (Li, 1977; Wilhelmi, 1982). The most prevalent bioassay measures the daily weight gain in hypophysectomized (hypoxed) rats given daily injections of hGH (Groesbeck and Parlow, 1987; USP Pharmacopeial Forum, 1990). The rats respond to exogenous GH (even GH from different species) as indicated by their increase in weight. While rat antibodies to the GH from other species do occur, they are not present in sufficient quantities to produce growth inhibition over the relatively short duration of the study. Typically, ten rats are used per group and the dosing and recording phase of the study takes 10 days. A reference standard and a blank are also run and usually two doses, 10 and 50 μg, are administered subcutaneously to each rat per day. The mean potency, P, of the sample relative to the Standard in USP Somatropin Units is defined as:

$$\text{P (USP Units per mg)} = \text{(RP Sample)} \times \text{(AP Standard)}$$

where RP Sample is the relative potency of the sample compared with the USP standard and AP Standard is the assigned potency of USP Somatropin, which has a potency of not less than 2 USP Somatropin Units per mg (USP Pharmacopeial Forum, 1990).

2.5. Chromatographic Methods

Chromatographic procedures for investigating proteins have evolved significantly during the 1980s with numerous advances in the technology of HPLC methods. Many improvements have occurred in pump technology, detectors, integration software, and other refinements. Also, new HPLC column technologies have resulted in a very large number of separation chemistries. The heightened interest in biotechnology, and the available source of ultrapure proteins, has provided a powerful catalyst for the development of new analytical HPLC methods. A number of myths have been debunked, including fears that denaturation was a natural consequence of exposure of proteins to organic solvents, or that the biological activity of proteins would be damaged upon contact with the stainless steel plumbing and fittings (necessary for the high-pressure systems in HPLC).

2.5.1. Reversed-Phase HPLC

Once the domain of small organic molecules, RP-HPLC has become a routine method for the analysis of a number of proteins (Hancock, 1984). Several different methods for the analysis of hGH, its variants and degradation products appear in the literature. In RP-HPLC, the analyte is loaded onto a hydrophobic column in a mixed polarity solvent, and eluted either with a gradient of decreasing polarity, or in an isocratic mode. The appropriate ion-pairing agent is added to the mobile phase, and elution of the analyte is typically monitored by a UV detector. Several groups have published methods for the analysis of hGH by RP-HPLC, and three methods appear in the USP Pharmacopeial Forum (1990).

The first of these methods is for Somatrem, utilizing a C_4 column with a gradient mobile phase. The mobile phase is composed of a water and acetonitrile mixture, with 0.1% trifluoroacetic acid (TFA) added as an ion-pairing agent, and achieves a final ratio of 30 : 70, water : acetonitrile. The flow rate is 1 ml/min, and the column temperature is 35°C. Detection is achieved by UV absorption at 214 nm. This method is based on one reported by Hancock (1986) who used it to study biosynthetic and pit-hGH preparations with regard to the presence of variants (Hancock *et al.*, 1988). It was shown by this method that biosynthetic hGH elutes in excess of 99% as the protein monomer, and is considerably more homogeneous than pit-hGH (Fig. 6).

This method has also been used to separate hGH from covalent dimers as well as from oxidized species (Pearlman *et al.*, 1987). Because it is run at a low pH, variants such as deamidated species are protonated, and therefore not resolved from the main hGH peak. Other chromatographic conditions

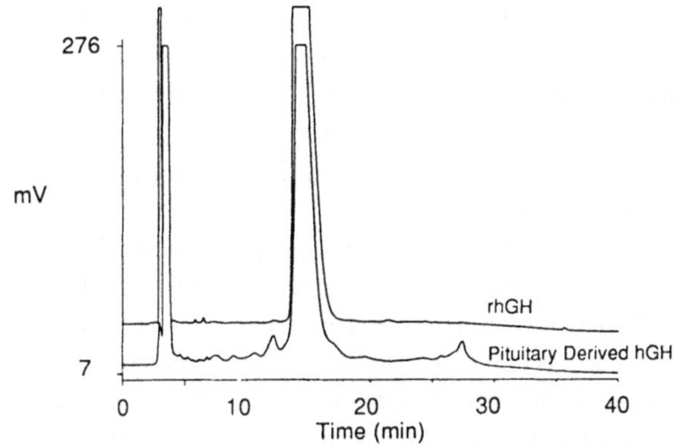

Figure 6. The reversed-phase comparison of biosynthetic hGH versus pit-hGH (Crescormon; Kabi). The aqueous mobile phase A was 0.1% TFA and the organic mobile phase B was acetonitrile containing 0.1% TFA. The 50-μg samples were chromatographed using a 0.45 × 15-cm Vydac C4 column by isocratic elution at 50% B for 10 min, followed by a linear gradient from 50% to 70% B in 20 min. The flow rate was 1 ml/min, the column temperature was maintained at 35°C, and the separations were monitored by UV absorption at 214 nm. After Hancock *et al.* (1988).

are necessary to resolve deamidated hGH from the main peak. This may be achieved at neutral pH, as described below for Somatropin.

The USP Pharmacopeial Forum method for Somatropin is an isocratic system using a mobile phase of 0.05 M Tris (pH 7.5), in 29% propanol/water, at 40°C. A 25-cm-long C_4 or C_8 column and a flow rate of 0.5 ml/min are generally used with UV detection at 220 nm. Under these conditions, hGH elutes at about 33 ± 3 min. Interestingly, this method is able to resolve Met-hGH from the methionyl-free hGH. It is also able to resolve mono- and di-deamidated forms of hGH, as described by Riggin *et al.* (1987). One of the problems in using this method is that a silica precolumn is necessary, since the high pH is detrimental to the solid support phase. Also, while this method is able to resolve both deamidated hGH and oxidized forms from the native hormone, the two types of modified species coelute. Thus, still another method is required to resolve these variants from each other. A third, gradient system is also described for Somatropin in the USP Pharmacopeial Forum (1990) using the same columns described above with a propanol/water system at neutral pH. This method is listed under "chromatographic purity" of hGH, as distinct from "chromatographic identity."

Welinder *et al.* (1987) used RP-HPLC to separate a number of hGH variants, but were unable to separate deamidated forms from the native form of hGH. These authors postulated that the lack of separation of the deamidated forms from the non-deamidated molecule was due to burial of the deamidated sites within the molecule. This explanation has been shown to be incorrect, since the method of Riggin *et al.* (1987) has been able to separate deamidated forms, and the sites of deamidation have been identified in several laboratories (Becker *et al.,* 1987; Canova-Davis *et al.,* 1988), and shown to be on the surface of the protein, based on the structure for porcine GH (Abdel-Meguid *et al.,* 1987).

Another RP-HPLC method run at low pH is described by Patience and Rees (1986). They used several HPLC methods to investigate variants occurring in pit-hGH samples, and were able to resolve an hGH variant with a smaller molecular mass (20 kDa), as well as other forms of the protein. They measured the immunoreactivity of the various fractions collected during a chromatographic run, and showed that different reactivities are elicited by dimeric, aggregated, and other species.

RP-HPLC is also used for the tryptic map analysis of hGH, as described by Kohr *et al.* (1982), and used in the USP Pharmacopeial Forum (1990). The tryptic map analysis involves incubating the protein (or a reduced and alkylated form of it) with trypsin to selectively cleave the protein at the COOH-terminal sides of lysine and arginine residues. In the native molecule this digestion results in the formation of 18 smaller peptides plus free lysine (Fig. 7), which may be separated by a shallow gradient on an RP-HPLC system. The resulting chromatogram, or tryptic map, is shown in Fig. 8. The tryptic map contains a wealth of information about the protein and is, in effect, a fingerprint of the molecule. The use of the tryptic digest, particularly for degraded samples of hGH, is described below as a means of obtaining information about the site of degradation in the protein.

Another form of HPLC which separates species according to their relative hydrophobicities is hydrophobic interaction chromatography (HIC). HIC employs a column of lower hydrophobic density than used for RP-HPLC. The samples are loaded in a mobile phase which contains a high concentration of an antichaotropic salt, and elution is effected by decreasing the concentration of this salt in the mobile phase. In HIC, separation of proteins is possible with little or no organic modifier in the mobile phase, thus resembling native conditions. Wu *et al.* (1990) describe the analysis of hGH by HIC. In this study, the authors examined a number of HIC columns for their abilities to resolve hGH from its variants, and showed that the TSK-phenyl column (Toyo Soda) was well suited in the analysis of hGH. They describe the optimization of a number of parameters, including temperature, salt type, and injection volume for the HIC method for hGH. The authors were able to resolve Somatropin from Somatrem, and cautioned about potential artifacts arising in HIC methods due to interactions between the sample and the loading buffer.

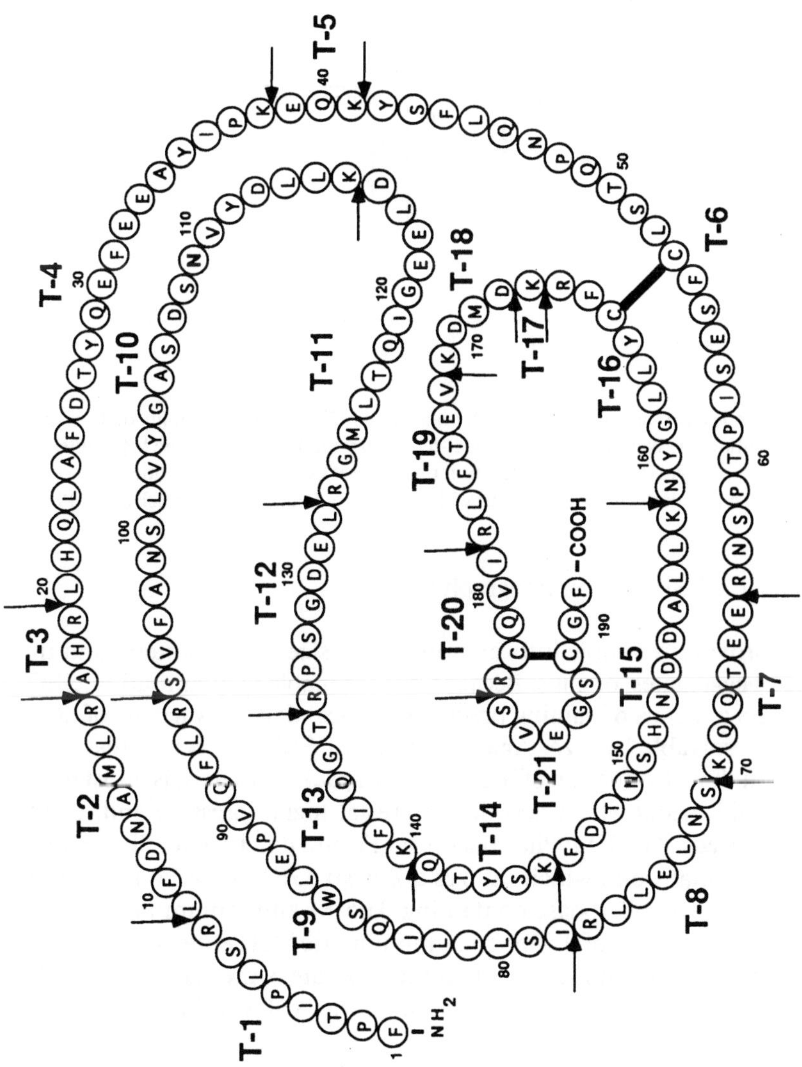

Figure 7. The primary structure of human growth hormone indicating the 21 theoretical peptides, including free lysine, produced by total tryptic digestion of the reduced–alkylated molecule. With digestion of the native molecule, 18 peptides plus free lysine will be produced due to disulfide linking of peptides T-6 to T-16 and T-20 to T-21.

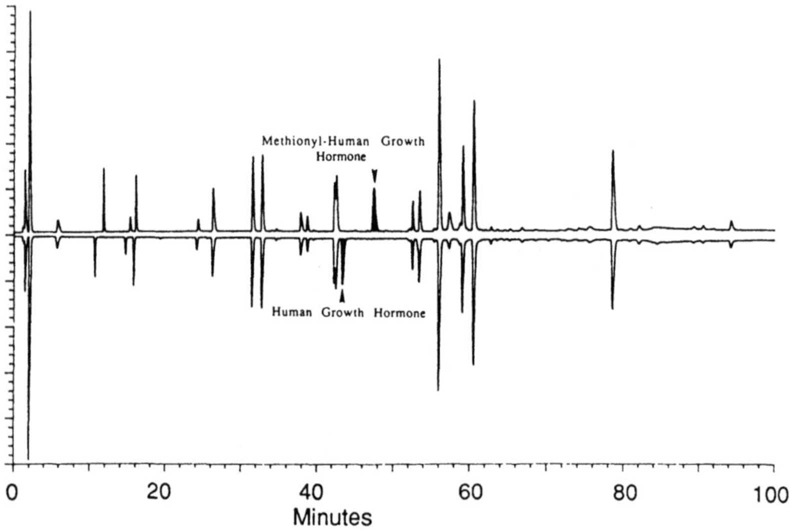

Figure 8. Tryptic map chromatograms of Met-hGH and hGH showing the ability of the tryptic map to differentiate the two materials. The NH_2-terminal peptide differences are highlighted by arrowheads. Note: Slight differences in retention times of peaks early in the peptide map are the result of HPLC pump performance problems with shallow gradients. After Garnick *et al.* (1988).

2.5.2. Size-Exclusion Chromatography (SEC)

SEC provides valuable information on the size of the protein, and the presence of aggregates, or clips. Because it is amenable to quantitation, SEC-HPLC offers a means of gaining precise analytical data on variants of hGH which are resolvable by size. Again, the use of SEC has been greatly expanded with the introduction of better column packing materials, most notably more rigid media which permit analyses at higher pressure and, therefore, faster speeds. The USP Pharmacopeial Forum describes an SEC-HPLC method for Somatropin, which utilizes a Zorbax GF250 (DuPont) column and an isocratic mobile phase containing 0.05 M ammonium bicarbonate buffer. Using a flow rate of 0.6 ml/min, monomeric hGH elutes between 750 and 1000 sec. The method suggests that the flow rate of mobile phase may be altered slightly in order to obtain the monomeric peak within this retention time window.

Riggin *et al.* (1988) describe an SEC-HPLC method, using a TSK G3000SW column (Toyo Soda) and a buffer of 0.025 M ammonium bicarbonate. A flow rate of 1 ml/min was used and detection was achieved using a UV monitor at 214 nm. A typical chromatogram is shown in Fig. 9, which depicts the elution times for monomer, dimer, and higher-order aggregates,

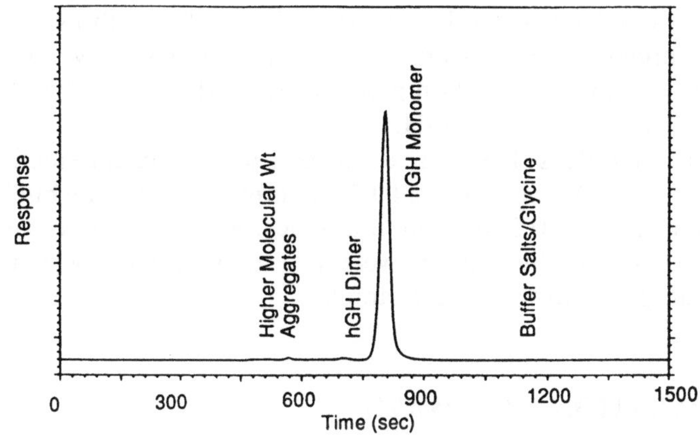

Figure 9. Size-exclusion HPLC of a sample of hGH under nondissociating conditions before final purification. Separation was performed using a TSK G3000SW (0.76 × 60 cm) column with a mobile phase of 25 mM ammonium bicarbonate at pH 7.6. Detection was at 214 nm, the injection volume was 20 μl, and the flow rate was 1 ml/min. From Riggin *et al.* (1988).

as well as the retention time of various components found in the formulation buffer. The authors show a strong correlation between the SEC method and two biopotency assays (rat weight gain and tibial cartilage width), with the higher precision of the SEC method being most noticeable. The authors argue for the use of mass, rather than biopotency units, as the measure for dosing of hGH.

2.5.3. Ion Exchange Chromatography

The utility of the ion exchange method resides in its ability to resolve species of different charges, often under native conditions. It is a flexible technique, because a wide variety of column types are available, including anionic and cationic matrices with either weak or strong chemistries. The choice of column type is dependent on the pI of the protein and the pH at which the analysis is to be performed. The protein is loaded onto the column at low ionic strength, where it binds, and elution is effected by means of a salt gradient of increasing ionic strength.

In the case of hGH, anion exchange chromatography has been investigated by Riggin *et al.* (1987) who describe an anion exchange method using a Pharmacia Mono Q column. The initial mobile phase is composed of 30% acetonitrile and 50 mM Tris at pH 8. Elution is effected by a gradient of the same mobile phase containing, in addition, 0.3 M NaCl. At a flow rate of 1 ml/min the gradient goes to 100% of the salt-containing buffer over 25 min, with the main hGH peak eluting at about 12 min. The authors were able to resolve deamidated species from native hGH using this assay, but were unable to resolve other degradation products which do not involve changes in net charge, such as oxidized hGH.

Patience and Rees (1986) also describe an anion exchange method using a TSK DEAE 5-PW column in a 0.05 M ammonium acetate buffer with a gradient formed with 0.5 M ammonium bicarbonate. In combination with SEC and RP-HPLC, the authors were able to characterize aggregates and deamidated species in their hGH samples.

3. DEGRADATION PATHWAYS

A number of published articles refer to the various ways in which hGH is degraded as well as describe the mechanisms and kinetics of its degradation. Early studies on the degradation of GH focused on ways of preserving the activity of the protein during extraction and isolation from the pituitary. More recently, with the advent of recombinant hGH, sufficient quantities have been available to study its degradation in detail.

In any discussion of protein degradation, it is important to determine whether degradation products possess any biological activity. Often analytical methods can detect changes occurring in a protein well before any change in biological effect is observed, since many degradation products possess full activity. Of interest, then, is the paper by Bristow and Jeffcoate (1992), in which ten different laboratories analyzed the same lot of rhGH. The activities of the samples were determined by a number of assays, including bioassays—hypophysectomized rat weight gain assay (eight labs) or hypophysectomized rat tibial width assay (three labs); an *in vitro* receptor assay (two labs); immunoassays (six labs); RP-HPLC (two labs); and SEC-HPLC (four labs).

In addition to measuring the potency of intact rhGH to determine a standard specific activity, the various groups also analyzed hGH that had been subjected to several stress conditions. Lyophilized samples of the protein were assayed either shortly after reconstitution, after exposure to proteolytic enzymes (to cleave the hormone) or hydrogen peroxide (to produce oxidized products), or after exposure to base (to form deamidated hormone).

Apart from the previously described higher potency seen in the reference sample (as compared with earlier values of the specific activity of hGH), the effect of modifications of the protein on measurements on potency were interesting. The results from the analyses of the degraded samples demonstrated that in general the bioassays were relatively insensitive to the degradation state of the molecule, whereas the RP-HPLC assay was shown to be the most sensitive gauge of biochemical integrity. In other words, the chromatographic assays could detect changes occurring in the protein with far more precision than the potency measurements; however, it may also be that these changes do not alter the pharmacologic activity of the protein. Such findings have also been reported by others and are detailed in the following section. Interestingly, mild proteolysis gave rise to a slightly increased potency of hGH in the *in vivo* bioassays, whereas it gave a marked decrease in "potency" in the chromatographic analyses. This study raises some important questions on the relative merits of analytical methods for hGH potency measurements.

3.1. Deamidation

Deamidation involves the conversion of the side chain in asparagine and glutamine residues to the carboxylate groups of aspartate and glutamate, respectively. It is a common reaction occurring in proteins in solution, with the generally accepted mechanism shown in Fig. 10. This mechanism in-

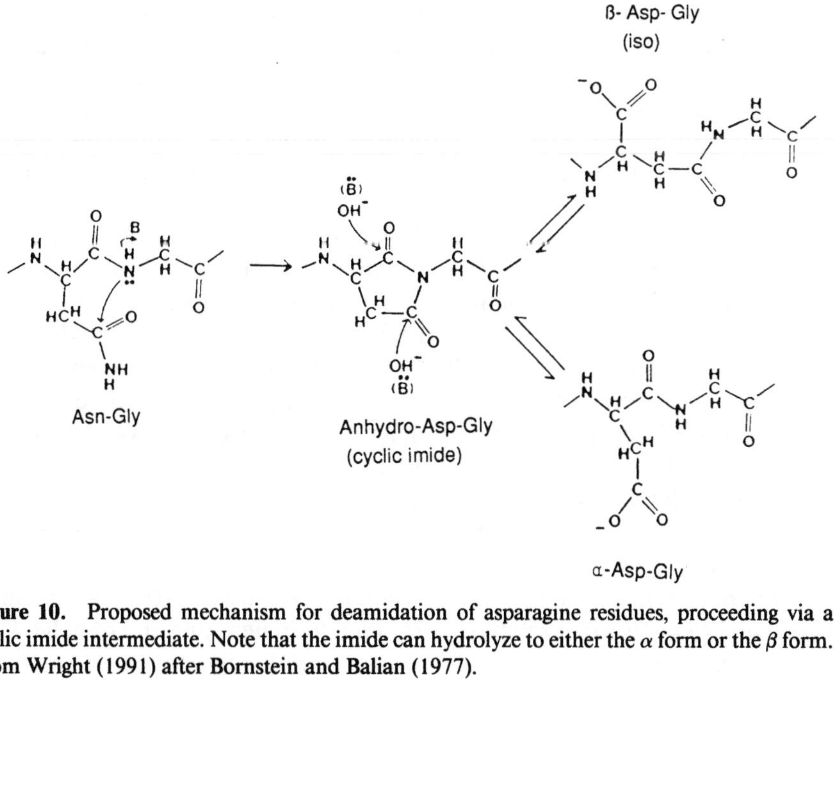

Figure 10. Proposed mechanism for deamidation of asparagine residues, proceeding via a cyclic imide intermediate. Note that the imide can hydrolyze to either the α form or the β form. From Wright (1991) after Bornstein and Balian (1977).

volves elimination of the -NH_2 group in asparagine to form a cyclic imide. The interesting feature of this intermediate state is that it can hydrolyze in either of two ways, yielding the aspartate residue (i.e., the α form) or the iso-peptide (the β form). Wright (1991) argues that while this mechanism has been demonstrated to occur in a number of proteins, it is not necessarily the only mechanism by which proteins deamidate. Robinson and Rudd (1974) describe the deamidation of asparagine and glutamine residues in a large number of small peptides. Also presented in this work is information on the role of pH, buffers, and temperature on the rate of this reaction. Recently, Wright (1991) has described a literature compilation of deamidation reactions for a large number of proteins. In this paper, Wright also discusses the mechanism of deamidation, first proposed by Bornstein and Balian (1977), which has subsequently been studied by others (McFadden and Clarke, 1982; Aswad, 1984).

hGH possesses 9 asparagine and 13 glutamine residues, so it is not surprising that it is susceptible to deamidation. Lewis *et al.* (1980) described the heterogeneous nature of pit-hGH and listed a number of variants present in isolates of the protein, including deamidated forms. In another study, Lewis *et al.* (1981) examined the effect of alkaline conditions on the deamidation of pit-hGH, demonstrating that deamidation altered the proteolytic cleavage of hGH. They also conclude that the principal site of deamidation in the presence of base was at Asn-152 as well as minor amounts at Gln-135, all presumably occurring during isolation procedures. These findings pose the question whether deamidated hGH is found in the pituitary, or whether it is always a consequence of the isolation procedures. Using electrophoretic procedures on freshly extracted pituitaries, under mild conditions, the authors concluded that deamidated forms of hGH are indeed present in the pituitary.

In contrast, more recent studies by Canova-Davis *et al.* (1988) identified Asp-149 as the principal site of deamidation for hGH stored under alkaline conditions, with only a minor amount of deamidation occurring at Asn-152. The deamidation site was identified after tryptic mapping of the degraded form, followed by sequence analysis. This finding has been independently confirmed in a study by Becker *et al.* (1988). In this study, after incubation of hGH in a pH 9 ammonium bicarbonate buffer, the sample was purified by anion exchange chromatography. The site of deamidation was confirmed by sequence analysis of the trypsin-digested material. The deamidated form behaved identically to native hGH on SDS-PAGE but, as expected, had a lower pI (4.9) than native hGH (pI ca. 5.3).

Examination of the three-dimensional structure of GH (Fig. 2) shows that Asn-149 occurs in a relatively extended loop near one end of helix 4. In fact, as described in Section 3.5, the entire peptide segment extending from approximately residue 130 to residue 153 exists as a flexible, exposed loop. Local conformation, and particularly this type of conformational flexibility, is thought to play a key role in promoting deamidation (Kossiakoff, 1988) and it is possible that Asn-149, being somewhat more exposed than other Asn residues in hGH, is more vulnerable to deamidation.

The mechanism of hGH deamidation at Asn-152 is also consistent with the previously described mechanism occurring via a cyclic imide structure (McFadden and Clarke, 1982). Evidence for this intermediate is obtained by examining the nature of the products of the deamidation reaction. Canova-Davis *et al.* (1988) found that approximately two-thirds of the product is isoAsp-152, with the remainder being Asp-152. Identity of the isoAsp was made during sequence analysis of the tryptic peptide referred to as T-15 (Fig. 7) which contains residues from Phe-146 to Lys-158. Becker *et al.* (1988) have obtained similar results. In further studies, Johnson *et al.* (1989) identified both isoAsp-149 and isoAsp-152 in deamidated hGH. Wright (1991) has discussed the deamidation reaction in detail in his review article, as well as presenting some challenging information on the mechanism and evolutionary pressure on deamidation as a posttranslational modification.

Becker *et al.* (1988) demonstrated that the deamidated form of hGH was equipotent to the native form of the protein in both the hypoxed rat weight gain assay and the tibial epiphyseal width bioassay. They also reported equal reactivity in an immunoassay compared with the native form of the hormone. These results are in good agreement with the work of Skottner *et al.* (1988) who studied the biological properties of deamidated forms of hGH isolated from the pituitary. These authors reported that two charged forms were separated from native hGH by IEF electrophoresis, and both were deamidated forms of hGH. The charged forms were shown to be essentially equipotent to native hGH as demonstrated in several bioassays, including the hypoxed rat weight gain and tibial growth assays.

Thus, it appears that although hGH does deamidate in solution upon storage, the deamidated forms have essentially the same degree of potency as the native hormone. Patel and Borchardt (1990) have commented that proteins could also undergo deamidation in the solid state, for example, upon storage as lyophilized dosage forms, and indeed, deamidation has been observed in freeze-dried preparations of GH (Pearlman and Oeswein, 1989; Pikal *et al.,* 1991).

3.2. Oxidation

A number of amino acid side chains in proteins are susceptible to oxidation, such as the methionine, tryptophan, histidine, and tyrosine residues

(Manning *et al.,* 1989). However, the major oxidative reaction of hGH is at the methionine residues, producing the corresponding sulfoxide, rather than proceeding to the sulfone derivative (Houghten *et al.,* 1977; The *et al.,* 1987; Becker *et al.,* 1988). Similar conclusions were made when hGH was oxidized in solution by hydrogen peroxide, or in the solid state under strong fluorescent light (Pearlman *et al.,* 1987).

hGH possesses three methionine residues: Met-14, Met-125, and Met-170. A number of studies have looked at the susceptibility of these groups to oxidation. The oxidation of hGH in solution was reported by Houghten *et al.* (1977), using hydrogen peroxide (H_2O_2). In this study, the authors describe reaction conditions employing 30% H_2O_2 at 23°C, with hGH at 5 mg/ml in Tris buffer at pH 8.5. The reaction was followed by quenching aliquots with catalase in order to follow the rate of reaction. Under these reaction conditions, the authors found that two of the three methionines in the protein were oxidized to the corresponding sulfoxides. The reaction product was identified as the di-methionine sulfoxide at positions 14 and 125. Met-170 did not react under these conditions, but could be oxidized if the protein was reacted with H_2O_2 under mild denaturing conditions in 8 M urea. This suggests that Met-170 either is sterically hindered, is interacting with another side chain, or perhaps is buried in the interior of the protein and hence unable to react with the peroxide.

Using several analytical techniques, Houghten and co-workers found that the partially oxidized form of the protein (i.e., methionyl sulfoxides at positions 14 and 125) behaved similarly to the native hormone, whereas the fully oxidized form showed a difference in its rate of digestion by trypsin, as well as in the degree of α-helical content as measured by CD. Similarly, in the rat tibia bioassay the partially oxidized hormone exhibited full bioactivity, whereas the fully oxidized form possessed only about one-fourth the activity of the native molecule.

In a similar study, Teh *et al.* (1987) used 0.2 M H_2O_2 at pH 8.0 to oxidize hGH at positions 14 and 125, again showing that the methionine at position 170 was resistant to oxidation under native conditions. The pseudo-first-order rate constants for the reactions at 4°C were reported to be $1.3 \times 10^{-4}\ sec^{-1}$ for oxidation at Met-14, and $4.8 \times 10^{-4}\ sec^{-1}$ for oxidation at Met-125. Oxidation at these two sites was shown to have no measurable effect on conformation of hGH as determined by CD and two-dimensional NMR, compared with native GH. Also, *in vitro* receptor binding showed only a minor reduction in potency of the oxidized forms when compared with native hGH. The authors comment that the lack of reactivity of Met-170 to oxidation supports the hypothesis that this group is buried within the protein, and is perhaps inaccessible to the action of peroxide.

Becker *et al.* (1988) employed 3% H_2O_2 at pH 7.5, at room temperature for 4 h to produce the methionine sulfoxide derivative of hGH. In this case, the reaction was quenched by adding free methionine, and the oxidized protein was purified in several steps including RP-HPLC. Identification of the oxidation product was made by fast atom bombardment mass spectrometry (FAB-MS) analysis of the tryptic peptide T-2. In contrast to the study of Teh *et al.* (1987), Becker *et al.* isolated the Met-14 sulfoxide derivative, rather than the Met-125 derivative, which was reported by Teh *et al.* to form at a faster rate. Becker *et al.* showed that the Met-14 sulfoxide derivative of hGH was fully active in the rat weight gain bioassay.

The oxidation of hGH in the solid state upon exposure to intense fluorescent light has been reported (Pearlman *et al.*, 1987). In this study, samples of hGH were lyophilized in a glycine/sodium phosphate buffer at pH 7.4. Vials of hGH (5 mg per vial) were placed in a light cabinet (1600 ft-candles) at 30°C, either wrapped in aluminum foil or uncovered. After 2 weeks the samples were removed and analyzed by RP-HPLC, using a C_4 column with an acetonitrile/TFA gradient as described in the USP Pharmacopeial Forum (1990, p. 1248). Chromatograms of the foil-wrapped control, and the light-exposed samples, are shown in Fig. 11A and B, respectively. Oxidized hGH formed in the sample exposed to light (Fig. 11B) and appears in the chromatogram as a peak with a retention time of 6 min, compared with the main hGH peak with a retention time of ca. 12 min. Identity of the oxidized peak was made by performing a tryptic digest followed by FAB-MS.

This observation was verified by oxidation of the tryptic peptide T-11 from native hGH with hydrogen peroxide. The product was run under the same HPLC conditions as for the tryptic map, and shown to be composed of two peaks, one of which eluted with the same retention time as the oxidized peak in the tryptic map. Further analysis of the two peaks showed that they consisted of different diastereomers of methionine sulfoxide, resulting from a new chiral center formed by oxidation of the methionine to the sulfoxide.

Becker *et al.* (1988) have observed oxidation of hGH as a degradation reaction occurring in solutions of hGH, and Pikal *et al.* (1991) recently reported the formation of oxidized hGH occurring upon storage of hGH in the freeze-dried state.

3.3. Reduction/Interchange of Disulfide Bonds

It is now generally accepted that the conformation of a protein is largely a result of noncovalent forces arising from complex interactions between the primary structure with itself and with the solvent. This raises an interesting question as to the specific role that disulfide bonds might play in either

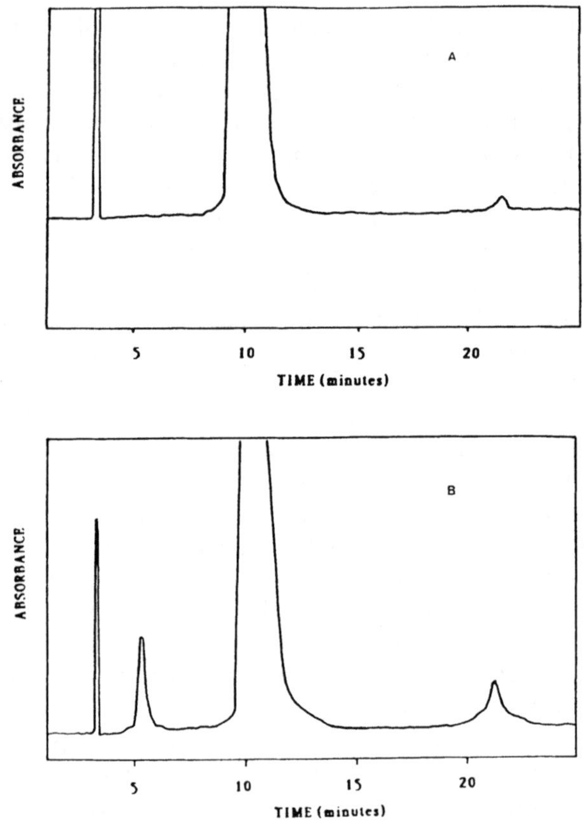

Figure 11. (A) Reversed-phase HPLC of a sample of freeze-dried hGH, covered in foil, after storage in a light cabinet at 1600 ft-candle intensity, and at 30°C for 14 days. The peak eluting at about 21 min corresponds to a dimeric form of hGH. (B) Reversed-phase HPLC of a sample of freeze-dried hGH, after storage in a light cabinet at 1600 ft-candle intensity, and at 30°C for 14 days. More dimer formed in this sample, and an earlier eluting peak (retention time of approximately 6 min) has also formed. This early eluting peak was isolated and shown to contain Met-125 sulfoxide hGH.

achieving or stabilizing these conformations by strategic placement of a few nonpeptidic links. In addition to serving as structural restraints, it may be that some disulfide bonds are intrinsically involved in the active site(s) of certain proteins, or serve some other, unknown function during the protein's "life cycle."

As shown in Fig. 1, the two disulfide bonds in hGH form a large, cross-linked loop comprising much of the middle third of the primary structure (Cys-53 to Cys-165), and a second, small cross-linked loop very near the COOH-terminus (Cys-182 to Cys-189). Both disulfides may be quantitatively reduced with dithiothreitol under mild conditions (0.1 M Tris-Cl, pH 8.2), and in the absence of any denaturants (Bewley *et al.*, 1969; Bewley and Li, 1970). In these studies, the gentle conditions, and in particular the lack of denaturants, are crucial for obtaining an unambiguous characterization of the reduced form of the protein, as well as to clearly understand the role of the disulfides in forming/stabilizing the native conformation. CD spectra indicate that the fully reduced protein does not undergo significant secondary or tertiary structural changes, and still remains in a fully folded form (Bewley and Li, 1970). Small changes in the side-chain CD spectrum have been shown to be due to chemical removal of the intrinsic CD of the disulfide bonds, and do not indicate a true conformational change (Bewley, 1977). The reduced form of the protein is extremely difficult to isolate in the absence of reducing agents due to rapid reoxidation, even under acid conditions (pH 3.0) where the protonated thiol groups should be much less reactive. This, and the fact that excellent yields of bioactive, reoxidized monomer can be obtained at surprisingly high protein concentrations (1–2 mg/ml, Li and Bewley, 1975), indicate that the reduced protein is still compactly folded, with the thiol pairs being held in proper proximity to form intramolecular disulfides rather than intermolecular disulfide-linked aggregates. It has also been shown that the COOH-terminal disulfide bond is much more readily reduced than the other, interior disulfide (Bewley, 1977; Brems *et al.*, 1990).

Alkylation of the thiol groups in reduced hGH with iodoacetamide produces a product which remains virtually identical to the native hormone in terms of conformation and biological profile (Bewley *et al.*, 1969; Aubert *et al.*, 1986; Brems *et al.*, 1990). The only significant difference is a slight increase in the rate of digestion by trypsin. In sharp contrast, alkylation with iodoacetic acid produces a derivative that displays no growth-promoting potency, and although its conformation does not appear to be highly altered, it is very rapidly digested by trypsin. The lack of bioactivity is assumed to reflect proteolytic degradation of the carboxylmethyl derivative following injection and systemic transport during *in vivo* bioassay. Apparently, the negative charges contributed by the carboxymethyl groups significantly destabilize the conformation to proteolytic attack, while the neutral carbamidomethyl derivative, though not quite as stable as the native protein, is still able to survive transport through the circulatory system and to function at the cellular level. These studies clearly indicate that while integrity of the disulfide bonds in hGH is not required to either produce the conformation of the native protein, or elicit biological activity, nevertheless, these structures do serve to provide additional physical restraints that stabilize the molecule.

Although the reoxidation of reduced hGH to produce good yields (>95%) of monomeric protein is highly efficient under laboratory conditions, the presence of up to 15% by weight of a disulfide-linked pit-hGH dimer has been reported (Lewis *et al.,* 1977). This covalent dimer shows greatly diminished immunologic and growth-promoting activities, but can be easily reduced and reoxidized to monomeric, native hGH. The authors suggest it may be formed in the pituitary either by disulfide interchange or by interchain bonding of a reduced form. It is of particular interest that the authors were able to prove that the half-cystine pairing pattern in the dimer was identical to the pairing in the native, monomeric form (i.e., both Cys-182 linked to a Cys-189 in the other monomer unit).

More recently, Lorenson and Jacobs (1987) have used extremely gentle procedures to isolate not only disulfide-linked dimers, but also high-molecular-weight disulfide-linked oligomers of GH from bovine and porcine GH secretory granules. These oligomers contain as many as 50 monomer units, and are readily converted to monomeric GH by treatment with reducing agents. The authors suggest that all GHs, including human, are stored within secretory granules of the pituitary somatotrophs, as biologically inactive, high-molecular-weight, disulfide-linked oligomers (presumably through the more easily reduced disulfide bond near the COOH-terminus). Bioactive GH is then released from these oligomers primarily as monomeric units by reduction with endogenous glutathione or other reducing agent. Autoxidation of the reduced monomers then proceeds quickly during or after secretion.

This is an interesting hypothesis and might be expanded on to explain why the homologous COOH-terminal disulfide loops in all species of GH studied to date, as well as those in the closely related prolactins and placental lactogens, are so highly conserved (Bewley and Li, 1975; Nicoll *et al.,* 1986). We have already mentioned that the disulfide bonds are not required for either formation of the native conformation or elicitation of biological activity. We might then wonder why nature has conserved this structure so assiduously? Could it be that, in addition to the increased stability toward proteolysis which these bonds obviously impart, the COOH-terminal disulfide is also centrally involved in the mechanism of storage and secretion of the hormone? That is to say, this particular structure has evolved to regulate another aspect of the molecule's overall life cycle that is entirely distinct from its interaction with a cellular receptor.

Table I shows partial sequences of GHs including the COOH-terminal disulfide bond (Nicoll *et al.,* 1986). A strongly conserved feature of this disulfide loop is the density of positively charged amino acids, i.e., Arg and Lys, which appear adjacent to Cys-182. In most GH species there are two, or even three such positive charges, and never a negative charge. The other three Cys residues (Cys-53, Cys-165, and Cys-189) in the GHs do not show any conservation of adjacent charges. This positively charged environment

Table I
COOH-Terminal Sequences of Nine Vertebrate Species of pit-GH[a]

			182								189		
Human	Val	Gln	Cys	**Arg**	—	Ser	Val	**Glu**	Gly	Ser	Cys	Gly	Phe-COOH
Bovine	Met	**Lys**	Cys	**Arg**	**Arg**	Phe	Gly	**Glu**	Ala	Ser	Cys	Ala	Phe-COOH
Ovine	Met	**Lys**	Cys	**Arg**	**Arg**	Phe	Gly	**Glu**	Ala	Ser	Cys	Ala	Phe-COOH
Equine	Met	**Lys**	Cys	**Arg**	**Arg**	Phe	Val	**Glu**	Ser	Ser	Cys	Ala	Phe-COOH
Porcine	Met	**Lys**	Cys	**Arg**	**Arg**	Phe	Val	**Glu**	Ser	Ser	Cys	Ala	Phe-COOH
Mouse	Met	**Lys**	Cys	**Arg**	**Arg**	Phe	Val	**Glu**	Ser	Ser	Cys	Ala	Phe-COOH
Rat	Met	**Lys**	Cys	**Arg**	**Arg**	Phe	Ala	**Glu**	Ser	Ser	Cys	Ala	Phe-COOH
Whale	Met	**Lys**	Cys	**Arg**	**Arg**	Phe	Val	**Glu**	Ser	Ser	Cys	Ala	Phe-COOH
Chicken	Met	**Lys**	Cys	**Arg**	**Arg**	Phe	Gly	**Glu**	Ser	Asn	Cys	Thr	Ile-COOH

[a] Modified from Nicoll *et al.* (1986). Cysteine residues are numbered as in pit-hGH. Charged residues are shown in boldface.

should lower the pK_a of the Cys-182 thiol, relative to the other thiol groups, promoting disulfide exchange at this site. This follows from the fact that virtually all thiol/disulfide interchange proceeds via a thiolate anion (Cecil, 1963), and Cys-182 would be expected to be the most reactive of the thiol groups in the GHs.

This increased reactivity, and the extremely high hGH concentrations estimated within the secretory granules (2.5–5 mM, equivalent to 50–100 mg/ml; Cunningham *et al.*, 1991) might well favor intermolecular exchange, and explain the presence of disulfide-linked dimers and oligomers isolated from the pituitary under exceptionally mild procedures. Presumably partial reduction of the dimeric or oligomeric storage form, catalyzed by endogenous reducing agent, and combined with rapid dilution of the reduced form, favors conversion to oxidized monomers in pituitary extracts produced by classical procedures. Although somewhat speculative, this proposed picture may add some additional understanding as to how the last 12 residues of the GH sequence might function, and why this particular design is so highly conserved.

3.4. Aggregation

Aggregation is a significant problem in the pharmaceutical development of proteins for several reasons. Small amounts of insoluble aggregates may become readily visible, affecting the pharmaceutical elegance of the product. With a parenteral product, this can render the substance unsuitable for use. A further, and far more serious, consequence is the implication of aggregated GH in promoting the presence of an antibody response to the protein (Moore, 1978; Rougeot *et al.*, 1991).

Aggregation of proteins has been studied by a number of groups, and many references have appeared (Thurow and Geisen, 1984; Moore, 1978; Manning *et al.*, 1989). Aggregation may be induced by several stress conditions, including heating (Brems *et al.*, 1990), freezing (Schwartz and Batt, 1973; Eckhardt *et al.*, 1991), or agitation (Hagenlocher and Pearlman, 1989; Pikal *et al.*, 1991). Aggregation of hGH may be either covalent, i.e., aggregates formed via covalent bonds such as disulfide bonds, ester or amide linkages, etc., or noncovalent, occurring via hydrophobic interaction or charge–charge complexation. In addition, aggregates may be either soluble (such as dimers, trimers, etc.) or insoluble. Insoluble forms are usually macroscopic, forming turbid solutions (Garcia-Rubino, 1989; Eckhardt *et al.*, 1991).

Quantitation of aggregation is often difficult, and compounded by the multiple forms in which aggregates may be found. The presence of insoluble aggregates poses a special problem in quantitation since even small amounts of such aggregates may make solutions opalescent, and the amount of light scattering is not (necessarily) in linear proportion to the mass of aggregate present. The amount of insoluble aggregate in a sample may be determined using a combination of UV spectroscopy, light scattering analysis, and centrifugation as described by Eckhardt *et al.* (1991). Native size-exclusion chromatography has been used successfully to quantitate total dimer and higher-order soluble aggregates occurring in hGH (Becker *et al.*, 1987), and RP-HPLC has been employed to measure covalent dimer (Pearlman and Nguyen, 1992). Noncovalent soluble aggregates such as dimers and higher-ordered species may also be measured by nondenaturing methods that avoid detergents or organic solvents. Typical sizing methods include native PAGE, size-exclusion chromatography, or analytical ultracentrifugation.

Becker *et al.* (1987) describe the characterization of a noncovalent dimer found during the purification of both pituitary and biosynthetic hGH. This dimer readily converts to monomer when run on an SDS gel, or in the RP-HPLC assay described by Riggin *et al.* (1987). Using a monoclonal antibody immunoassay, the authors could distinguish the dimeric from the monomeric hGH, implying that some epitopes were buried in the dimeric form. Also, the authors found the dimer to have about one-third the potency of monomeric hGH in the rat weight gain bioassay.

Eckhardt *et al.* (1991) studied the effects of freezing hGH solutions on the formation of both covalent and noncovalent aggregates. Using light scattering as well as SEC-HPLC, the authors studied the effects of a number of parameters on the aggregation profile of hGH. They concluded that the formation of soluble (noncovalent) aggregates was largely unaffected by the rate of freezing, whereas greater particulate formation (insoluble aggregates) occurred when solutions were frozen more rapidly. The authors also found that a solution of hGH at pH 7.8 produced significantly less aggregate upon freezing than a solution at pH 7.4.

Schwartz and Batt (1973) reported that ^{125}I-labeled hGH was susceptible to aggregation upon freezing and thawing solutions of the protein. Aggregation was measured using a Sephadex G-100 (Pharmacia) gel filtration system. Rapid freezing was unable to prevent formation of the aggregate, and this aggregate could not be dissociated in 5 M guanidine. Also, the aggregated protein had diminished binding affinity for hGH antibody, implying that epitopes toward the antibody may be buried in the aggregate. Interestingly, Livesey *et al.* (1980) found that hGH samples in plasma were able to withstand five freeze–thaw cycles without loss in immunoreactivity. This study has important implications since many plasma samples require frozen storage prior to analysis.

Also, hGH was seen to form dimers when stored as a lyophilized powder, in the presence of strong fluorescent light (Pearlman *et al.*, 1987). In Fig. 11, another peak appeared at approximately 21 min. retention time in addition to the early eluting oxidized hGH. This peak was found to be dimeric hGH which appears to a higher extent in the light-exposed sample. The dimer eluting at 21 min was heterogeneous, and it was made up of both reducible and nonreducible species. In order to assess any differences in dimer composition between the light-exposed and heated-treated samples, analysis of the dimer content of the samples by SEC under native conditions and (denaturing) RP-HPLC was performed (Fig. 12). It can be seen that the total amount of dimer formed in lyophilized hGH, upon exposure to light, is greater than in samples exposed to 30°C alone. Also, the majority of dimer formed upon exposure to light is nondissociable under RP-HPLC conditions, and is most likely covalent in nature. Contrasted to this is the observation that the dimer formed in the samples kept protected from light is mostly dissociable, and hence noncovalent in nature.

Agitation has been shown to readily induce aggregation in hGH solutions (Hagenlocher and Pearlman, 1989). In this study, the authors exposed hGH solutions in test tubes to a defined agitation protocol, and measured the amount of visible aggregate formed by light scattering. It was postulated that increased particulates formed in solution due to surface denaturation of the protein, followed by aggregation of the unfolded molecules. A number of excipients were screened for their ability to suppress aggregation, including cyclodextrins and cellulose derivatives, as shown in Fig. 13. While a number of these agents are effective, note the large concentration (20%) of cyclodex-

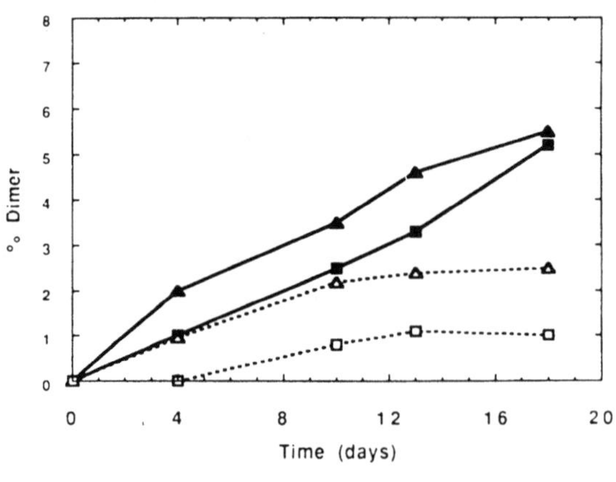

Figure 12. A plot of the percent dimer formed on storage over time, for freeze-dried hGH samples stored in the presence (solid lines) or absence (dashed lines) of light. The light source produced approximately 1600 ft-candles, and the cabinet maintains a temperature of 30°C. The amount of dimer was measured both by size-exclusion HPLC (a nondissociating method) providing the total amount of dimer present (triangles), and by reversed-phase HPLC (a dissociating method) to measure the amount of covalent dimer present (squares).

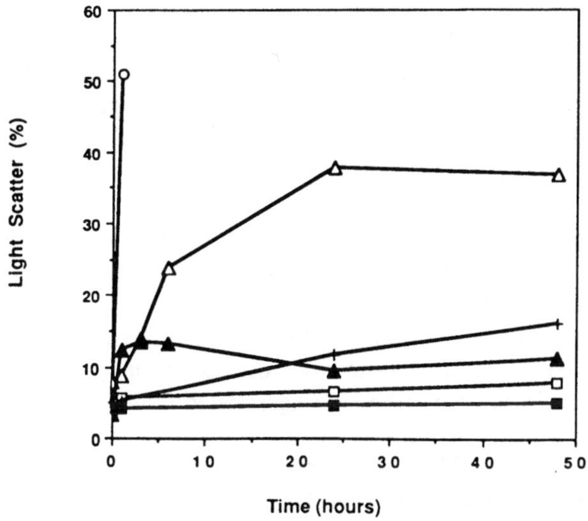

Figure 13. A plot of the amount of aggregation occurring in hGH solutions (determined as the % absorbance at 278 nm due to scattered light) containing different modified celluloses, and shaken for varying times. The control (○) contains hGH at 1.0 mg/ml, dissolved in 5 mM phosphate buffer at pH 7.4; all other solutions contain the control buffer plus one of the following modified celluloses: ■, 0.5% hydroxypropylmethylcellulose (HPMC); □, 2% HPMC; △, 0.1% methylcellulose 400 (MC); ▲, 0.5% MC; +, 20% hydroxypropyl-β-cyclodextrin.

trin required to reduce aggregation, relative to the amount of methylcellulose or hydroxypropylmethylcellulose that produced the same result. This protective effect against aggregation was postulated to be due to a specific interaction between the protein and the agent, rather than due to an increase in solution viscosity. Surfactants, particularly the nonionic type such as polysorbates and poloxamers, have also been shown to be quite effective in preventing surface-induced aggregation of hGH (Pearlman and Oeswein, 1989) as well as other proteins such as insulin (Thurow and Geisen, 1984).

It has also been reported that small changes in buffer concentrations may alter the amount of aggregate formation of hGH in solution (Pearlman and Nguyen, 1992). In this study, formulations of hGH at 2 mg/ml were prepared at the same pH, but with different concentrations of phosphate buffer. The samples were shaken for various times, and the amount of absorbance due to light scattering was determined by a UV scan. Interestingly, an optimum concentration of sodium phosphate buffer of about 5 mM, produced less aggregate compared with buffers at concentrations of 2.5, 10, or 20 mM. Such interactions are quite complex, and empirical observation and screening are the only ways to ascertain such information.

The clinical significance of aggregated hGH is discussed in several papers. Many of the early preparations of pit-hGH were found to contain, among other variants, significant amounts of aggregated material (Moore, 1978; Moore and Leppert, 1980). These papers describe an increased incidence of antibodies occurring in patients treated with pit-hGH which had higher levels of aggregate present. However, it is clear from the literature that these early, and highly heterogeneous preparations also contained a number of other variants of hGH in addition to aggregates, and caution should be used in ascribing the antibodies formed solely to the presence of aggregated hGH (Skyler *et al.,* 1977; Lewis *et al.,* 1980).

Holmstrom and Fholenhag (1975) have characterized a number of preparations of hGH used in the clinic, and concluded that preparations purified from frozen pituitaries contained less aggregated material than those derived from pituitaries stored in acetone. Rougeot *et al.* (1991) also speculated on the role of aggregated hGH in the formation of antibodies in children administered recombinant forms of hGH. Using a sensitive assay directed against hGH antibodies, they measured the presence of anti-hGH in children given the hormone over a 2-year period. The authors observed antibody levels occurring in three of the six preparations, with an incidence of between 7 and 22%. The authors concluded that the higher antibody levels were due to increased levels of aggregated hGH in certain preparations, even though the antibody levels were quite low, and all of the products showed extremely high purity.

We should now recognize that the overly simple term "aggregate" has historically been inappropriately used to describe a highly complex (and still largely uninvestigated) mixture of molecular forms, produced by an unknown number of different mechanisms. The individual components of these complex mixtures undoubtedly display differences in biological profile, as well as in physicochemical properties. To expect all "aggregates" to be equivalent in any particular respect would be as unwise as to state that all proteins are equivalent. The truth regarding which of these aggregated forms (if any) may or may not be immunogenic is completely uncertain at present.

3.5. Proteolysis/Hydrolysis

Exhaustive digestion of hGH with proteolytic enzymes such as trypsin will produce at least 18 individual peptides, plus free Lys (see Fig. 7), and completely abolish all biological activity (Li, 1957). In contrast, early reports (Li, 1957, 1962; Li and Samuelsson, 1965) indicated that partial digests, using trypsin, chymotrypsin, or pepsin, still show significant biological activity and suggested the presence of an "active core" in the molecule. Unfortunately, it is extremely difficult to assess the validity of these experiments since it is likely that such unfractionated digests are highly heterogeneous, and may still contain considerable undigested protein.

More recent work conclusively shows that under the proper conditions, limited digestion with serine proteases such as plasmin (Li and Graf, 1974; Reagan *et al.,* 1975), thrombin (Mills *et al.,* 1980; Graf *et al.,* 1981), and trypsin (Graf *et al.,* 1982) will cleave hGH at various sites, all of which lie between residues 133 and 149 of the hGH sequence. Enzymatically modified products which have been purified from these digests show various degrees of proteolytic attack. Some are merely clipped at a single site (thrombin) while others are missing as many as 10 residues (plasmin, trypsin). In all cases, the two large fragments composed of approximately the NH_2-terminal 130 residues, and the COOH-terminal 50 residues, are held together by the disulfide bond linking Cys-53 to Cys-165. All such variants have been shown to retain full growth-promoting activity. In at least one report, substantial increases in activity have been claimed (Lewis *et al.,* 1971), but this interesting result has not been generally reproducible. The two fragments may be separated and purified after reduction and alkylation of the linking disulfide bond. Very low levels ($<10\%$) of biological activity have been reported for the larger of these two pieces, but nothing of clinical importance.

It is most interesting that the purified, reduced–alkylated fragments prepared from these digests can often be recombined noncovalently (a reaction known as complementation) to form a stable complex that displays essentially full biological potency, and a native hGH-like conformation (Li and Bewley, 1975, 1976; Graf *et al.,* 1981, 1982; Reagan *et al.,* 1981). This again demonstrates that disulfide bonds are not required either for the formation of the native conformation, or for eliciting growth-promoting activity. In the case of the plasmin-modified form, the purified NH_2-terminal, 134-residue fragment has been shown to retain a globular protein-like structure with a considerable content of α-helix, while the smaller COOH-terminal fragment (141–191) exists by itself as a random coil (Li and Bewley, 1975, 1976). On complementation, the random coil peptide appears to fold spontaneously into an α-helix, while "docking" with the larger piece. This process can now be better understood by reference to the three-dimensional structure of GH (Abdel-Meguid *et al.,* 1987) shown in Fig. 2.

In the native protein, the first 134 residues from the NH_2-terminus contain the first three helical sections and the complex loop structure at the bottom of the molecule. It is undoubtedly important that the helical content of this portion is fully retained in the isolated fragment, as is some of its tertiary structure (Li and Bewley, 1975, 1976). Again, referring to the intact molecule, the COOH-terminal 51 residues appear as a small bit of loop structure, and all of helix 4. Apparently, residues 141–191 can form helix 4 only in the presence of the larger fragment, and it is the folding and docking of helix 4 with the other helices, and the loop structure, that forms the stable noncovalent complex and returns full biological potency.

In addition to the plasmin-modified variant, another variant of hGH exhibiting an unusual proteolytic "clip" between Phe-139 and Lys-140 has been isolated from pituitary extracts by Singh *et al.* (1974). Another unusual clip has been demonstrated to occur between Thr-142 and Tyr-143 in recombinant hGH isolated from *E. coli* fermentations (Canova-Davis *et al.*, 1990). Both modified forms show full biological activity, and again, the clips occur within the same sequence segment as produced by limited *in vitro* digests.

Once again, it is the three-dimensional structure of GH which helps us understand why all of these sites of initial proteolytic attack might occur in such a localized region of the protein sequence. The long, ribbonlike structure spanning residues 130 to 153, and seen lying across the middle of the molecule in Fig. 2, is described (Abdel-Meguid *et al.*, 1987) as one of the most extended, and flexible segments of the protein, and appears to be highly exposed to the solvent. These are precisely the local conformational requirements (in addition to side-chain specificity requirements) which dictate relative proteolytic susceptibility (Markus, 1965), and most certainly explain why all of these initial clips occur within a 16-residue segment of the peptide backbone. The fact that 6 residues can be removed from this segment without producing a significant conformational change (Li and Bewley, 1975, 1976) is consistent with a relatively weak set of secondary interactions between this segment and the rest of the molecule, in complete agreement with the x-ray data.

3.6. Shear

Shear has often been implicated in the aggregation of proteins, for example when protein solutions are shaken. Since hGH is especially vulnerable to agitation, forming noticeable amounts of precipitate, the potential for damage due to shear was naturally considered (Hagenlocher and Pearlman, 1989). The potential for shear damage during pharmaceutical processing has been addressed by Hsu *et al.* (1988). The authors demonstrated that the energies due to shear forces which might be encountered during processing, for example during sterile filtration, were roughly an order of magnitude less than those involved in hydrogen bonding, and therefore much lower than that necessary to disrupt the tertiary structure of proteins. Thus, the authors concluded that shear was not responsible for the aggregation (sometimes referred to as protein shedding) seen during processing of hGH solutions. It is more likely that this phenomenon is related to protein–surface interactions, similar to those described by Thurow and Geisen (1984) for insulin. Pikal *et al.* (1991) also concluded that shear was not responsible for hGH aggregation, by studying hGH solutions pumped through capillary tubes at shear rates approximating those encountered during sterile filtration. The authors found no difference in the amount of soluble or insoluble aggregates occurring after passage through the tubing, as compared with control solutions. They also ascribed the aggregation seen in hGH solutions to surface-induced denaturation.

4. STABILITY PROFILE OF hGH

From the preceding sections it is apparent that hGH is susceptible to a variety of degradation processes. Many of these have been well defined by the excellent methods available for the separation, isolation, and characterization of individual degradation products. While there is a wealth of information on the biochemical aspects of hGH degradation, very little has appeared in the literature from a pharmaceutical perspective.

In order to preserve the chemical and biological activities of the protein, commercial preparations of hGH are supplied as freeze-dried formulations, stored at 2–8°C. Thus, any description of the stability profile of hGH needs to address both the stability over long-term storage in the freeze-dried form (i.e., the shelf life) as well as the stability in solution after reconstitution (the utilization time). Generally, stability in the freeze-dried form results in a shelf life on the order of 2 years at 2–8°C. Reconstitution is performed with sterile water for injection, which must contain a preservative if multiple injections are to be made from the vial. Typical preservatives include benzyl alcohol and *m*-cresol. Once reconstituted, the solution is stable for about 2 weeks at 2–8°C. Care should be exercised to avoid shaking the reconstituted solution, since the protein is susceptible to aggregation. Also, reconstituted vials may not be refrozen, since the preservative will adversely interact with the protein during freezing, resulting in severe aggregation.

4.1. Solution Stability

Deamidation is the most prevalent degradation reaction hGH undergoes in neutral or alkaline solution. Lewis *et al.* (1970) studied hGH deamidation in solution, using IEF electrophoresis in tube gels. Quantitation was achieved by cutting out the fixed and Coomassie-stained bands, dissolving them in base, and measuring their absorbance. The authors found the pseudo-first-order rate constant for deamidation of hGH at 37°C to be 5 $\times 10^{-7}$ sec^{-1} at pH 7.5 to 8. Verification that the reaction rates measured by IEF were indeed due to deamidation was obtained by also measuring the rate of ammonia production during the reaction. Increases in temperature, pH, or ionic strength all increased the rate of deamidation, and the presence of urea not only increased the rate, but also resulted in a larger number of degradation products. The authors also concluded that the measured deamidation was nonenzymatic.

Similar studies have been performed using IEF slab gels, and quantitation of the deamidated bands was achieved using a scanning densitometer (Pearlman and Nguyen, 1989). In this study, solutions of hGH at 2 mg/ml were prepared over the pH range 6 to 8, and stored at 25 and 40°C. At appropriate times, a sample was removed and run on a narrow-range (pH 4.5–6.0) IEF gel. Figure 14 shows the result of this limited deamidation rate study. At 40°C and pH 7.4, 7.6, and 8.0, the pseudo-first-order rate constants for deamidation were observed to be 6.4×10^{-2}, 6.8×10^{-2}, and 7.2 $\times 10^{-2}$ $days^{-1}$, respectively. These rate constants translate to 7.4×10^{-7}, 7.9 $\times 10^{-7}$, and 8.3×10^{-7} sec^{-1}, in excellent agreement with the data of Lewis *et al.* (1970) quoted above.

Oxidation of hGH has been observed to occur in solution by the action of hydrogen peroxide as described in Section 3.2. To date, no studies have appeared in the literature describing the oxidation of hGH occurring in solution upon storage, i.e., without oxidants. Since the fermentation process for the production of recombinant proteins is performed under oxidizing conditions, it is not surprising that oxidized hGH is formed during processing.

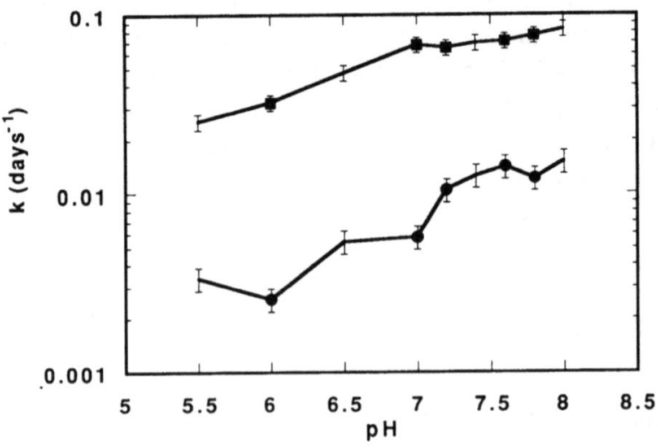

Figure 14. Plot of the first-order rate constants in $days^{-1}$ for deamidation of hGH in solution as a function of pH at 25°C (●) and 40°C (■). The rate constants were obtained from the slopes of the disappearance of hGH versus time, when solutions of hGH were stored at various pH values. Measurements of the amount of hGH and deamidated hGH were made by scanning IEF gels.

Becker *et al.* (1988) describe the potential for oxidation during isolation and purification of the protein.

Aggregation of hGH in solution has been discussed above, and it is a reaction with serious pharmaceutical consequences. Shaking of hGH solutions results in the progressive formation of both soluble and insoluble aggregates. A number of excipients, such as surfactants, methylcellulose, and cyclodextrins, have been shown to reduce or prevent this phenomenon (Hagenlocher and Pearlman, 1989). Freezing of hGH solutions is another means of forming aggregates, as described by Eckhardt *et al.* (1991). Particular care must be taken to avoid freezing reconstituted solutions of hGH which contain a preservative, such as benzyl alcohol, *m*-cresol, or phenol. Freezing in the presence of these agents appears to cause significant aggregation, including precipitation of the protein.

4.2. Stability in the Solid State

Very little has been reported regarding the stability of hGH formulations in the solid state, or on the rationale for solid-state formulation design. A patent application (Pearlman and Oeswein, 1989) and a recent paper (Pikal *et al.*, 1991) describe the optimization of freeze-dried formulations of hGH. In the paper by Pikal *et al.*, the authors examine the effects of excipients, pH, temperature, formulation design, and residual moisture on the characterization and stability of lyophilized formulations of hGH (Table II). The loss in hGH main-peak, as detected by isocratic RP-HPLC, was found to follow first-order kinetics over the degradation range followed. They describe the major degradation reactions occurring upon storage in the freeze-dried state as being deamidation, oxidation, and aggregation. Oxidation was exacerbated when the vial headspace was purged with 90% oxygen, indicat-

Table II
Freeze-Dried Formulations of hGH Used to Study the Effect of Excipients on the Stability of hGH as Described by Pikal *et al.* (1991)[a]

Excipient code	Excipient weight ratio hGH : mannitol : glycine : dextran	Crystallinity (x-ray diffraction)
A	1 : 0 : 0 : 0	Amorphous
B	1 : 1 : 0 : 0	Glycine crystals
C	1 : 0 : 5 : 0	Mannitol crystals
D	1 : 1 : 5 : 0	Mannitol crystals
Dextran 40	1 : 0 : 0 : 6	Amorphous

[a] Ratios of protein to excipient were calculated on a weight basis. Samples also contained 0.227 mg (1.69 mmole) $Na_2HPO_4 \cdot 7H_2O$ at pH 7.4.

ing that molecular oxygen can play a role in solid-state methionine oxidation. The amount of soluble aggregates measured by SEC after reconstitution was observed on storage of the freeze-dried samples at 40°C over 4 weeks. The formulation components were found to affect the amount of aggregate formed, with the least amount occurring in a formulation composed of a mixture of glycine and mannitol in a ratio of 1 : 1 (w/w). The enhanced stability of hGH in the glycine/mannitol formulation was believed to be due to a lyoprotectant effect of the glycine in the amorphous protein phase, as distinct from the crystalline mannitol phase. Significant aggregation (6%) occurred in the hGH sample which was freeze-dried in the absence of any excipient. However, protection from aggregation in the excipient-containing dextran 40 showed four times as much aggregate as the one with glycine/mannitol.

Similar findings have been described for the stabilization of hGH formulations on freeze-drying by Pearlman and Oeswein (1989). In this study, the stabilizing effect of glycine on the aggregation of hGH in the freeze-dried state was evaluated. Samples of hGH were freeze-dried from several buffers containing sodium phosphate (5 mM, pH 7.4) and the mannitol : glycine excipient ratios shown in Table III. Vials of each formulation were freeze-dried and stored at 40°C for 4 weeks, after which samples were reconstituted and the amount of dimer measured by nondissociating SEC. In this study, the authors found a molar ratio of glycine : hGH (100 : 1) which uniquely produced minimum aggregate formation relative to either the 10 : 1 or 1000 : 1 ratios (Fig. 15). Also, the 100 : 1 glycine : hGH system was superior to either glycine alone or mannitol alone (Fig. 16). Thus, it can be seen that relatively minor changes in excipient concentrations may cause measurable changes in the stability profile of hGH, further supporting the need for carefully designed formulation studies.

Table III

Freeze-Dried Formulations Used to Optimize the Excipient System for hGH as Described by Pearlman and Oeswein (1989)[a]

Excipient	Excipient molar ratio hGH : glycine : mannitol	Excipient weight ratio hGH : glycine : mannitol
Mannitol	1 : 0 : 1100	1 : 0 : 8
Glycine	1 : 5544 : 0	1 : 18.9 : 0
10 Gly	1 : 10 : 1100	1 : 0.034 : 8
100 Gly	1 : 100 : 1100	1 : 0.341 : 8
1000 Gly	1 : 1000 : 1100	1 : 3.41 : 8

[a] Ratios of hGH to excipient were calculated on a molar basis, and are also listed on a weight basis. All samples were buffered with 5 mM sodium phosphate buffer at pH 7.4.

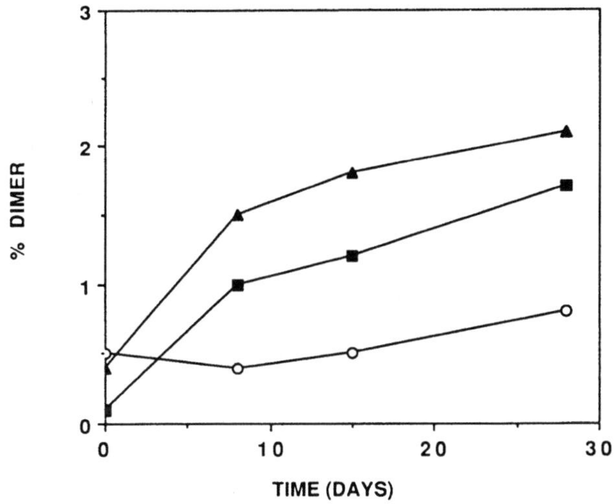

Figure 15. Plot of the percent dimer, as measured by a size-exclusion HPLC assay, for freeze-dried samples of hGH, as a function of storage time at 40°C. In these formulations, a constant amount of mannitol was used, and the ratio of glycine to hGH was varied, as shown in Table III. Molar ratios of glycine : hGH— ▲, 10 : 1; ■, 1000 : 1; ○, 100 : 1.

5. FORMULATIONS

Some of the following information has been obtained from published prescribing information for commercially available hGH preparations in the United States (Physicians Desk Reference, 1992). Recombinant derived hGH is available in the United States as Somatrem (Protropin®; Genentech, Inc.) and as Somatropin (Humatrope®; Eli Lilly & Co.). Protropin is available in 5- and 10-mg vial configurations, and Humatrope is sold in a 5-mg vial. Patients may reconstitute the vials with varying amounts of special diluent provided with each vial. In the case of Protropin, this diluent is bacteriostatic water for injection (BWFI), benzyl alcohol preserved. Dilution with other types of BWFI, e.g., parabens preserved, results in severe precipitation of the protein, and thus patients must exercise care in reconstitution and handling of the reconstituted vials. The diluent provided with Humatrope is a special diluent containing *m*-cresol and glycerol. Typical dilution concentrations range from 1 mg/ml to 5 mg/ml, and the dilution is determined by the physician in order to permit accurate dosing of the smallest injection volume.

Outside the United States, several other companies market Somatropin, including Biotechnology General (BioTropin®, Eskatrope®), Kabi (Genotro-

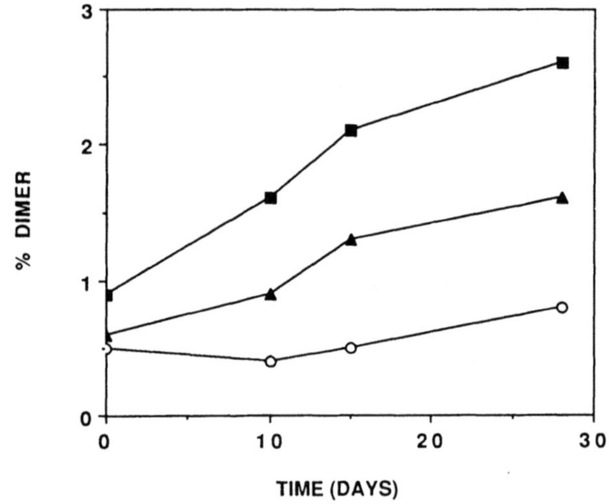

Figure 16. Plot of the percent dimer, as measured by a size-exclusion HPLC assay, for freeze-dried samples of hGH, as a function of storage time at 40°C. Squares represent a mannitol formulation, triangles a glycine formulation, and circles a mixed mannitol/glycine formulation, where the molar ratio of glycine : hGH is 100 : 1, as described in Table III.

pin®), NovoNordisk (Norditropin®), Serono (Saizen®), and Sumitomo (Genotropin, licensed from Kabi). The quantity of hGH listed in these vials is based upon IU rather than weight, and various dosage amounts are available ranging from 4 IU to 16 IU. All forms are presented as freeze-dried products for reconstitution. A summary of the formulations, derived from the package inserts of these products, is presented in Table IV. Although in most instances injections are made with a tuberculin-type syringe, in Europe, "pen delivery systems" have been used for the administration of hGH. One form of this device is provided with a dual-chamber vial, containing both lyophilized hGH and diluent (KabiPen®; Kabi). Reconstitution of the protein is effected by movement of a barrier between the powder and diluent, thus avoiding external transfer of the diluent by the patient.

6. CONCLUSIONS

This chapter has described the history, background, and clinical uses of hGH. We have also presented a survey of the analytical methods used for characterization of the protein, as well as a description of the degradation products found under a variety of conditions. A discussion of the stability

Table IV
Commercial hGH Preparations: Listed Contents per Vial[a]

Protropin (somatrem for injection), Genentech, Inc.	
Somatrem	5.0 mg (~13 IU)
Mannitol	40.0 mg
Sodium phosphates (as monobasic sodium phosphate 0.1 mg and dibasic sodium phosphate 1.6 mg)	1.7 mg
Phosphoric acid may be used for pH adjustment. The diluent is BWFI, containing 0.9% benzyl alcohol. Upon reconstitution, the pH is ~7.8. The utilization time upon reconstitution is 14 days when stored at 2–8°C.	
Humatrope (somatropin for injection), Eli Lilly & Co.	
Somatropin	5.0 mg (~13 IU)
Mannitol	25.0 mg
Glycine	5.0 mg
Dibasic sodium phosphate	1.13 mg
Phosphoric acid and/or sodium hydroxide may have been added at the time of manufacture. The diluent contains 0.3% *m*-cresol as a preservative and 1.7% glycerin to render the solution isotonic. Upon reconstitution, the pH is ~7.5. The utilization time upon reconstitution is 14 days when stored at 2–8°C.	
Saizen (somatropin), Serono S.A.	
Somatropin	4.0 IU
D-Mannitol	20.0 mg
Disodium phosphate	2.2 mg
Monosodium phosphate	0.35 mg
Sodium chloride	1.0 mg
The diluent is isotonic saline, which is not preserved; thus, the contents should be used immediately.	
Genotropin (somatropin), Kabi	
Somatropin	16 IU
Mannitol	41.0 mg
Glycine	2.0 mg
Dibasic sodium phosphate	0.26 mg
Disodium phosphate	0.28 mg
The diluent contains 0.3% *m*-cresol. The utilization time upon reconstitution is 3 weeks when stored at 2–8°C.	
Norditropin (somatropin), Novo Nordisk	
Somatropin	12.0 IU
Glycine	20–25 mg
Sodium bicarbonate	2.5–3.1 mg
Mannitol	2.0–2.5 mg
The diluent contains 0.9% benzyl alcohol. The utilization time upon reconstitution is 14 days when stored at 2–8°C.	

[a] Protropin and Humatrope are available in the United States, the other preparations are available in Europe, Japan and the rest of the world.

profile for formulations of hGH both in solution and in the solid state is included along with a listing of some of the commercially available products.

hGH is one of the more extensively studied proteins, and we have been fortunate to have such a wide and varied literature to draw upon. Most references are of a biochemical or biophysical nature, along with a large number of clinical papers. However, more recently, with the advent of commercially available recombinant hGH, a number of papers written from a pharmaceutical perspective have appeared. It should be apparent from this chapter that the key to successful protein formulation is a solid analytical foundation, on which one may confidently base the characterization, stability, and ultimately the formulation of the agent.

ACKNOWLEDGMENTS

The authors gratefully acknowledge the fine technical assistance of their colleagues, James Oeswein, Barbara O'Connor, and Tue Nguyen, and express their thanks to Mike Powell, Armin Ramel, and Jessica Burdman for critical reading of the manuscript.

REFERENCES

Abdel-Meguid, S. S., Shieh, H. S., Smith, W. W., Dayringer, H. E., Violand, B. N., and Bentle, L. A., 1987, Three dimensional structure of a genetically engineered variant of porcine growth hormone, *Proc. Natl. Acad. Sci. USA* **84**:6434.

Aloj, S., and Edelhoch, H., 1972, The molecular properties of human growth hormone, *J. Biol. Chem.* **247**:1146.

Aswad, D., 1984, Stoichiometric methylation of porcine adrenocorticotropin by protein carboxy methyltransferase requires deamidation of asparagine 25, *J. Biol. Chem.* **259**:10714.

Aubert, M. L., Bewley, T. A., Grumbach, M. M., Kaplan, S. L., and Li, C. H., 1986, Structure–function studies on human growth hormone. Evidence that tertiary structure is essential for biological activity, *Int. J. Pept. Protein Res.* **28**:45.

Basuyaux, B., Paolucci, F., Clavies, C., Hervaud, E., Pau, B., and Peyrouset, A., 1987, Production and characterization of monoclonal antibodies to human growth hormone, *Hybridoma* **6**:423.

Beck, J. C., McGarry, E. E., Dryenfurth, I., and Venning, E. H., 1957, Metabolic effects of human and monkey growth hormone in man, *Science* **125**:884.

Becker, G. W., Bowsher, R. R., Mackellar, W. C., Poor, M. L., Tackitt, P. M., and Riggin, R. M., 1987, Chemical, physical, and biological characterization of a dimeric form of biosynthetic human growth hormone, *Biotechnol. Appl. Biochem.* **9**:478.

Becker, G. W., Tackitt, P. M., Bromer, W. W., Lefeber, D. S., and Riggin, R. M., 1988, Isolation and characterization of a sulfoxide and a desamido derivative of biosynthetic human growth hormone, *Biotechnol. Appl. Biochem.* **10**:326.

Bennett, L. L., Weinberger, H., Escamilla, R., Margen, S., Li, C. H., and Evans, H. M., 1950, Failure of hypophyseal growth hormone to produce nitrogen storage in a girl with hypophyseal dwarfism, *J. Clin. Endocrinol.* **10**:492.

Bewley, T. A., 1977, Optical activity of disulfide bonds in proteins. 1. Studies on plasmin modified human somatotropin, *Biochemistry* **16**:209.

Bewley, T. A., 1979, Circular dichroism of pituitary hormones, *Recent Prog. Horm. Res.* **35**:155.

Bewley, T. A., and Li, C. H., 1970, Human pituitary growth hormone. XXII. The reduction and reoxidation of the hormone, *Arch. Biochem. Biophys.* **138**:338.

Bewley, T. A., and Li, C. H., 1972, Circular dichroism studies on human pituitary growth hormone and ovine pituitary lactogenic hormone, *Biochemistry* **11**:884.

Bewley, T. A., and Li, C. H., 1975, The chemistry of human pituitary growth hormone, *Adv. Enzymol.* **42**:73.

Bewley, T. A., and Li, C. H., 1984, Conformational comparison of human pituitary growth hormone and human chorionic somatomammotropin (human placental lactogen) by second-order absorption spectroscopy, *Arch. Biochem. Biophys.* **233**:219.

Bewley, T. A., and Li, C. H., 1986, The conformation of monkey pituitary somatotropin, *Arch. Biochem. Biophys.* **248**:646.

Bewley, T. A., Brovetto-Cruz, J., and Li, C. H., 1969, Human pituitary growth hormone. Physicochemical investigations of the native and reduced-alkylated protein, *Biochemistry* **8**:4701.

Bornstein, P., and Balian, G., 1977, Cleavage at asn-gly bonds with hydroxylamine, *Methods Enzymol.* **47**:132.

Brems, D. N., Brown, P. L., and Becker, G. W., 1990, Equilibrium denaturation of human growth hormone and its cysteine-modified forms, *J. Biol. Chem.* **265**:5504.

Bristow, A. F., and Jeffcoate, S. L., 1992, Analysis of therapeutic growth hormone preparations: Report of an interlaboratory collaborative study on growth hormone assay methodologies, *Biologicals* **20**:221.

Brown, P., Gajdusek, C. C., Gibbs, C. J., and Asher, D. M., 1985, Potential epidemic of Creutzfeldt–Jakob disease from human growth hormone therapy, *N. Engl. J. Med.* **313**:728.

Canova-Davis, E., Chloupek, R. C., Baldonado, I. P., Battersby, J. E., Spellman, M. W., Basa, L. J., O'Connor, B. H., Pearlman, R., Quan, C., Chakel, J. A., Stults, J. T., and Hancock, W. S., 1988, Analysis by FAB-MS and LC of proteins produced by either biosynthetic or chemical techniques, *Am. Biotechnol. Lab.* **6**:8.

Canova-Davis, E., Baldonado, I. P., Moore, J. A., Rudman, C. G., Bennet, W. F., and Hancock, W. S., 1990, Properties of a cleaved two-chain form of recombinant human growth hormone, *Int. J. Pep. Protein Res.* **35**:17.

Cecil, R., 1963, Intramolecular bonds in proteins. I. The role of sulfur in proteins. in: *The Proteins,* 2nd ed., Vol. I (H. Neurath, ed.), Academic Press, New York, pp. 379–476.

Cohen, F. E., and Kuntz, I. D., 1987, Prediction of the three-dimensional structure of human growth hormone, *Proteins* **2**:162.

Cunningham, B. C., and Wells, J. A., 1989, High-resolution epitope mapping of hGH–receptor interactions by alanine-scanning mutagenesis, *Science* **244**:1081.

Cunningham, B. C., Mulkerrin, M. G., and Wells, J. A., 1991, Dimerization of human growth hormone by zinc, *Science* **253**:545.

Daughaday, W. F., 1985, The anterior pituitary, in: *Williams' Textbook of Endocrinology,* 7th ed. (J. Wilson and D. W. Foster, eds.), Saunders, Philadelphia, pp. 568–613.

Eckhardt, B. E., Oeswein, J. Q., and Bewley, T. A., 1991, Effect of freezing on aggregation of human growth hormone, *Pharm. Res.* **8**:1360.

Evans, H. M., and Long, J. A., 1921, The effect of the anterior lobe administered intraperitoneally upon growth, maturity, and oestrous cycles of the rat, *Anat. Rec.* **21**:62.

Evans, W. S., Faria, A. C., Christiansen, E., Ho, K. Y., Weiss, J., Rogol, A. D., Johnson, M. L., Blizzard, R. M. Veldhuis, J. D., and Thorner, M. O., 1987, Impact of intensive venous sampling on characterization of pulsatile GH release, *Am. J. Physiol.* **252**:549.

Garcia-Rubino, L. H., 1989, Characterization of proteins during aggregation using turbidimetry, *Chem. Eng. Commun.* **80**:193.

Garnick, R. L., Ross, M. J., and Dumee, C. P., 1988, Analysis of recombinant biologicals, in: *Encyclopedia of Pharmaceutical Technology,* Vol. 1 (J. Swarbrick and J. C. Boylan, eds.), Dekker, New York, pp. 253–313.

Goeddel, D. V., Heyneker, H. L., Hozumi, T., Arentzen, R., Hakura, K., Yansura, D. G., Ross, M. J., Miozzari, G., Crea, R., and Seeburg, P. M., 1979, Direct expression in Escherichia coli of a DNA sequence coding for human growth hormone, *Nature* **281**:544.

Graf, L., Li, C. H., Cheng, C. H. K., and Jibson, M. D., 1981, Two contiguous thrombin fragments of human somatotropin form a functionally active recombinant, but the two homologous fragments from sheep hormone do not, *Biochemistry* **20**:7251.

Graf, L., Li, C. H., and Jibson, M. D., 1982, Human somatotropin; selective removal with trypsin of residues 135–145 from the hormone molecule with no loss of biological activity, *J. Biol. Chem.* **257**:2365.

Groesbeck, M. D., and Parlow, A. F., 1987, Highly improved precision of the hypophysectomized female rat body weight gain bioassay for growth hormone by increased frequency of injections, avoidance of antibody formation, and other simple modifications, *Endocrinology* **120**:2582.

Hagenlocher, M., and Pearlman, R., 1989, Use of a substituted cyclodextrin for stabilization of human growth hormone, *Pharm. Res. Suppl.* **6**:S30.

Hancock, W. S., (ed.) 1984, *Handbook of HPLC for the Separation of Amino Acids, Peptides and Proteins,* CRC Press, Boca Raton, Fla.

Hancock, W. S., 1986, Significance of purity in the manufacture of recombinant-DNA-derived proteins, *Chromatogr. Forum* **1**:57.

Hancock, W. S., Canova-Davis, E., Chloupek, R. C., Wu, S. L., Baldonado, I. P., Battersby, J. E., Spellman, M. W., Basa, L. J., and Chakel, J. A., 1988, The characterization of degradation products of recombinant human growth hormone, in: *Therapeutic Peptides and Proteins: Assessing the New Technologies* (D. Marshak and D. Liu, eds.), Cold Spring Harbor Laboratory, Cold Spring Harbor, N.Y., pp. 95–118.

Hashida, S., Ishikawa, E., Kato, Y., Imura, H., Mohri, Z., and Murakami, Y., 1987, Human growth hormone (hGH) in urine and its correlation to serum hGH examined by a highly sensitive sandwich enzyme immunoassay, *Clin. Chim. Acta* **162**:229.

Hindmarsh, P. C., and Brook, C. G., 1987, Effect of growth hormone on short normal children, *Br. Med. J. Clin. Res.* **295**:573.

Hintz, R. L., Rosenfeld, R. G., Wilson, D. M., Bennett, A., Fenno, J., McClellan, B., and Swift, R., 1982, Biosynthetic methionyl-human growth hormone is biologically active in adult humans, *Lancet* **1**:1276.

Holladay, L. A., Hammonds, R. G., and Puett, D., 1974, Growth hormone conformation and conformational equilibria, *Biochemistry* **13**:1653.

Holmstrom, B., and Fholenhag, K., 1975, Characterization of human growth hormone preparations for the treatment of pituitary dwarfism: A comparison of concurrently used batches, *J. Clin. Endocrinol. Metab.* **40**:856.

Holzwarth, G., and Doty, P., 1965, The ultraviolet circular dichroism of polypeptides, *J. Am. Chem. Soc.* **178**:350.

Hormone Drugs, 1982, *Proceedings of the FDA–USP Workshop on Drug and Reference Standards for Insulins, Somatropins and Thyroid-Axis Hormones* (J. L. Gueriguian, E. D. Bransome, and A. S. Outschoorn, eds.), United States Pharmacopeial Convention, Rockville, Md.

Houghten, R. A., Glaser, C. B., and Li, C. H., 1977, Human somatotropin reaction with hydrogen peroxide, *Arch. Biochem. Biophys.* **178**:350.

Hsu, C. C., Pearlman, R., and Curley, J. C., 1988, Some factors causing protein denaturation and aggregate formation in pharmaceutical processing, *Pharm. Res. Suppl.* **5**:S34.

Johnson, B. A., Shirokawa, J. M., Hancock, W. S., Spellman, M. W., Basa, L. J., and Aswad, D. W., 1989, Formation of isoaspartate at two distinct sites during in vitro aging of human growth hormone, *J. Biol. Chem.* **264**:14262.

Jones, A. J. S., and O'Connor, J. V., 1982, Chemical characterization of methionyl human growth hormone, in: *Hormone Drugs, Proceedings of the FDA–USP Workshop on Drug and Reference Standards for Insulins, Somatropins and Thyroid-Axis Hormones* (J. L. Gueriguian, E. D. Bransome, and A. S. Outschoorn, eds.), United States Pharmacopeial Convention, Rockville, Md., pp. 335–351.

Jorgensen, J. O. L., 1991, Human growth hormone replacement therapy: Pharmacological and clinical aspects, *Endocr. Rev.* **12**:189.

Jorgensen, J. O. L., Flyvbjerg, A., Lauritzen, T., Orskov, H., and Christiansen, J. S., 1988, Subcutaneous degradation of biosynthetic human growth hormone in growth hormone deficient patients, *Acta Endocrinol. (Copenhagen)* **118**:154.

Jorgensen, K. D., 1987, Comparison of the pharmacological properties of pituitary and biosynthetic human growth hormone. Demonstration of antinatriuretic/antidiuretic and barbital sleep effects of human growth hormone in rats, *Acta Endocrinol. (Copenhagen)* **114**:124.

Kawauchi, H., Bewley, T. A., and Li, C. H., 1976, Alkaline titrations of human somatotropin, human chorio-mammotropin and ovine prolactin by circular dichroism and fluorescence, *Biochim. Biophys. Acta* **446:**525.

Kohr, W. J., Keck, R., and Harkins, R. N., 1982, Characterization of intact and trypsin-digested biosynthetic human growth hormone by high-pressure liquid chromatography, *Anal. Biochem.* **122:**348.

Kossiakoff, A. A., 1988, Tertiary structure is a principal determinant to protein deamidation, *Science* **240:**191.

Langley, K. E., Lai, P. H., Wypych, J., Everrett, R. R., Berg, T. F., Krabill, L. F., Davis, J. M., and Souza, L. M., 1987, Recombinant-DNA-derived bovine growth hormone from *Escherichia coli* 2. Biochemical, biophysical, immunological and biological comparison with pituitary hormone, *Eur. J. Biochem.* **163:**323.

Larhammar, H., Sjoholm, I., Skoog, B., and Wichman, A., 1985, Comparative conformational analysis of human somatotropin and biosynthetic methionyl human somatotropin by spectroscopic titrations, *Int. J. Pharm.* **23:**13.

Lewis, U. J., Cheever, E. V., and Hopkins, W. C., 1970, Kinetic study of the deamidation of growth hormone and prolactin, *Biochim. Biophys. Acta* **214:**498.

Lewis, U. J., Singh, R. N. P., and Seavey, B. K., 1971, Human prolactin: Isolation and some properties, *Biochem. Biophys. Res. Commun.* **44:**1169.

Lewis, U. J., Peterson, S. M., Bonewald, L. F., Seavey, B. K., and VanderLaan, W. P., 1977, An interchain disulfide dimer of human growth hormone, *J. Biol. Chem.* **252:**3697.

Lewis, U. J., Singh, R. N. P., Tutweiler, G. F., Sigel, M. B., VanderLaan, E. F., and VanderLaan, W. P., 1980, Human growth hormone: A complex of proteins, *Recent Prog. Horm. Res.* **36:**477.

Lewis, U. J., Singh, R. N. P., Bonewald, L. F., and Seavey, B. K., 1981, Altered proteolytic cleavage of human growth hormone as a result of deamidation, *J. Biol. Chem.* **256:**11645.

Li, C. H., 1957, Properties of and structural investigations on growth hormones isolated from bovine, monkey and human pituitary glands, *Fed. Proc.* **16:**775.

Li, C. H., 1962, Preliminary investigations on the action of pepsin on human pituitary growth hormone, *J. Gen. Physiol.* **45:**169.

Li, C. H., 1977, Bioassay of pituitary growth hormone, in: *Hormonal Proteins and Peptides* (C. H. Li, ed.), Academic Press, New York, pp. 2–41.

Li, C. H., 1969, Aspects of the comparative chemistry of human pituitary growth hormone and chorionic somatomammotrophin, in: *Growth and Growth Hormone, Proceedings of the Second International Symposium, Milan* (A. Pecile and E. E. Muller, eds.), Excerpta Medica International Congress Series No. 244, pp. 17–23.

Li, C. H., and Bewley, T. A., 1975, Studies on plasmin-modified growth hormone and its fragments, in: *Growth Hormone and Related Peptides,* Proceedings of the IIIrd International Symposium, Milan, Italy, Excerpta Medica, Amsterdam, pp. 14–32.

Li, C. H., and Bewley, T. A., 1976, Human pituitary growth hormone: Restoration of full biological activity by noncovalent interaction of two fragments of the hormone, *Proc. Natl. Acad. Sci. USA* **73:**1476.

Li, C. H., and Evans, H. M., 1944, The isolation of pituitary growth hormone, *Science* **99:**183.

Li, C. H., and Graf, L., 1974, Human pituitary growth hormone: Isolation and properties of two biologically active fragments from plasmin digests, *Proc. Natl. Acad. Sci. USA* **71:**1197.

Li, C. H., and Papkoff, H., 1956, Preparation and properties of growth hormone from human and monkey pituitary glands, *Science* **124:**1038.

Li, C. H., and Samuelsson, G., 1965, Human pituitary growth hormone. XI. Rate of hydrolysis by trypsin, chymotrypsin, and pepsin; effect of trypsin on the biological activity, *Mol. Pharmacol.* **1:**47.

Li, C. H., Liu, W. K., and Dixon, J. S., 1962, Human pituitary growth hormone. XII. Modified procedure for isolation and NH_2-terminal amino acid sequence, *Arch. Biochem. Biophys. Suppl.* **1:**327.

Li, C. H., Chung, D., Lahm, H.-W., and Stein, S., 1986, The primary structure of monkey pituitary somatotropin, *Arch. Biochem. Biophys.* **245:**287.

Li, C. H., Bewley, T. A., Chung, D., and Oosthuizen, M. M., 1987, Elephant growth hormone. Isolation and characterization, *Int. J. Pept. Protein Res.* **29:**62.

Livesey, J. H., Hodgkinson, S. C., Roud, H. R., and Donald, R. A., 1980, Effect of time, temperature and freezing on the stability of immunoreactive LH, FSH, TSH, growth hormone, prolactin and insulin in plasma, *Clin. Biochem.* **13:**151.

Lorenson, M. Y., and Jacobs, L. S., 1987, Osmotic pressure regulation of prolactin and growth hormone release from bovine secretory granules, *Endocrinology* **120:**365.

McFadden, P. N., and Clarke, S., 1982, Methylation at D-aspartyl residues in erythrocytes: Possible step in the repair of aged membrane proteins, *Proc. Natl. Acad. Sci. USA* **79:**2460.

Maddaiah, V. T., Collipp, P. J., Sharma, R. K., Chen, S. Y., and Thomas, J., 1972, Fluorescence studies of human growth hormone, *Biochim. Biophys. Acta* **263:**133.

Manning, M. C., Patel, K., and Borchardt, R. T., 1989, Stability of protein pharmaceuticals, *Pharm. Res.* **6:**903.

Markus, G., 1965, Protein substrate conformation and proteolysis, *Proc. Natl. Acad. Sci. USA* **54:**253.

Mills, J. B., Kostyo, J. L., Reagan, C. R., Wagner, S. A., Moseley, M. H., and Wilhelmi, A. E., 1980, Fragments of human growth hormone produced by digestion with thrombin: Chemistry and biological properties, *Endocrinology* **107:**391.

Milner, R. D. G., and Flodh, H. (eds.), 1985, *Immunological Aspects of Human Growth Hormone,* Oxford Medical Education Services.

Milner, R. D., Barnes, N. D., Buckler, J. M., Carson, D. J., Hadden, D. R., Hughes, I. A., Johnston, D. I., Parkin, J. M., Price, D. A., Rayner, P. H., Savage, D. C. L., Savage, M. O., Smith, C. S., and Swift, P. G. F., 1987, United Kingdom multicentre clinical trial of somatrem, *Arch. Dis. Child.* **62:**776.

Moore, J. A., Pletcher, A., and Ross, M. J., 1986, Absorption enhancement of growth hormone from the gastrointestinal tract of rats, *Int. J. Pharm.* **34:**35.

Moore, W. V., 1978, The role of aggregated human growth hormone in the therapy of human growth hormone deficient children, *J. Clin. Endocrinol. Metab.* **46**:20.

Moore, W. V., and Leppert, P., 1980, Role of aggregated human growth hormone in development of antibodies to human growth hormone, *J. Clin. Endocrinol. Metab.* **51**:691.

Nederman, T., and Sjodin, L., 1987, The quantitation of human growth hormone by a radioreceptor assay using an established human cell line, *J. Biol. Stand.* **15**:199.

Nicoll, C. S., Mayer, G. L., and Russell, S. M., 1986, Structural features of prolactins and growth hormones related to their biological activity, *Endocr. Rev.* **7**:169.

Okada, Y., Taira, K., Takano, K., and Hizuka, N., 1987, A case report of growth attenuation during methionyl human growth hormone treatment, *Endocrinol. Jpn.* **34**:621.

Patel, K., and Borchardt, R. T., 1990, Deamidation of asparaginyl residues in proteins: A potential pathway for chemical degradation of proteins in lyophilized dosage forms, *J. Parenteral Sci. Tech.* **44**:300.

Patience, R. L., and Rees, L. H., 1986, Comparison of reversed-phase and anion exchange high-performance liquid chromatography for the analysis of human growth hormones, *J. Chromatogr.* **352**:241.

Pearlman, R., and Nguyen, T. H., 1989, Formulation strategies for recombinant proteins: Human growth hormone and tissue plasminogen activator, in: *Therapeutic Peptides and Proteins: Formulation, Delivery and Targeting* (D. Marshak and D. Liu, eds.), Cold Spring Harbor Laboratory, Cold Spring Harbor, N.Y., pp. 23–31.

Pearlman, R., and Nguyen, T. H., 1991, Analysis of protein drugs, in: *Peptide and Protein Drug Delivery* (V. H. Lee, ed.), Dekker, New York, pp. 247–301.

Pearlman, R., and Nguyen, T., 1992, Pharmaceutics of protein drugs, *J. Pharm. Pharmacol.* **44**:(Suppl.):178.

Pearlman, R., and Oeswein, J. Q., 1989, Human growth hormone formulation, European Patent Publication PCT WO 89/09614.

Pearlman, R., O'Connor, B. H., and Brehm, S. L., 1987, Formulation of recombinant derived human growth hormone, 194th National Meeting of the American Chemical Society, New Orleans.

Physicians Desk Reference, 46th ed., 1992, Medical Economics Company, Montvale, pp. 1049 and 1226.

Pikal, M. J., Dellerman, K. M., Roy, M. L., and Riggin, R. M., 1991, The effects of formulation variables on the stability of freeze-dried human growth hormone, *Pharm. Res.* **8**:427.

Preece, M. A., and Tanner, J. M., 1976, Treatment of growth hormone deficiency, in: *Proceedings of the International Congress of Paediatric Endocrinology,* Bologna, Italy, Serono, **173**:94.

Raben, M. S., 1957, Preparation of growth hormone from pituitaries of man and monkey, *Science* **125**:883.

Rappaport, R., 1987, Preliminary results of authentic recombinant somatropin treatment in human growth hormone deficient children. French Collaborative Study, *Acta Paediatr. Scand. Suppl.* **331**:25.

Reagan, C. R., Mills, J. B., Kostyo, J. L., and Wilhelmi, A. E., 1975, Biological properties of plasmin digests of S-carbamidomethylated human growth hormone, *Proc. Natl. Acad. Sci. USA* **72**:1684.

Reagan, C. R., Kostyo, J. L., Mills, J. B., Gennick, S. E., Messina, J. L., Wagner, S. A., and Wilhelmi, A. E., 1981, Recombination of fragments of human growth hormone: Altered activity profile of the recombinant molecule, *Endocrinology* **109**:1663.

Riggin, R. M., Dorulla, G. K., and Miner, D. J., 1987, A reversed-phase high-performance liquid chromatographic method for characterization of biosynthetic human growth hormone, *Anal. Biochem.* **167**:199.

Riggin, R. M., Sharr, C. J., Dorulla, G. K., Lefeber, D. S., and Miner, D. J., 1988, High-performance size-exclusion chromatographic determination of the potency of biosynthetic human growth hormone products, *J. Chromatogr.* **435**:307.

Robinson, A. B., and Rudd, C. L., 1974, Deamidation of glutaminyl and asparaginyl residues in peptides and proteins, *Curr. Top. Cell Regul.* **8**:247.

Roongta, V., Powers, R., Jones, C., Beakage, M. J., Shields, J. E., and Gorenstein, D. G., 1989, Solution conformation of a synthetic fragment of human pituitary growth hormone. Two-dimensional NMR of an alpha-helical dimer, *Biochemistry* **28**:1048.

Roos, P., Fevold, H. R., and Gemsell, C. A., 1963, Preparation of human growth hormone by gel filtration, *Biochim. Biophys. Acta* **74**:523.

Rougeot, C., Marchand, P., Dray, F., Job, J. C., Pierson, M., Ponte, C., Rochiccoili, P., and Rappaport, R., 1991, Comparative study of biosynthetic human growth hormone immunogenicity in growth hormone deficient children, *Horm. Res.* **35**:76.

Saxena, B. B., and Henneman, P. H., 1966, Isolation and properties of the electrophoretic components of human growth hormone by Sephadex-gel filtration and preparative polyacrylamide-gel electrophoresis, *Biochem. J.* **100**:711.

Schalch, D. S., and Parker, M. L., 1964, A sensitive double antibody immunoassay for growth hormone in plasma, *Nature* **203**:1141.

Schwartz, P. L., and Batt, M., 1973, The aggregation of ^{125}I human growth hormone in response to freezing and thawing, *Endocrinology* **92**:1795.

Sieradzki, J., Stanuch, H., Golda, W., and Szybinski, Z., 1987, Chronobiological analysis of changes in circadian rhythm of anterior pituitary hormone level during glycemia normalization by means of biostator, *Horm. Metab. Res.* **19**:208.

Singh, R. N. P., Seavey, B. K., and Lewis, U. J., 1974, Heterogeneity of human growth hormone, *Endocr. Res. Commun.* **1**:449.

Skottner, A., Forsman, A., Skoog, B., Kostyo, J. L., Cameron, C. M., Adamafio, N. A., Thorngren, K. G., and Hagerman, M., 1988, Biological characterization of charge isomers of human growth hormone, *Acta. Endocrinol. (Copenhagen)* **118**:14.

Skyler, J. S., Baumann, G., and Chrambach, A., 1977, A catalogue of isohormones of human growth hormone based upon quantitative polyacrylamide gel electrophoresis, *Acta Endocrinol. (Copenhagen) Suppl.* **212**:5.

Teh, L. C., Murphy, L. J., Huq, N. L., Surus, A. S., Friesen, H. G., Lazarus, L., and Chapman, G. E., 1987, Methionine oxidation in human growth hormone and human chorionic somatomammotropin. Effects on receptor binding and biological activities, *J. Biol. Chem.* **262**:6472.

Thurow, H., and Geisen, K., 1984, Stabilisation of dissolved proteins against denaturation at hydrophobic interfaces, *Diabetologia* **27**:212.

USP Pharmacopeial Forum, Nov.–Dec. 1990, The United States Pharmacopeial Convention, Rockville, Md., pp. 1246–1264.

Wallace, A. L. C., and Ferguson, K. A., 1961, Preparation of human growth hormone, *J. Endocrinol.* **23**:285.

Welinder, B. S., Sorensen, H. H., and Hansen, B., 1987, Reversed-phase high-performance liquid chromatography of human growth hormone, *J. Chromatogr.* **398**:309.

Wetlaufer, D. B., 1962, Ultraviolet spectra of proteins and amino acids, *Adv. Protein Chem.* **17**:303.

WHO Technical Report, 1975, Serial Number 565, 26th Report of the World Health Organization Expert Committee on Biological Standardization, WHO, Geneva, pp. 27–28.

Wilhelmi, A. E., 1961, Fractionation of human pituitary glands, *Can. J. Biochem.* **39**:1659.

Wilhelmi, A. E., 1982, Structure and function in growth hormone, in: *Hormone Drugs, Proceedings of the FDA–USP Workshop on Drug and Reference Standards for Insulins, Somatropins and Thyroid-Axis Hormones* (J. L. Gueriguian, E. D. Bransome, and A. S. Outschoorn, eds.), United States Pharmacopeial Convention, Rockville, Md., pp. 369–381.

Winder, A. F., and Gent, W. L. G., 1971, Correction of light-scattering errors in spectrophotometric protein determination, *Biopolymers* **10**:1243.

Wright, H. T., 1991, Sequence and structure determinants of the nonenzymatic deamidation of asparagine and glutamine residues in proteins, *Protein Eng.* **4**:283.

Wu, S. L., Hancock, W. S., Pavlu, B., and Gellerfors, P., 1990, Application of high-performance hydrophobic-interaction chromatography to the characterization of recombinant DNA-derived human growth hormone, *J. Chromatogr.* **500**:595.

Chapter 2

Characterization and Formulation Considerations for Recombinantly Derived Bovine Somatotropin

Stephen R. Davio and Michael J. Hageman

1. INTRODUCTION

1.1. bSt Commercial Applications

Bovine somatotropin (bSt), also known as bovine growth hormone, is produced and released in the bovine anterior pituitary gland. bSt regulates growth in the bovine (Isaksson *et al.*, 1985). Exogenously administered bSt increases the rate of growth in steers (Moseley *et al.*, 1992) and increases milk yield in dairy cows (Machlin, 1973; Peel *et al.*, 1981, 1983). The ability of bSt to increase milk yield is the primary impetus for its commercial development.

1.2. bSt Primary Structure

bSt from bovine pituitary gland is a polypeptide hormone of 191 amino acids and molecular weight 21,812. Amino acid sequences for bSt which

---

Stephen R. Davio and Michael J. Hageman • Drug Delivery R&D—Pharmaceutics, Upjohn Laboratories, The Upjohn Company, Kalamazoo, Michigan 49007.

Stability and Characterization of Protein and Peptide Drugs: Case Histories, edited by Y. John Wang and Rodney Pearlman. Plenum Press, New York, 1993.

differed slightly were published almost simultaneously by Santome *et al.* (1973) and Wallis (1973). Miller *et al.* (1980) confirmed the Wallis sequence (Fig. 1) using DNA sequencing techniques on cDNA generated from mRNA. The amino acid composition of bSt reflects a high content of hydrophobic amino acids. There is no associated lipid or carbohydrate. Pituitary-derived bSt has been reported to show sequence heterogeneity at the NH_2-terminus where approximately one-half of the pituitary bSt molecules contain a full-length sequence beginning with Ala-Phe-Pro-Ala-Met-. The other half of the molecules begins with Phe-Pro-Ala-Met- (Li and Ash, 1953; Santome *et al.*, 1973; Wallis, 1973). This heterogeneity is likely due to variability in the removal of a prehormone leader sequence (Lingappa *et al.*, 1977). Another site of sequence heterogeneity is at position 127 where an allelic variation results in either Leu or Val with an approximate 2 : 1 ratio. bSt has four cysteine residues which are all involved in disulfide linkages. These are Cys_{53}–Cys_{164} and Cys_{181}–Cys_{189}.

```
                                       10                                      20
Ala-Phe-Pro-Ala-Met-Ser-Leu-Ser-Gly-Leu-Phe-Ala-Asn-Ala-Val-Leu-Arg-Ala-Gln-His-
------------------------------ T1 ------------------------------- ↑ -----------
                                       30                                      40
Leu-His-Gln-Leu-Ala-Ala-Asp-Thr-Phe-Lys-Glu-Phe-Glu-Arg-Thr-Tyr-Ile-Pro-Glu-Gly-
----------- T2 ---------------------- ↑ ---- T3 ------↑ --------------- T4 ----
                                       50                                      60
Gln-Arg-Tyr-Ser-Ile-Gln-Asn-Thr-Gln-Val-Ala-Phe-Cys-Phe-Ser-Glu-Thr-Ile-Pro-Ala-
------- ↑ --------------------------------------- T5 -------------------------
                                       70                                      80
Pro-Thr-Gly-Lys-Asn-Glu-Ala-Gln-Gln-Lys-Ser-Asp-Leu-Glu-Leu-Leu-Arg-Ile-Ser-Leu-
--------------- ↑ --------- T6 --------- ↑ ---------- T7 ------------ ↑ -----------
                                       90                                     100
Leu-Leu-Ile-Gln-Ser-Trp-Leu-Gly-Pro-Leu-Gln-Phe-Leu-Ser-Arg-Val-Phe-Thr-Asn-Ser-
---------------------------- T8 --------------------------- ↑ ------------------
                                      110                                     120
Leu-Val-Phe-Gly-Thr-Ser-Asp-Arg-Val-Tyr-Glu-Lys-Leu-Lys-Asp-Leu-Glu-Glu-Gly-Ile-
-------- T9 ------------------- ↑ --- T10 ----- ↑ - T11 - ↑ ------------- T12 -----
                                      130                                     140
Leu-Ala-Leu-Met-Arg-Glu-Leu-Glu-Asp-Gly-Thr-Pro-Arg-Ala-Gly-Gln-Ile-Leu-Lys-Gln-
------------------ ↑ -------------- T13 ---------- ↑ --------- T14 -------- ↑ ---
                                      150                                     160
Thr-Tyr-Asp-Lys-Phe-Asp-Thr-Asn-Met-Arg-Ser-Asp-Asp-Ala-Leu-Leu-Lys-Asn-Tyr-Gly-
--- T15 ------- ↑ ---------- T16 ------- ↑ ---------- T17 ----------- ↑ ----------
                                      170                                     180
Leu-Leu-Ser-Cys-Phe-Arg-Lys-Asp-Leu-His-Lys-Thr-Glu-Thr-Tyr-Leu-Arg-Val-Met-Lys-
------- T18 ----------- ↑T19↑ --- T20 ---- ↑ ----------  T21 ------- ↑ --- T22 -- ↑
                                      190
Cys-Arg-Arg-Phe-Gly-Glu-Ala-Ser-Cys-Ala-Phe
- T23 - ↑T24↑ --------- T25 ---------------
```

Figure 1. Amino acid sequence of pituitary-derived bovine somatotropin as determined by Wallis (1973) and confirmed by DNA sequencing techniques by Miller *et al.* (1980). This is also the sequence of Somavubove, which is a recombinantly derived bovine somatotropin. Tryptic peptides are designated T1 through T25.

1.3. rbSt Production

bSt is being produced for commercial purposes using *Escherichia coli* bacteria which have been modified by recombinant DNA technology to overexpress large quantities of bSt (Seeburg *et al.*, 1983; Schoner *et al.*, 1984; Langley *et al.*, 1987a; Wingfield *et al.*, 1987; Calcott *et al.*, 1988). The "recombinantly derived" bSt (rbSt) accumulates in the bacteria as insoluble cytoplasmic granules or inclusion bodies which are visible by light microscopy (Schoner *et al.*, 1985). The following summarizes a detailed procedure provided by Langley *et al.* (1987a) for isolation of active rbSt from inclusion bodies. Inclusion bodies are harvested by cell lysis followed by centrifugation to obtain the inclusion bodies in the pellet. The pellet is extracted to remove lipid, endotoxin, nucleic acid, and protein contaminants. The extracted pellet is then dissolved in high concentrations of denaturant like guanidine hydrochloride and disulfide bonds are allowed to re-form by air oxidation. A size exclusion chromatography step of the guanidine-solubilized rbSt is performed to remove disulfide-linked aggregates. The denaturant is then slowly removed to allow refolding of rbSt to a native conformation.

Other procedures for isolating rbSt from inclusion bodies have appeared (e.g., see the method of Mitchell, 1991). However, all forms of rbSt which derive from *E. coli* will utilize a refolding step. Much of the characterization work to be discussed in Section 2 was conducted to confirm that refolded rbSt has assumed a native, biologically active conformation. This is particularly true of data pertaining to rbst secondary and tertiary structure.

One report has appeared of bSt produced in cultured murine fibroblasts (Leung *et al.*, 1986). bSt produced and secreted from these mammalian cells should be native in conformation and should not require refolding.

1.4. bSt Predicted Conformational Structure

A detailed three-dimensional structure of bSt has not been determined by x-ray crystallography. An x-ray crystallographic structure has been determined for porcine somatotropin (pSt) which has Met substituted for the normal NH_2-terminal Ala residue (Abdel-Meguid *et al.*, 1987). pSt and bSt differ at 18 positions in their sequences, and many of these differences are conservative substitutions (Seeburg *et al.*, 1983). Thus, it is likely that the Met-pSt structure will be analogous to that of bSt. Carlacci *et al.* (1991) developed a predicted structure of bSt utilizing the published information on the Met-pSt structure, the bSt sequence, and physical, chemical, and biological principles. A schematic drawing of this bSt structure is shown in Fig. 2. The predicted bSt structure resembles that of so-called four-helix bundle proteins (Abdel-Meguid *et al.*, 1987; Kaumaya *et al.*, 1990). The four helices of bSt are in a closely packed bundle which is left-hand twisted. Adjacent helices are antiparallel. The helical segments correspond to residues 8–34, 75–96, 106–128, and 153–183 in the sequence. A long loop region connects helices 1 and 2; another long loop region connects helices 3 and 4.

Hydrophobic interactions play a major role in stabilizing the core of the helix bundle and in stabilizing close association of the loop regions with the helix bundle. Loop–bundle association accounts for a large component of the stability of bSt structure. Overall 77% of hydrophobic amino acid side chains are either buried or partly buried inside the bSt molecule (Carlacci *et al.*, 1991).

2. rbSt STRUCTURAL CHARACTERIZATION

2.1. rbSt Primary Structure (Sequence)

Various forms of rbSt having slightly different sequences have been presented in the literature. One recombinant form of bSt has been produced in *E. coli* whose sequence is virtually identical to that of pituitary bSt shown in Fig. 1 (Langley *et al.*, 1987a). Much of the information presented in this

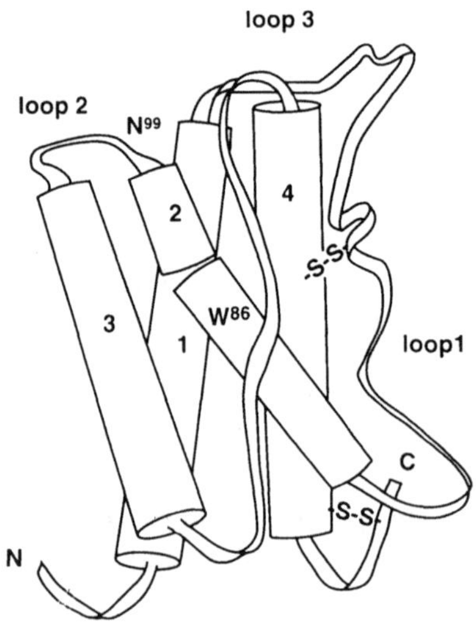

Figure 2. Schematic diagram of bSt sketched from the predicted structure of bSt by Carlacci *et al.* (1991). Helices are numbered 1 through 4. The approximate location of Trp_{86}, an amino acid important in monitoring tertiary structure spectroscopically, is shown. Also shown is the approximate location of Asn_{99}, which is an important residue in the stability of rbSt.

chapter pertains specifically to this protein which we refer to as rbSt. This protein produced by a specific process has the generic name Somavubove and is under development by The Upjohn Company.

The Monsanto Company is developing an rbSt analogue beginning with the sequence Met-Phe_2-Pro_3-Ala_4-Met_5- (Wingfield *et al.*, 1987; Bogosian *et al.*, 1989). The remainder of this methionyl-bSt (Met-bSt) sequence is identical to that shown in Fig. 1. This protein has the generic name Sometribove (Fleeger, 1991). Eli Lilly & Company are developing a bSt analogue which has the NH_2-terminal sequence Met-Phe-Pro-Leu-Asp-Asp-Asp-Asp-Lys- preceding the normal bSt sequence starting at Phe_2. This bSt analogue has the generic name Somidobove (Fleeger, 1991). American Cyanamid is developing a bSt analogue with the sequence Met-Asp-Gln- preceding the normal bSt sequence starting at Phe_2. This analogue has the generic name Somagrebove (Fleeger, 1991). The NH_2-terminal sequences of these analogues and how they compare with the sequence of native bSt are summarized in Table I.

An analogue of bSt has been biosynthesized in *E. coli* with the NH_2-terminal eight amino acids removed and the Gly at position 9 changed to a Ser which is the NH_2-terminal residue (Wingfield *et al.*, 1987). This analogue, called Delta-9 by the authors, is claimed to show a higher expression level in recombinant *E. coli* than did Met-bSt under the same conditions.

In all of the above examples, the rbSt's had single NH_2-terminal sequences, i.e., they were not heterogeneous as was bSt isolated from pituitary glands. Biosynthesis of bSt in mammalian cells (Leung *et al.*, 1987) produced bSt which was heterogeneous at the NH_2-terminus. Roughly 51% of molecules were full-length (i.e., started at position 1), 30% started at position

Table I
A Summary of the NH_2-Terminal Sequences for rbSt and rbSt Analogues under Development

Generic name	Sequence Summary[a,b]	Company
Somavubove	Ala_1-Phe_2-Pro_3-. . .	The Upjohn Company
Sometribove	**Met**-Phe_2-Pro_3-. . .	Monsanto
Somidobove	**Met-Phe-Pro-Leu-Asp-Asp-Asp-Asp-Lys**-Phe_2-Pro_3-. . .	Eli Lilly
Somagrebove	**Met-Asp-Gln**-Phe_2-Pro_3-. . .	American Cyanamid

[a] Amino acid residues are numbered according to sequence of native bSt as determined by Wallis (1973).
[b] These sequences were obtained from Fleeger (1991). Sequences which differ from bSt are indicated in boldface.

2, and 16% started at position 3. This heterogeneity may also result from variability in the removal of a prehormone leader sequence (Lingappa *et al.*, 1977; Leung *et al.*, 1987).

2.1.1. SDS-PAGE

SDS-PAGE has been extensively used in rbSt characterization to indicate purity and apparent molecular weight (Langley *et al.*, 1987b; Wood *et al.*, 1989). rbSt usually runs as a single sharp band on SDS-PAGE with a possible trace band indicating the presence of covalent dimer. This dimer is probably not disulfide bonded since it is present after reduction with β-mercaptoethanol. The apparent molecular weight of reduced rbSt is 22,000–23,000 (determined by comparison of band position relative to molecular weight standards). Unreduced bSt migrates faster than reduced bSt. This is common for disulfide-linked proteins and is probably due to a smaller hydrodynamic radius of the protein when disulfide bonds are intact.

2.1.2. Isoelectric Focusing

Isoelectric focusing (IEF) has been used to characterize the charge heterogeneity of rbSt preparations and to compare rbSt preparations with bSt derived from pituitary extracts. IEF of rbSt shows a single major isoelectric component at pI 8.5 (Langley *et al.*, 1987b) which corresponds with the major isoelectric component of pituitary bSt. rbSt also shows bands at pI 7.3 and 6.5 which may be isoforms resulting from deamidation or acetylation of lysine residues (Harbour *et al.*, 1991). Met-bSt shows a single isoelectric band around a pI of 8.0 which is near the isoelectric points of two isoforms of pituitary-derived bSt (Wood *et al.*, 1989).* The two isoforms having reported pI's of 8.1 and 7.9 are the forms of bSt with NH_2-terminal Ala and Phe, respectively (Ellis *et al.*, 1972). Thus, NH_2-terminal differences in sequences can result in different pI values. This is demonstrated clearly by the Delta-9 analogue of rbSt which has a pI of 7.4 (Wingfield *et al.*, 1987). The low pI of this analogue is not explained by any difference in the number of ionizable groups between this analogue and native rbSt. Somidobove with the four additional acidic groups added to the NH_2-terminus (see Table I)

* Differences in apparent pI's of pituitary bSt in the work of Langley *et al.* (1987b) and others may reflect differences in methodology, standards used, preparations of pituitary bSt, etc.

has the lowest pI of the rbSt analogues. Its isoelectric pH is reportedly 5.5 (Ferguson *et al.*, 1988).

2.1.3. Reversed-Phase High-Performance Liquid Chromatography (RP-HPLC)

RP-HPLC has been used to determine the purity of rbSt preparations and to monitor the stability of the rbSt sequence. RP-HPLC methods typically use C_4, C_8, or C_{18} bonded phase large-pore silica columns; rbSt elution is accomplished using increasing linear gradients of acetonitrile containing trifluoroacetic acid. rbSt typically appears as a single major symmetrical peak with one or more smaller early eluting and possibly several late-eluting peaks (see Fig. 3 for a sample RP-HPLC chromatogram). Langley *et al.* (1987b) showed that the main RP-HPLC component of rbSt chromatographs identically to that of pituitary-derived bSt. The early eluting peaks in the RP-HPLC chromatogram have been identified as isoforms of bSt resulting from deamidation of Asn_{99} (Harbour *et al.*, 1989). These isoforms will be discussed further in the section dealing with chemical modifications of the primary sequence. Violand *et al.* (1990) observed an RP-HPLC chromatogram equivalent to that of Fig. 3 with Met-bSt and also made similar identifications on their early eluting peaks. Bogosian *et al.* (1989) identified a late-

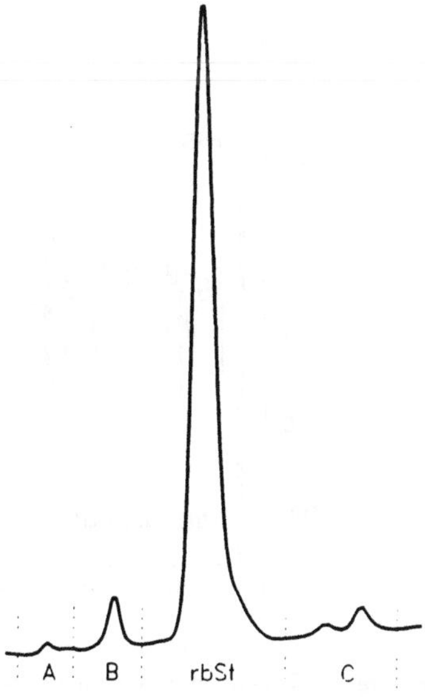

Figure 3. RP-HPLC chromatogram of rbSt and reaction by-products. rbSt is the large main component. Early eluting components A and B have been identified as reaction products resulting from chemical degradation at Asn_{99}. Later-eluting components designated "C" have not been identified.

eluting peak in the Met-bSt chromatogram as due to a sequence beginning with formyl-Met-Phe-Pro-.

2.1.4. Tryptic Mapping

The most sensitive use of RP-HPLC for detection of changes in rbSt sequence has been in tryptic mapping. In this method rbSt is digested by trypsin into peptide fragments which are then separated and quantitated by RP-HPLC. The tryptic peptides are small enough to interact fully with the RP-HPLC stationary phase. Hence, the likelihood of RP-HPLC insensitivity to changes in protein structure is substantially reduced. Two papers have appeared describing tryptic mapping of rbSt (Hartman *et al.,* 1986; Dougherty *et al.,* 1990). Other papers demonstrate the utility of tryptic maps without exploring the details of the assay methodology (Bogosian *et al.,* 1989; Wood *et al.,* 1989; Violand *et al.,* 1990).

Based on the sequence of rbSt (shown in Fig. 1), 25 peptides should result from a complete trypsin digestion of disulfide-reduced rbSt. Twenty-three peptides should result from disulfide-intact rbSt. The peptides that should result from trypsin digestion are designated T1 through T25 in Fig. 1. Tryptic maps of nonreduced rbSt and pituitary bSt are shown in Fig. 4 (John J. Dougherty *et al.,* unpublished data). Each peak in the chromatogram was identified in the rbSt sequence by thermospray mass spectrometry or NH_2-terminal sequencing using a gas-phase analyzer. The maps have a few

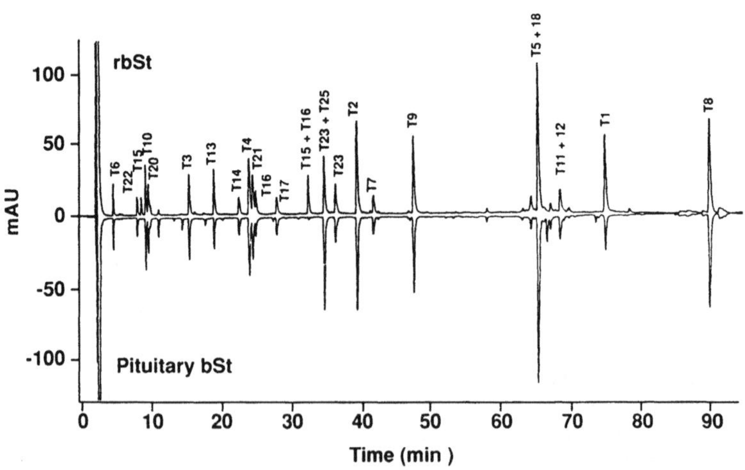

Figure 4. Tryptic map of rbSt and pituitary-derived bSt (unpublished data of J. J. Dougherty, used with permission).

subtle but distinct differences. The T1 peptide in pituitary-derived bSt is smaller than in rbSt due to NH_2-terminal sequence heterogeneity in the pituitary material. Also, T13 is smaller in pituitary bSt because of the allelic variation at position 126. Tryptic peptide mapping has been used to identify the location of sequence alterations in degradation products seen as early eluting RP-HPLC peaks (Harbour *et al.*, 1989; Violand *et al.*, 1990). Wood *et al.* (1989) used peptide mapping to detect isoaspartic acid at position 129 (normally Asp) in highly purified pituitary bSt. Bogosian *et al.* (1989) observed changes in the tryptic maps of bSt which had norleucine substituted in two positions for methionine. These studies show that tryptic peptide mapping is an extremely sensitive technique for the detection of small changes in bSt primary structure.

2.1.5. Raman Spectroscopy

Raman spectroscopy has been used to monitor the integrity of the disulfide bonds in rbSt (Havel *et al.*, 1989). The S–S stretching band typically occurs at 500 cm^{-1} in proteins, while the S–H stretching due to a reduced disulfide bond occurs at about 2550 cm^{-1}. A small band associated with the stretching of the two disulfide bonds in rbSt was detected at 508 cm^{-1}. This band sits within a large envelope of non-disulfide-related Raman bands. Chemical treatments to reduce and irreversibly modify rbSt disulfide bonds eliminate the Raman band at 508 cm^{-1}. Also, γ irradiation, a possible means of terminally sterilizing rbSt, causes significant decreases in the disulfide stretching band consistent with disulfide bond disruption (Havel *et al.*, 1989).

2.2. rbSt Secondary Structure

2.2.1. Far-UV Circular Dichroism Spectroscopy (Far-UV CD)

Far-UV CD allows a look directly at the secondary structure of rbSt (see Fig. 5). Langley *et al.* (1987b) have shown that rbSt prepared by refolding from a guanidine hydrochloride solution has a far-UV CD spectrum identical to that of pituitary-derived bSt. This comparison provides evidence that rbSt can acquire native secondary structure by refolding from a guanidine hydrochloride solution.

The far-UV CD spectrum with its characteristic minima at 208 and 222 nm and maximum at 193 nm resembles that of other proteins known to have

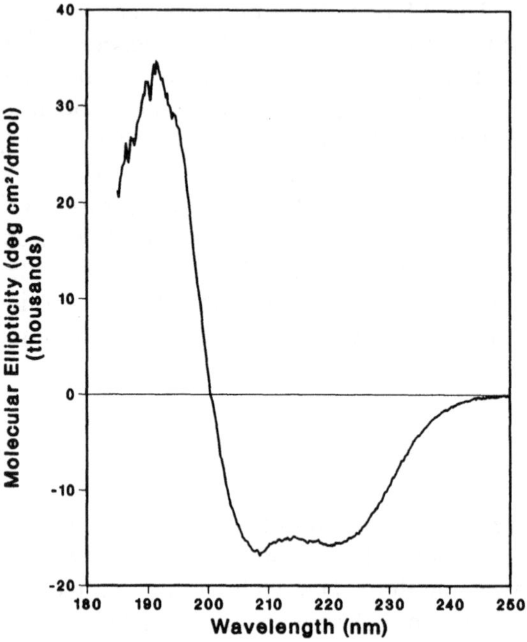

Figure 5. Far-UV circular dichroism spectrum of rbSt at pH 9.5.

a high content of α-helix. The far-UV CD spectrum has been used to estimate a secondary structure composition of 55% α-helix, 10% β-pleated sheet, and 35% random structures (Sonenberg and Beychok, 1971).

2.2.2. Fourier Transform Infrared Spectroscopy (FT-IR)

FT-IR has also provided an estimate of secondary structure composition (Havel *et al.*, 1989). This method uses special deconvolution methods to separate and integrate overlapping amide I infrared absorption bands associated with α-helix, β-pleated sheet, and random structures. The deconvolution analysis has indicated the relative composition of secondary structures in rbSt to be 45% α-helix, 15% β-sheet, and 40% random structure. FT-IR experiments have also indicated that α-helical content is dependent on the integrity of the two disulfide bonds in rbSt. Reduction of these disulfides reduces the measure of helical content from 45% to 30%. FT-IR has compared secondary structures of rbSt in solution, as a lyophilized solid, compressed pellet, and in oil suspension. These studies indicate that the secondary structure of rbSt, particularly the α-helical content, remains unchanged regardless of its medium (Havel *et al.*, 1989).

2.3. rbSt Tertiary Structure

2.3.1 Near-UV Absorption Spectroscopy

UV absorption spectroscopy obtained at wavelengths of 250–320 nm is one of several spectroscopic techniques used to monitor rbSt tertiary structure. The UV absorbance spectrum of rbSt is typical for proteins and shows an absorption maximum around 278 nm. Unfolding of rbSt using a denaturant (studies discussed below) shifts the absorbance spectrum to shorter wavelengths which indicates that the single Trp is located in a nonpolar environment in the native molecule and is transferred to a more polar aqueous environment in the denatured state.

Second derivatives of UV absorbance spectra have been used to provide information about the Tyr residues (six of them) in the rbSt tertiary structure (Havel *et al.,* 1986, 1989). Tyr can be monitored by changes in second-derivative UV absorbance spectra from 276 to 278 nm (Terada *et al.,* 1984). Unfolding of rbSt using a denaturant causes the Tyr second-derivative absorbance peaks to shift to a shorter wavelength, indicating that the Tyr environment also changes from relatively nonpolar in the native molecule to more polar upon unfolding (Havel *et al.,* 1986).

2.3.2. Near-UV CD Spectroscopy

The near-UV CD spectrum, shown in Fig. 6, provides a fingerprint which is indicative of the native tertiary structure of rbSt. The individual absorption bands have been assigned by Holladay *et al.* (1974) to specific electronic transitions of bSt chromophores. The very existence of a near-UV CD spectrum indicates that rbSt chromophores detect a highly chiral environment which indicates that rbSt has a definite tertiary structure. Langley *et al.* (1987b) have used near-UV CD to show that the spectrum of rbSt refolded from a guanidine hydrochloride solution is identical to pituitary-derived bSt. This provides evidence that rbSt can be refolded to assume native tertiary structure.

2.3.3. Fluorescence Spectroscopy

Fluorescence spectroscopy has also been used to provide information on rbSt tertiary structure. Fluorescence emission wavelength and quantum yield can both provide information about the local Trp environment. The

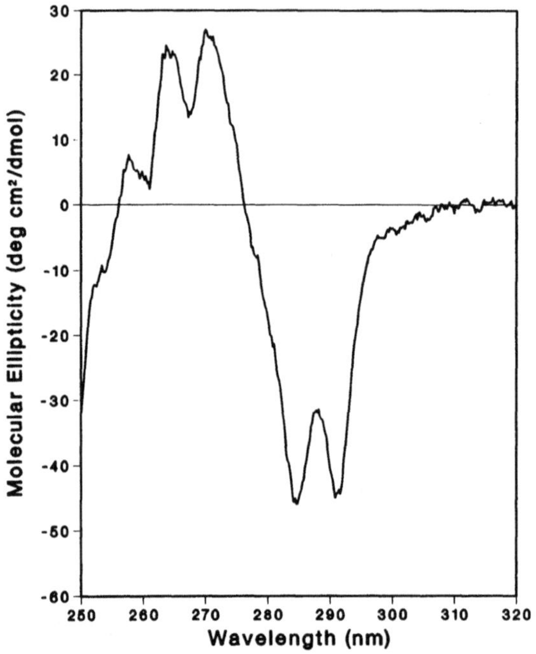

Figure 6. Near-UV circular dichroism spectrum of rbSt at pH 9.5.

rbSt emission spectrum for the single Trp occurs with a maximum at 335 nm, indicating that the Trp is in a nonpolar environment (Lakowicz, 1983). The quantum yield is quite low for rbSt in its native state, indicating a high level of internal quenching possibly due to the close proximity of a charged group(s) to Trp in the native structure (Burger *et al.*, 1966; Brems *et al.*, 1985; Holzman *et al.*, 1990). This quenching interaction is disrupted by certain changes in pH and by increasing concentrations of denaturants (both to be discussed below).

2.4. rbSt Quaternary Structure

Pituitary bSt has been shown to aggregate into dimers and possibly higher-order oligomers in solution (Dellacha *et al.*, 1968). Size exclusion (SE) HPLC has been used to monitor oligomer formation of rbSt. Methods developed by Stodola *et al.* (1986) included both denaturing mobile phases (6 M urea and 0.1% SDS) and nondenaturing mobile phases (sodium phosphate buffer, pH 6.8). With both types of mobile phases, rbSt chromatographs with the same retention time as pituitary-derived bSt, indicating an equivalent hydrodynamic radius for both. Also, monomer rbSt is separated from dimer, trimer, and oligomer peaks. Monomer rbSt chromatographs with an apparent molecular weight of around 25,000 relative to molecular weight standards. rbSt oligomer peaks are present when using mobile phases containing SDS or guanidine hydrochloride, which indicates that the oligomers are covalently linked and not hydrophobic aggregates.

2.5. rbSt Bioassay Characterization

2.5.1. Rat Growth Bioassay

The hypophysectomized rat weight gain assay (Marx *et al.,* 1942) has been used to assess the biological potency of various rbSt preparations. Rats with pituitary glands removed are administered daily injections of rbSt solution for 10 days. Weights are recorded daily. rbSt potency is determined based on the rate of body weight gain above the mean rate of animals injected with a buffer control solution. Langley *et al.* (1987b) found their rbSt refolded from guanidine hydrochloride solution to be slightly more potent than a sample of pituitary-derived material. These authors concluded that their rbSt was folded in the native, biologically active, conformation. Wingfield *et al.* (1987) also found that their Met-bSt and their Delta-9 analogue of rbSt, both produced in *E. coli,* had growth-promoting activity equal to that of pituitary bSt. rbSt produced in mammalian cells had similar growth-promoting activity as pituitary bSt (Leung *et al.,* 1987). Harbour *et al.* (1989) found that an isoform of rbSt consisting of an isoaspartic acid at position 99 was slightly more potent than native rbSt in the rat growth bioassay.

2.5.2. Radioreceptor Assay

A radioreceptor assay of Haro *et al.* (1984) has also been used to characterize rbSt (Wood *et al.,* 1989). Bovine liver microsomal fractions contain membrane receptors for bSt which bind ^{125}I-labeled bSt with specificity. Concentrations of rbSt ranging over several orders of magnitude are combined with constant amounts of liver microsomal preparation and ^{125}I-labeled bSt in individual assay tubes. After an equilibration period, the amount of ^{125}I-labeled bSt displaced from receptor binding was determined as a function of experimental bSt concentration. Wood *et al.* (1989) found that Met-bSt displaced ^{125}I-labeled bSt in a manner superimposable with pituitary bSt, indicating equal affinity for the bSt receptor. Therefore, the conformations of the two proteins as detected by the receptor are identical. Leung *et al.* (1987) reached this same conclusion for their rbSt produced in mammalian cells. Wingfield *et al.* (1987) found that their Delta-9 analogue of rbSt has affinity equal to pituitary bSt for the bSt receptor. This means that the NH_2-terminal eight amino acids are not critical for binding since these residues are missing in the Delta-9 analogue.

3. rbSt CONFORMATIONAL STABILITY

3.1. The Effect of Solution pH on rbSt Conformation

The effect of pH on the conformation of rbSt has been studied by Holzman *et al.* (1990) using several spectroscopic techniques. Fluorescence spectroscopy, which is sensitive to changes in tertiary structure, showed that neutral-to-basic pH titrations cause essentially no change in the Trp emission spectrum. However, a neutral-to-acidic pH titration results in an approximate fourfold increase in the fluorescence intensity occurring with a transition midpoint around pH 4.7 (Fig. 7). The increase in fluorescence intensity occurs with no shift in the wavelength of the emission maximum. These changes signal the occurrence of a conformational change which disrupts the interaction in native rbSt between Trp and nearby quenchers. The acid-induced conformational change does not result in complete un-

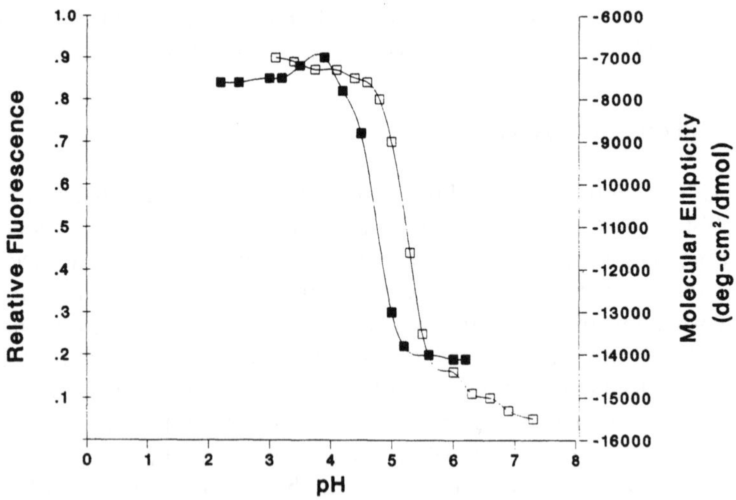

Figure 7. Effect of pH on rbSt intrinsic fluorescence (■) and molecular ellipticity at 222 nm (□). Data replotted from Holzman *et al.* (1990).

folding since this would result in exposure of Trp to water and, consequently, a 15-nm redshift in the Trp emission maximum.

The partial unfolding induced by acid pH is an extremely cooperative process as indicated by the steepness of the change in fluorescence as the pH is decreased. The transition fits the profile expected if it is triggered by the protonation of two acid groups simultaneously. This same transition was also indicated by changes in UV absorbance spectra (Holzman *et al.*, 1990).

Far-UV CD at 222 nm was used to monitor the effects of acid pH on the secondary structure of rbSt, specifically the α-helical content (Holzman *et al.*, 1990). An acid-induced transition resulted in a 50% loss of native rbSt helical content occurring at a pH of ~5.3 (Fig. 7). This transition in secondary structure clearly does not coincide with the transitions in tertiary structure seen by fluorescence and UV absorbance spectroscopy. These results indicate that acid-induced unfolding of rbSt involves intermediates resulting from an uncoupling of the tertiary and secondary levels of rbSt structure.

3.2. The Effect of Denaturants on rbSt Conformation

Denaturation studies on rbSt using urea and guanidine hydrochloride (Gdn-HCl) have provided additional information on the conformational properties of rbSt. Edelhoch and Burger (1966) studied urea denaturation of pituitary bSt and found that the relative fluorescence of bSt at neutral pH increases with increasing concentrations of urea. Holzman *et al.* (1990) found that urea denaturation of rbSt results in an increase in fluorescence and no change in the Trp emission maximum until the urea concentration exceeds 9 M. Since the increase in fluorescence ("dequenching") does not coincide with the shift in emission maximum, it is clear that rbSt unfolding by urea is not "two-state" and involves intermediates.

Brems *et al.* (1985) used a variety of techniques to study Gdn-HCl denaturation of rbSt. They found that Gdn-HCl-induced unfolding of rbSt and pituitary-derived bSt were virtually superimposable as monitored by UV absorbance and fluorescence spectroscopy, indicating that the two forms of bSt have virtually identical conformational stability. Brems *et al.* (1985) found that the rbSt Gdn-HCl denaturation curves obtained by UV absorbance (290 nm), fluorescence, and CD (222 nm) spectroscopy were not superimposable, indicating that rbSt unfolding is not a two-state process and involves intermediates. The loss of tertiary structure as indicated by UV absorbance and fluorescence spectra occurred at lower concentrations of Gdn-HCl than did the loss of α-helical structure as indicated by CD at 222 nm. This suggests that the intermediate in rbSt denaturation might lack native tertiary structure but retain at least some of its native secondary structure. Brems *et al.* (1985) proposed that rbSt folding might be consistent with the framework model of protein folding (Kim and Baldwin, 1982) in which hydrogen-bonded secondary structure develops before tertiary structure. This hypothesis is further strengthened by kinetic studies on rbSt folding which show that the rate of helix formation as measured by CD (222 nm) is significantly faster than the formation of native tertiary structure as measured by UV absorbance (290 nm) (Brems *et al.*, 1987).

Unfolding intermediates of rbSt show a propensity to self-associate. This is the conclusion reached by Havel *et al.* (1986) and Brems *et al.* (1986) in rbSt denaturation studies at protein concentrations around 1 mg/ml. The self-associated intermediate possesses a characteristic peak in the near-UV CD at 300 nm which is clearly distinct from anything in the spectra of the native or denatured forms of rbSt. The associated intermediate has a larger Stokes radius than monomeric native or denatured rbSt which is detectable by SE-HPLC. These methods have shown that the associated intermediate occurs only under partially denaturing conditions. Complete denaturation due to high concentrations of denaturant disrupts the associated intermediate. The associated intermediate is concentration dependent and exists at rbSt concentrations greater than 0.2 mg/ml.

Formation of the associated intermediate results in a change in Trp environment but not in Tyr environment (Havel *et al.*, 1986). This is a conclusion drawn from second derivatives of UV absorbance spectra. This suggests that the interaction resulting in formation of the associated intermediate involves specific structural components of rbSt. This is further suggested by the observation that a peptide with a sequence corresponding to the rbSt sequence 96–133 diminishes formation of the associated intermediate. This observation supports the hypothesis that an amphiphilic helix from residues 110–127 mediates formation of the associated intermediate. The hypothesis further states that under partially denaturing conditions this amphiphilic helix is disrupted from its native-state intramolecular associations and made available for intermolecular associations (Brems *et al.*, 1986).

4. rbSt PHYSICAL PROPERTIES

4.1. Solubility Properties

Solubility measurements on rbSt have been made using saturated solutions prepared by dialysis, pH jump, and dissolution of solid rbSt; results by these methods agree reasonably well (see Table II; Hageman *et al.*, 1990).

Table II
Comparison of Apparent Solubilities of rbSt Measured by Different Techniques

Method for inducing a saturated solution, pH 7 (μ = 0.055)	Measured solubility
Acid-to-base pH jump	4.5 mg/ml
Base-to-acid pH jump	4.4 mg/ml
Dissolution of solid rbSt to saturation	3.0 mg/ml
Dialysis	5.6 mg/ml
Ultrafiltration (μ = 0.12)	12.5 mg/ml

Ultrafiltration studies result in slightly higher solubility determinations, possibly resulting from supersaturation. Temperature has been noted to have very little effect on rbSt solubility (Hageman *et al.,* unpublished observations, 1992a). Solubilities as measured by the dissolution method ranged from 8.1 mg/ml at 5°C to 11.7 mg/ml at 37°C in pH 7.4 low-ionic-strength buffers.

rbSt solubility is extremely dependent on pH as indicated by Fig. 8 (Hageman, 1987). The pH of minimum solubility is between 7 and 8 in low-ionic-strength solution and depends on whether the insolubility was induced by an acid-to-base pH jump or vice versa. Ionic strength has a profound effect on the rbSt pH–solubility profile as shown in Fig. 9. Increasing ionic strength shifts the pH of minimum solubility to lower pH values (Hageman, 1987). The shift in the pH–solubility relationship suggests a specific interaction of anions with the rbSt molecule, perhaps resulting in a

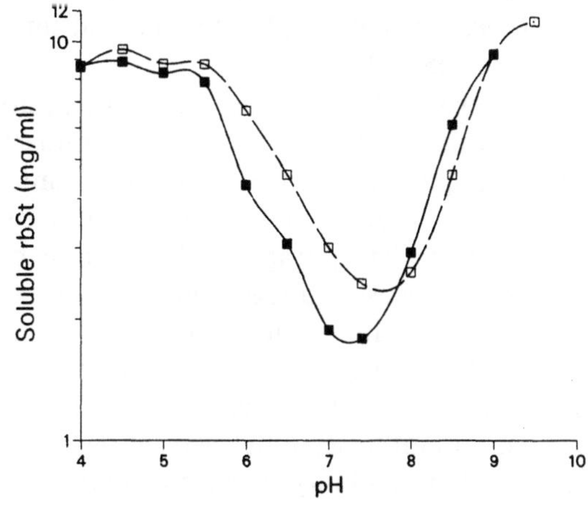

Figure 8. Solubility of rbSt in water at low ionic strength (μ < 0.008). Final pH values were achieved by adjusting pH from basic solutions or acid solutions as indicated (■, base to acid; □, acid to base) (Hageman, 1987).

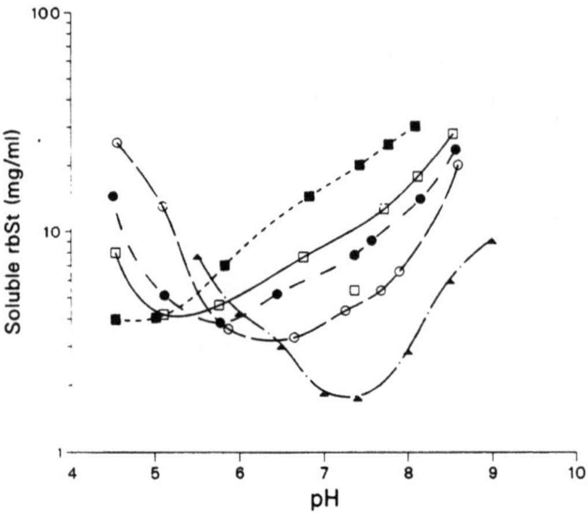

Figure 9. Effect of ionic strength on rbSt solubility–pH profiles. Ionic strength (μ) was increased with NaCl (■, μ = 0.24; □, 0.12; ●, 0.072; ○, 0.024; ▲, <0.008) (Hageman, 1987).

masking of electrostatic charges and a lowering of the "apparent" isoelectric pH. An alternative explanation is that specific ionic interactions or increasing ionic strength induce a shift in the conformational equilibrium previously mentioned to occur at acidic pH conditions, thus affecting the apparent solubility of rbSt.

At specific pH conditions, increasing ionic strength can have opposite effects on rbSt solubility. This is predictable from the data of Fig. 9. Increasing ionic strength increased the rbSt solubility at pH 7 but decreased the solubility at pH 5. RbSt solubility at pH 5 in solutions of equal ionic strength was relatively higher in solutions containing Cl^-, lower in solutions containing ClO_3^-, and lower still in solutions containing SO_4^{2-}. This result is apparently not due to a "salting out" phenomenon since the relative effect of these anions on rbSt solubility is not consistent with the Hofmeister series. These unexpected results imply the potential importance of specific ion interactions on rbSt solubility.

Interpretation of these apparent solubility measurements should be made with caution due to the likely impact of solution conditions on the chemical potential of the solid. Also, conformational changes which have been noted to occur under acidic conditions could result in a multicomponent system where the apparent solubilities obtained may involve a heterogeneous solid phase. Furthermore, as noted with rbSt at low pH values, there may be significant precipitation which is actually a result of conformational instability leading to hydrophobic aggregation. These aggregates are usually irreversible under the conditions of the solubility assessment. In other words, it is important to separate aggregation/precipitation phenomena due to extremely low solubility of the hydrophobic aggregate from precipitation of native protein at concentrations exceeding its solubility, the latter being easily reversible.

4.2. Dissolution Properties

The dissolution properties of rbSt would be expected to have a significant impact on the development of formulations for this protein, especially sustained-release formulations. The dissolution properties of rbSt have been studied by Hageman *et al.* (1992a) using the rotating disk dissolution system. The dissolution rate is expressed as a flux of protein, J, into the dissolution medium. Based on experience with nonprotein organic molecules, flux (dissolution) is expected to follow the equation (Ritschel, 1973)

$$J = \frac{DAC_s}{h} \tag{1}$$

where D is the apparent diffusion coefficient, A is the surface area of the disk, C_s is the saturation solubility of the drug, and h is the thickness of the unstirred aqueous diffusion layer.

Typical rotating disk data show a constant flux with time. Moreover, the flux is typically proportional to the solubility under experimental conditions where all other variables are constant. This was not observed for rbSt dissolution (Fig. 10; the flux is the slope at any point on these curves). rbSt shows a rapid initial dissolution followed by a much slower but constant dissolution of protein. For an interesting comparison, the dissolution of recombinantly

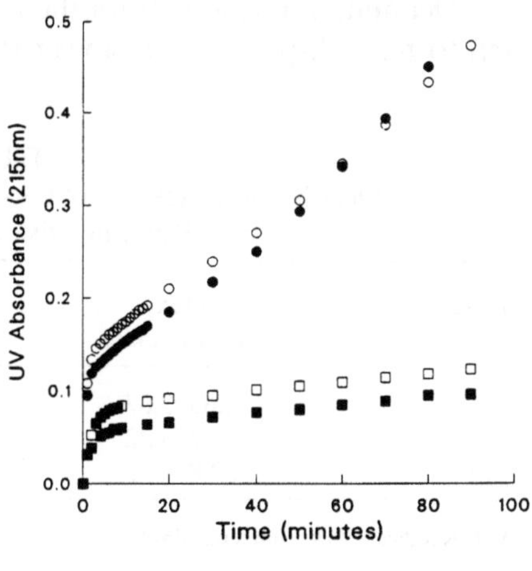

Figure 10. Dissolution of sodium salts of rbSt (squares) and recombinantly derived porcine somatotropin (circles) at 37°C and pH 7.4. The data represent the appearance of protein in the dissolution media. Closed and open symbols represent replicate runs (data replotted from Hageman *et al.*, 1992a).

derived porcine somatotropin (rpSt) is also shown in Fig. 10. rpSt also shows the biphasic dissolution. The slower rates of dissolution were used to calculate fluxes for both proteins. The rate of rpSt dissolution was almost ten times faster than that of rbSt, whereas the solubility of rpSt was only four times greater than that of rbSt. Thus, the relationship of Eq. (1) does not hold rigorously across a comparison of these two similar proteins.

A similar test of Eq. (1) was conducted by comparing the dissolution rates of rbSt under pH conditions where the solubility of rbSt is known. According to Eq. (1), the ratios of J/C_s should be constant. These data are given in Table III and clearly indicate that these ratios are not constant. In short, dissolution data with rbSt do not fit standard models for rotating disk dissolution data. This is probably because some of the assumptions associated with rotating disk dissolution do not apply for rbSt. Changes in the diffusion coefficient of rbSt as it passes through its unstirred boundary layer, either due to a viscosity gradient or due to association–dissociation equilibria, could cause deviations from the typical rotating disk dissolution profile (Kuu *et al.*, 1989). Models used to describe the dissolution of macromolecular polymers might better describe dissolution data for rbSt and other proteins (Kuneida and Shinoda, 1975; Ouano *et al.*, 1977).

5. PHYSICAL/CHEMICAL STABILITY

5.1. Solution Stability

5.1.1. Deamidation Reactions

Deamidation results from the hydrolysis of side chain amide groups of Asn to form Asp and, at a slower rate, Gln to form Glu (Manning *et al.*,

Table III
Dissolution Rates for rbSt Sodium Salts at 37°C as Obtained by a Rotating Disk Apparatus (600 rpm)[a]

Media pH	Flux (μg/min-cm^2)	C_s (mg/ml)	J/C_s (cm/min)
6.4	1.63–2.07	7.4	$2.2–2.8 \times 10^{-4}$
7.4	4.22–5.22	9.0	$4.7–5.8 \times 10^{-4}$
8.0	39.8–47.5	18.0	$2.2–2.6 \times 10^{-3}$

[a] Values indicate 95% confidence limits.

1989, and references cited therein). Deamidation of bSt was first studied by Lewis *et al.* (1970) using pituitary-derived bSt. The rate of deamidation was studied as a function of pH, temperature, and ionic strength using IEF to monitor the loss of the main isoelectric species from its isoionic pH of ~8.5 (Langley *et al.*, 1987b). The rates of deamidation at pH values ranging from 2 to 7 (37°C) were slow ($k < 0.1 \times 10^{-6}$ sec^{-1}). This rate increased steadily with increasing pH above 7 to a value of $k = 1.1 \times 10^{-6}$ sec^{-1} at pH 10 (37°C). Temperature increased the rate of deamidation; the reaction rate at pH 10 and 37°C was tenfold greater than its value at 5°C. Ionic strength increased the rate of deamidation; addition of 1 M $(NH_4)_2SO_4$ at pH 10 increased the reaction rate by fourfold. Denaturation with urea increased the rate of deamidation at pH 7, indicating a greater stability of the folded form of bSt toward deamidation than the unfolded form. This is consistent with studies using ribonuclease A (Wearne and Creighton, 1989). Urea-induced unfolding of bSt also changed its IEF pattern by creating more acid bands, indicating a larger variety of deamidated species resulting from deamidation at many sites.

Deamidation of Asn residues in proteins probably proceeds through a cyclic succinimide intermediate (Geiger and Clarke, 1987, and references cited therein). Succinimide intermediates of Asn hydrolyze to isoaspartic acid (isoAsp) or to the classical product of deamidation, Asp. A third by-product of Asn degradation is a cleaved product on the COOH-terminal side of the Asn. Harbour *et al.* (1989) have observed the isoAsp and cleaved products resulting from chemical reaction of rbSt at Asn_{99}. The cleaved form of rbSt consists of two chains held together by a disulfide bond with Asp (formerly Asn_{99}) at the COOH-terminus of one chain and Ser_{100} at the NH_2-terminus of the other chain. The cleaved form of rbSt appears as peak A in the RP-HPLC chromatogram shown in Fig. 3 while the isoAsp form of rbSt appears as peak B in this chromatogram. The deamidated form of rbSt with Asp at position 99 coelutes with normal rbSt under the RP-HPLC conditions used for this chromatogram.

Violand *et al.* (1990) have observed the same bSt isoforms resulting from deamidation at Asn_{99} of Met-bSt and have drawn conclusions about the relative rates of formation of each product. A comparison of appropriate peaks in the Met-bSt chromatogram indicates that the concentration of the cleaved form is one-tenth that of the isoAsp form. The relative quantities of the $isoAsp_{99}$ and Asp_{99} forms can only be determined by RP-HPLC of trypsin-digested Met-bSt. The tryptic peptides containing residues 96–108 with isoAsp, Asp, or Asn at position 99 can be resolved by RP-HPLC. This analysis of Met-bSt degraded at pH 10 showed that the $isoAsp_{99}$ form occurs at a rate threefold faster than does the Asp_{99} form. This ratio is consistent with what is observed with model peptides (Geiger and Clarke, 1987).

Instability at Asn_{99} also occurs in pSt (Violand *et al.*, 1990), suggesting that the pSt three-dimensional structure might help explain the reason for this site of instability in both bSt and pSt. The Asn_{99} site of pSt occurs in a flexible region of the structure between helices (Violand *et al.*, 1990; Abdel-Meguid *et al.*, 1987). The Asn_{99} site in the bSt predicted structure of Carlacci *et al.* (1991) occurs in a short region between helices (see Fig. 2). Conformational flexibility is probably a requirement for formation of the succinimide intermediate prior to deamidation (Wearne and Creighton, 1989).

The instability at Asn_{99} has stimulated efforts to enhance the stability of rbSt by replacing the residue at position 99 with other amino acids (Lyle *et al.*, 1990). Replacement of Asn_{99} with Asp, Ser, Gly, Pro, or Glu resulted in significantly improved aqueous stability at pH 10 as assessed by RP-HPLC and SDS-PAGE. The Asp, Ser, and Gly analogues were reported to be equipotent with native rbSt in enhancing milk yield in dairy cows (Garlick *et al.*, 1989).

Sites of bSt deamidation other than Asn_{99} were identified by Secchi *et al.* (1986) using pituitary-derived bSt. Deamidations of Asn_{13}, Asn_{148}, and Gln_{140} were identified using tryptic mapping. There was no indication of deamidation at Asn_{99} from these studies. Differences in the sites of deamidation observed by Secchi *et al.* (1986) and others may derive from differences in the sources of bSt and the methods of purification.

5.1.2. Oligomerization Reactions

Covalent cross-linking of rbSt to dimers and oligomers is another chemical reaction of rbSt in solution. These reactions have been studied by Hageman *et al.* (1992a) using SE-HPLC and RP-HPLC. Figure 11 shows the rate

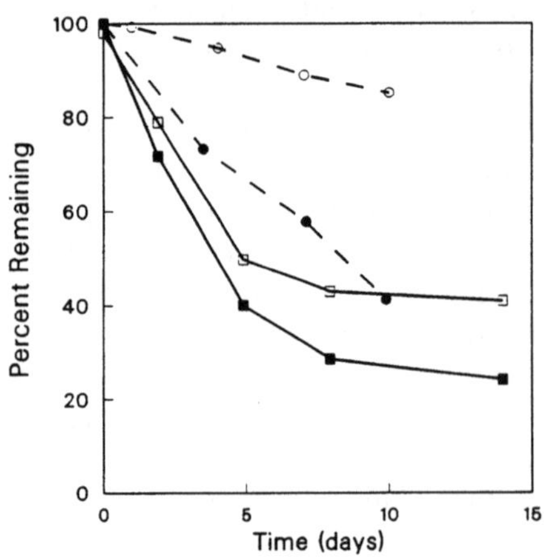

Figure 11. rbSt degradation as detected by RP-HPLC (●, ■) and SE-HPLC (○, □). Dashed lines indicate degradation of rbSt (10 mg/ml) in solution at pH 9.8 at 30°C. Solid lines indicate degradation of lyophilized sodium salts of rbSt at 96% relative vapor pressure, and 30°C (data replotted from Hageman *et al.*, 1992a).

of monomer loss due to covalent cross-linking reactions from a solution of 10 mg/ml, pH 9.8, and 30°C. These data are provided by SE-HPLC. Also shown is the rate of loss of RP-HPLC main component which results from the sum of covalent cross-linking reactions and hydrolysis reactions including several reactions just discussed involving Asn_{99}. The cross-linking reactions result in an approximately 15% decrease of rbSt monomer by SE-HPLC over a 10-day incubation period. The loss of RP-HPLC main component was about 60% during this same period. Therefore, cross-linking reactions account for approximately 25% of the total rbSt degradation under these conditions. The majority of these cross-linked degradation products are not reducible by β-mercaptoethanol, indicating that they are probably not disulfide-linked. One possibility is that they are oligomers cross-linked by a condensation of a Lys ϵ-amine and a carboxyl side chain of either Asp or Glu. This reaction remains to be confirmed, however. The formation of reducible dimers and oligomers, cross-linked via disulfide linkages, has been noted for pSt at very high concentrations (Shieh *et al.*, 1990). These products are likely to form for rbSt as well under similar solution conditions.

Typical of many proteins, rbSt undergoes significant agitation-induced precipitation (M. J. Hageman, unpublished observations). As has been demonstrated for human somatotropin (Pikal *et al.*, 1991), this is likely due to denaturation at the air/water interface with subsequent aggregation and not shear-induced disruption of the conformational structure. At neutral pH, especially at elevated temperature (i.e., 35–55°C) and high protein concentration, the predominant pathway of loss from aqueous solution becomes irreversible precipitation of hydrophobic aggregates. These aggregates are noncovalent and can be dissolved in the presence of denaturant.

5.2. Solid-State Stability

The relative occurrence of cross-linking and hydrolysis reactions have been determined on lyophilized rbSt (Hageman *et al.*, 1992a) using SE-HPLC and RP-HPLC analyses of reconstituted samples (Fig. 11). The fraction of total degradation products consisting of covalent dimers increased to 80–90% with the remaining 10–20% consisting of deamidation and chain cleavage products. Again, the covalent dimers were found to be predominantly nonreducible by β-mercaptoethanol. Both the dimerization and deamidation/cleavage reactions increased proportionately in rate with exposure of lyophilized rbSt to increasing relative humidity (data not shown). Both classes of degradation reactions were sensitive to the pH of the rbSt solution before lyophilization and increased in rate by a factor of about 10 in going from pH 4.7 to pH 9.6. The fractional loss due to dimerization remains constant at around 80% of the total degradation, independent of pH.

5.3. Absorption of Water in the Solid State

This topic has been extensively covered by Hageman *et al.* (1992b) who show that rbSt absorbs water according to a sigmoidal-shaped isotherm (Fig. 12). This type of isotherm is fairly typical of proteins (Kuntz and Kauzmann, 1974). The rbSt isotherm is described by mathematical equations which yield "monolayer" values of 5–8 g H_2O/100 g protein. Absorption isotherms for rbSt show evidence for penetration of water into the solid. Thus, the amount of water absorbed is independent of particle surface area. Lyophilized and spray-dried rbSt, despite having quite different surface areas, absorb similar amounts of water.

Lyophilized formulations of rbSt incubated at relative humidities from 10 to 43% resulted in water contents ranging from 1.5 to 5%. This was shown to affect the stability of the rbSt as monitored by RP-HPLC which detects both cross-linking and hydrolysis reactions. rbSt equilibrated at 43% relative humidity, its vial capped and then transferred to 47°C, degraded at a rate roughly tenfold higher than rbSt equilibrated at 10% relative humidity.

6. FORMULATIONS

Formulation approaches for bSt and analogues have been driven primarily by market needs for a long-acting preparation. Formulation efforts

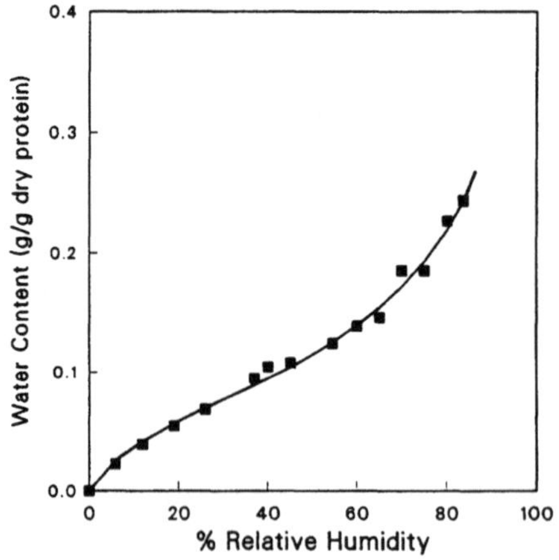

Figure 12. Water sorption isotherm for rbSt as a function of relative humidity at room temperature. The solid line is a nonlinear least-squares best fit to the data as described by Hageman *et al.* (1992b).

have been limited by the inherent instability of bSt in an aqueous environment, by the relatively high dose of protein required (i.e., 10–30 mg/day), and by limits on the cost of goods for the final product. As a consequence, the most common formulation approaches have been those involving the dispersion of bSt in oils of low cost.

The use of less-soluble metal ion salts of somatotropins with viscosity-inducing agents such as salts of fatty acids (i.e., aluminum stearate and calcium stearate) can further retard release of bSt from an oil suspension (Mitchell, 1986; Bramley *et al.*, 1989). The data of Mitchell (1986) show that aluminum stearate prolongs the *in vivo* release of zinc-bSt from an oil suspension. Bramley *et al.* (1989) have also included dextrans or celluloses with their oil suspensions to provide elevated serum levels of bSt in sheep for up to 21 days following a 320-mg dose. The celluloses and dextrans either were added directly into the oil suspension or were incorporated via granulation with the bSt prior to suspension in the oil. The addition of glycerides to oil suspensions provided for somewhat improved rat growth enhancement over 7–14 days in the rat growth assay as compared with rats receiving an unmodified oil suspension of bSt (Thakkar and Harrison, 1987). Cady *et al.* (1989) studied the utility of bSt incorporated into glyceryl tristearate microspheres suspended in oil vehicles, and demonstrated elevated plasma levels of bSt and enhanced milk production in cows following a biweekly dose of 525 mg

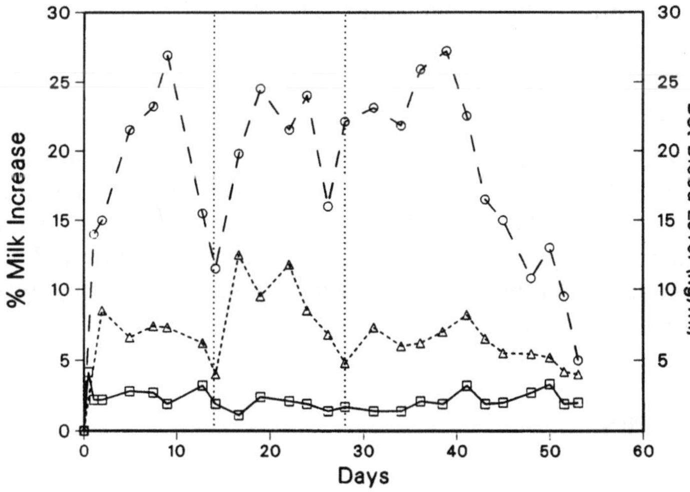

Figure 13. Plasma levels (△) and percent milk yield increase (○) for cows treated on days 0, 14, and 28 with bSt/glyceryl tristearate microspheres suspended in an oleaginous vehicle. Cows received 525 mg bSt dose at each injection. Plasma levels (□) for untreated cows are also shown. Data replotted from Cady *et al.* (1989).

bSt (Fig. 13). The use of waxes in oil suspensions has also resulted in elevated serum levels of bSt for up to 28 days in lactating Holstein cows with a bSt dose of 500–900 mg (Ferguson *et al.,* 1987).

Sustained release from matrices comprised of polymers such as polylactic acids, polycaprolactones, or ethyl vinyl acetate has been studied (Shalati and Viswanathan, 1988) and found to be limited in its ability to prolong the release of bSt. The coating of bSt/polymer matrices with microporous polyethylene as a controlling membrane was used to obtain *in vitro* release rates out to at least 14 days (Sivaramakrishnan *et al.,* 1989). Implantation of bSt pellets into dairy cows at 14-day intervals resulted in elevated serum levels and 15–40% increases in milk yield over controls (Azain *et al.,* 1988). This study was conducted over 56 days. bSt pellets contained little or no excipient. Several aqueous systems have been demonstrated to provide some sustained release. These include associated complexes of bSt with polysaccharides (Cady *et al.,* 1986) and multiple water/oil emulsions (Tyle and Cady, 1990). Ultimately, the utility of these aqueous preparations is likely to be limited by stability problems.

The prevailing formulation approach seems to be that of oil dispersions of protein with various additives or waxes resulting in elevated plasma levels for 1 to 4 weeks. The development of longer-duration delivery systems for the somatotropins may be primarily limited by the stability of the protein within the delivery systems subsequent to implantation. High protein concentration, an aqueous environment, and elevated temperature are deleterious to the integrity of the protein prior to its release from the system (Hageman *et al.,* 1992a,b).

ACKNOWLEDGMENTS

We gratefully acknowledge the help of Julianne Bauer, Kathryn Kienle, and Margaret McShane for preparing figures; Drs. Robert Garlick, James Lauderdale, and Michael Moseley for their assistance in finding specific information; and the editors of this volume for their patience. Lastly, we warmly thank our families for their generous sacrifice of time during the preparation of this chapter.

REFERENCES

Abdel-Meguid, S. S., Shieh, H.-S., Smith, W. W., Dayringer, H. E., Violand, B. N., and Bentle, L. A., 1987, Three-dimensional structure of a genetically engineered variant of porcine growth hormone, *Proc. Natl. Acad. Sci. USA* **84:**6434–6437.

Azain, M. J., Kasser, T. R., Eigenberg, K. E., and Sabacky, M. J., 1988, Somatotropin prolonged release, European Patent Application, EP 0 283 458 A2.

Bogosian, G., Violand, B. N., Dorward-King, E. J., Workman, W. E., Jung, P. E., and Kane, J. F., 1989, Biosynthesis and incorporation into protein of norleucine by *Escherichia coli, J. Biol. Chem.* **264**:531–539.

Bramley, M. R., Carter, A. B., and Dunwell, D. W., 1989, Somatotropin formulations, European Patent Application, EP 0 314 421 A1.

Brems, D. N., Plaisted, S. M., Havel, H. A., Kauffman, E. W., Stodola, J. D., Eaton, L. C., and White, R. D., 1985, Equilibrium denaturation of pituitary- and recombinantly-derived bovine growth hormone, *Biochemistry* **24**:7662–7668.

Brems, D. N., Plaisted, S. M., Kauffman, E. W., and Havel, H. A., 1986, Characterization of an associated equilibrium folding intermediate of bovine growth hormone, *Biochemistry* **25**:6539–6543.

Brems, D. N., Plaisted, S. M., Dougherty, J. J., and Holzman, T. F., 1987. The kinetics of bovine growth hormone folding are consistent with a framework model, *J. Biol. Chem.* **262**:2590–2596.

Burger, H. G., Edelhoch, H., and Condliffe, P. G., 1966, The properties of bovine growth hormone, *J. Biol. Chem.* **241**:449–457.

Cady, S. M., Fishbein, R., and Schroder, U., 1986, Water dispersible and water soluble carbohydrate polymer compositions for parenteral administration, European Patent Application, EP 0 193 917 A2.

Cady, S. M., Steber, W. D., and Fishbein, R., 1989, Development of a sustained release delivery system for bovine somatotropin, *Proc. Int. Symp. Control. Rel. Bioact. Mater.* **16**:22–23.

Calcott, P. H., Kane, J. F., Krivi, G. G., and Bogosian, G., 1988, Parameters affecting production of bovine somatotropin in *Escherichia coli* fermentations, *J. Ind. Microbiol.* **29**:257–266.

Carlacci, L., Chou, K.-C., and Maggiora, G. M., 1991, A heuristic approach to predicting the tertiary structure of bovine somatotropin, *Biochemistry* **30**:4389–4398.

Dellacha, J. M., Santome, J. A., and Paladini, A. C., 1968, Physicochemical and structural studies on bovine growth hormone, *Ann. N.Y. Acad. Sci.* **148**:313–327.

Dougherty, J. J., Jr., Snyder, L. M., Sinclair, R. L., and Robins, R. H., 1990, High-performance tryptic mapping of recombinant bovine somatotropin, *Anal. Biochem.* **190**:7–20.

Edelhoch, H., and Burger, H. G., 1966, The properties of bovine growth hormone. II. Effects of urea, *J. Biol. Chem.* **241**:458–463.

Ellis, S., Lorenson, M., Grindeland, R. E., and Callahan, P. X., 1972, Separation of the phenylalanyl and alanyl chains of bovine and ovine growth hormones by electrofocusing, in: *Growth and Growth Hormones* (A. Pecile and E. E. Muller, eds.), Excerpta Medica, Amsterdam, pp. 55–67.

Ferguson, T. H., Harrison, R. G., and Moore, D. L., 1987, Injectable sustained release formulation, European Patent Application, EP 0 211 691 A2.

Ferguson, T. H., Thompson, W. W., Moore, D. L., and Rodewald, J. M., 1988, Chemical and physical characterization of somidobove, a recombinantly produced bovine somatotropin, *Proc. Int. Symp. Control. Rel. Bioact. Mater.* **15**:55a–55b.

Fleeger, C. A., (ed.), 1991, *USAN and the USP Dictionary of Drug Names,* United States Pharmacopeial Convention, Rockville, Md.

Garlick, R. L., Lyle, S. B., and Mott, J. E., 1989, Somatotropin analogues, International patent application number PCT/US89/05447.

Geiger, T., and Clarke, S., 1987, Deamidation, isomerization, and racemization at asparaginyl and aspartyl residues in peptides, *J. Biol. Chem.* **262**:785–794.

Hageman, M. J., 1987, Preformulation solubility studies on proteins: Recombinant bovine somatotropin, paper presented at the ACS National Meeting, New Orleans.

Hageman, M. J., Possert, P. L., Bauer, J. M., and Darrington, R. T., 1990. Preformulation studies oriented toward sustained delivery of recombinant somatotropins, Paper presented at the ACS National Meeting, Washington, D.C.

Hageman, M. J., Bauer, J. M., Possert, P. L., and Darrington, R. T., 1992a, Preformulation studies oriented toward sustained delivery of recombinant somatotropins, *J. Agric. Food Chem.* **40**:348–355.

Hageman, M. J., Possert, P. L., and Bauer, J. M., 1992b, Prediction and characterization of the water sorption isotherm for bovine somatotropin, *J. Agric. Food Chem.* **40**:342–347.

Harbour, G. C., Hoogerheide, J. G., Lyle, S. B., Olsen, R. F., Charlton, M. D., Haak, W. J., and Garlick, R. L., 1989, Alkaline incubation of recombinant bovine somatotropin leads to isoaspartic acid formation and chain cleavage at position 99, Paper presented at the Third Symposium of the Protein Society, Seattle.

Harbour, G. C., Lyle, S. B., Hoogerheide, J. G., Crow, F. W., and Garlick, R. L., 1991, N-ϵ-acetylation can occur at lysines residues 157, 167, 171, and 180 of recombinant bovine somatotropin, Paper presented at the Fifth Symposium of the Protein Society, Baltimore.

Haro, L. S., Collier, R. J., and Talamantes, F. J., 1984, Homologous somatotropin radioreceptor assay utilizing recombinant bovine growth hormone, *Mol. Cell. Endocrinol.* **38**:109–116.

Hartman, P. A., Stodola, J. D., Harbour, G. C., and Hoogerheide, J. G., 1986, Reversed-phase high-performance liquid chromatography peptide mapping of bovine somatotropin, *J. Chromatogr.* **360**:385–395.

Havel, H. A., Kauffman, E. W., Plaisted, S. M., and Brems, D. N., 1986, Reversible self-association of bovine growth hormone during equilibrium unfolding, *Biochemistry* **25**:6533–6538.

Havel, H. A., Chao, R. S., Haskell, R. J., and Thamann, T. J., 1989, Investigations of protein structure with optical spectroscopy: Bovine growth hormone, *Anal. Chem.* **61**:642–650.

Holladay, L. A., Hammonds, R. G., Jr., and Puett, D., 1974, Growth hormone conformation and conformational equilibria, *Biochemistry* **13**:1653–1661.

Holzman, T. F., Dougherty, J. J., Jr., Brems, D. N., and MacKenzie, N. E., 1990, pH-induced conformational states of bovine growth hormone, *Biochemistry* **29:**1255–1261.

Isaksson, O. G. P., Eden, S., and Jansson, J.-O., 1985, Mode of action of pituitary growth hormone on target cells, *Annu. Rev. Physiol.* **47:**483–499.

Kaumaya, P. T. P., Berndt, K. D., Heidorn, D. B., Trewhella, J., Kezdy, F. C., and Goldberg, E., 1990, Synthesis and biophysical characterization of engineered topographic immunogenic determinants of $\alpha\alpha$ topology, *Biochemistry* **29:**13–23.

Kim, P. S., and Baldwin, R. L., 1982, Specific intermediates in the folding reactions of small proteins and the mechanism of protein folding, *Annu. Rev. Biochem.* **51:**459–489.

Kuneida, H., and Shinoda, K., 1975, Dissolution mechanism of water-soluble polymers, in: *Colloidal Dispersions and Micellar Behavior* (K. L. Mittal, ed.), American Chemical Society, Washington, D.C.

Kuntz, I. D., Jr., and Kauzmann, W., 1974, Hydration of proteins and polypeptides, *Adv. Protein Chem.* **28:**239–345.

Kuu, W. Y., Prisco, M. R., Wood, R. W., and Roseman, T. J., 1989, Studies of dissolution behavior of highly soluble drugs using a rotating disk, *Int. J. Pharm.* **55:**77–89.

Lakowicz, J. R., 1983, *Principles of Fluorescence Spectroscopy,* Plenum Press, New York, pp. 342–379.

Langley, K. E., Berg, T. F., Strickland, T. W., Fenton, D. M., Boone, T. C., and Wypych, J., 1987a, Recombinant-DNA-derived bovine growth hormone from *Escherichia coli.* 1. Demonstration that the hormone is expressed in reduced form, and isolation of the hormone in oxidized, native form, *Eur. J. Biochem.* **163:**313–321.

Langley, K. E., Lai, P.-H., Wypych, J., Everett, R. R., Berg, T. F., Krabill, L. F., Davis, J. M., and Souza, L. M., 1987b, Recombinant-DNA-derived bovine growth hormone from *Escherichia coli.* 2. Biochemical, biophysical, immunological and biological comparison with pituitary hormone, *Eur. J. Biochem.* **163:**323–330.

Leung, F. C., Jones, B., Steelman, S. L., Rosenblum, C. I., and Kopchick, J. J., 1986, Purification and physiochemical properties of a recombinant bovine growth hormone produced by cultured murine fibroblasts, *Endocrinology* **119:** 1489–1496.

Lewis, U. L., Cheever, E. V., and Hopkins, W. C., 1970, Kinetic study of the deamidation of growth hormone and prolactin, *Biochim. Biophys. Acta* **214:**498–508.

Li, C. H., and Ash, L., 1953, The nitrogen terminal end-groups of hypophyseal growth hormone, *J. Biol. Chem.* **203:**419–424.

Lingappa, V. R., Devillers-Thiery, A., and Blobel, G., 1977, Nascent prehormones are intermediate in the biosynthesis of authentic bovine growth hormone and prolactin, *Proc. Natl. Acad. Sci. USA* **74:**2432–2436.

Lyle, S. B., Mott, J. E., Trepod, C. M., Hoogerheide, J. G., Fairbanks, M. B., Campbell, C. M., and Garlick, R. L., 1990, Replacement of asparagine-99 in recombinant bovine somatotropin leads to enhanced aqueous stability, Paper presented at the Fourth Symposium of the Protein Society, San Diego.

Machlin, L. J., 1973, Effect of growth hormone on milk production and feed utilization in dairy cows, *J. Dairy Sci.* **56**:575–580.

Manning, M. C., Patel, K., and Borchardt, R. T., 1989, Stability of protein pharmaceuticals, *Pharm. Res.* **6**:903–918.

Marx, W., Simpson, M. E., and Evans, K. M., 1942, Bioassay of the growth hormone of the anterior pituitary, *Endocrinology* **30**:1–10.

Miller, W. L., Martial, J. A., and Baxter, J. D., 1980, Molecular cloning of DNA complementary to bovine growth hormone mRNA, *J. Biol. Chem.* **255**:7521–7524.

Mitchell, J. W., 1986, Prolonged release of biologically active polypeptides, European Patent Application, EP 0 177 478 A2.

Mitchell, J. W., 1991, Prolonged release of biologically active somatotropin, U.S. Patent 5,013,713.

Moseley, M., Paulissen, J. B., Goodwin, M. C., Alaniz, G. R., and Claflin, W. H., 1992, Recombinant bovine somatotropin improves growth performance in finishing beef steers, *J. Anim. Sci.* **70**:412–425.

Ouano, A. C., Tu, Y. O., and Carothers, J. A., 1977, Dynamics of polymer dissolution, in: *Structure–Solubility Relationships in Polymers* (F. W. Harris, ed.), Academic Press, New York, pp. 11–20.

Peel, C. J., Bauman, D. E., Gorewit, R. C., and Sniffen, C. J., 1981, Effect of exogenous growth hormone on lactational performance in high yielding dairy cows, *J. Nutr.* **111**:1662–1671.

Peel, C. J., Fronk, T. J., Bauman, D. E., and Gorewit, R. C., 1983, Effect of exogenous growth hormone in early and late lactation on lactational performance of dairy cows, *J. Dairy Sci.* **66**:776–782.

Pikal, M. J., Dellerman, K. M., Roy, M. L., and Riggin, R. M., 1991, The effects of formulation variables on the stability of freeze-dried human growth hormone, *Pharm. Res.* **8**:427–436.

Ritschel, W. A., 1973, Parenteral dosage forms with prolonged action, in: *Drug Design* (E. J. Amiens, ed.), Academic Press, New York, pp. 75–92.

Santome, J. A., Dellacha, J. M., Paladini, A. C., Pena, C., Biscoglio, M. J., Daurat, S. T., Poskus, E., and Wolfenstein, C. E. M., 1973, Primary structure of bovine growth hormone, *Eur. J. Biochem.* **37**:164–170.

Schoner, B. E., Hsiung, H. M., Belagaje, R. M., Mayne, N. G., and Schoner, R. G., 1984, Role of mRNA translational efficiency in bovine growth hormone expression in *Escherichia coli, Proc. Natl. Acad. Sci. USA* **81**:5403–5407.

Schoner, R. G., Ellis, L. F., and Schoner, B. E., 1985, Isolation and purification of protein granules from Escherichia coli cells overproducing bovine growth hormone, *Bio/Technology* **1985**:151–155.

Secchi, C., Biondi, P. A., Negri, A., Borroni, R., and Ronchi, S., 1986, Detection of desamido forms of purified bovine growth hormone, *Int. J. Pept. Protein Res.* **28**:298–306.

Seeburg, P. H., Sias, S., Adelman, J., de Boer, H. A., Hayflick, J., Jhurani, P., Goeddel, D. V., and Heyneker, H. L., 1983, Efficient bacterial expression of bovine and porcine growth hormones, *DNA* **2**:37–45.

Shalati, M. D., and Viswanathan, R., 1988, Sustained release implant and method for preparing same, US Patent 4,761,289.

Shieh, H.-M., Buckwalter, B. L., Cady, S. M., Chaudhuri, A. K., and Johnson, D. F., 1990, Enhancement of PST stability by chemical modifications of cysteine residues, Paper presented at the 200th ACS National Meeting, Washington, D.C.

Sivaramakrishnan, K. N., Rahn, S. L., Moore, B. M., and O'Neil, J., 1989, Sustained release of bovine somatotropin from implants, *Proc. Int. Symp. Control. Rel. Bioact. Mater.* **16**:14–15.

Sonenberg, M., and Beychok, S., 1971, Circular dichroism studies of biologically active growth hormone preparations, *Biochim. Biophys. Acta* **229**:88–101.

Stodola, J. D., Walker, J. S., Dame, P. W., and Eaton, L. C., 1986, High-performance size-exclusion chromatography of bovine somatotropin, *J. Chromatogr.* **357**:423–428.

Terada, H., Inoue, Y., and Ichikawa, T., 1984, Second derivative spectral properties of tryptophan and tyrosine residues in proteins. Effects of guanidine hydrochloride and dodecyl sulfate on the residues in lysozyme, ribonuclease and serum albumin, *Chem. Pharm. Bull.* **32**:585–590.

Thakkar, A. L., and Harrison, R. G., 1987, Injectable semi-solid formulations, European Patent Application, EP 0 213 851 A2.

Tyle, P., and Cady, S. M., 1990, Sustained release multiple emulsions for bovine somatotropin delivery, *Proc. Int. Symp. Control. Rel. Bioact. Mater.* **17**:49–50.

Violand, B. N., Schlittler, M. R., Toren, P. C., and Siegel, N. R., 1990, Formation of isoaspartate 99 in bovine and porcine somatotropins, *J. Prot. Chem.* **9**:109–117.

Wallis, M., 1973, The primary structure of bovine growth hormone, *FEBS Lett.* **35**:11–14.

Wearne, S. J., and Creighton, T. E., 1989, Effect of protein conformation on rate of deamidation: Ribonuclease A, *Proteins: Structure, Function, and Genetics* **5**:8–12.

Wingfield, P. T., Graber, P., Buell, G., Rose, K., Simona, M. G., and Burleigh, B. D., 1987, Preparation and characterization of bovine growth hormones produced in recombinant *Escherichia coli*, *Biochem. J.* **243**:829–839.

Wood, D. C., Salsgiver, W. J., Kasser, T. R., Lange, G. W., Rowold, E., Violand, B. N., Johnson, A., Leimgruber, R. M., Parr, G. R., Siegel, N. R., Kimack, N. M., Smith, C. E., Zobel, J. F., Ganguli, S. M., Garbow, J. R., Bild, G., and Krivi, G. G., 1989, Purification and characterization of pituitary bovine somatotropin, *J. Biol. Chem.* **264**:14741–14747.

Chapter 3

Stability Characterization and Formulation Development of Alteplase, a Recombinant Tissue Plasminogen Activator

Tue H. Nguyen and Carole Ward

1. INTRODUCTION

Plasminogen activators (PA) are endogenous serine proteases involved in a cascade of events leading to the dissolution of a blood clot. These proteins are classified in two distinct groups: urokinase-type plasminogen activator (u-PA) and tissue-type plasminogen activator (t-PA).

Alteplase is a tissue plasminogen activator produced by recombinant DNA technology (rt-PA). It is a glycoprotein synthesized by expressing the complementary DNA (cDNA) for natural human t-PA in Chinese hamster ovary (CHO) cells. It is produced and marketed under the brand name of Activase for use in the management of acute myocardial infarction and pulmonary embolism.

In this chapter, the biochemistry and pharmacology of alteplase are briefly reviewed, analytical methods specific for the characterization of alteplase are described, and stability studies leading to the development of Activase are summarized.

Tue H. Nguyen and Carole Ward • Department of Pharmaceutical Research and Development, Genentech, Inc., South San Francisco, California 94080.

Stability and Characterization of Protein and Peptide Drugs: Case Histories, edited by Y. John Wang and Rodney Pearlman. Plenum Press, New York, 1993.

1.1. The Fibrinolytic System

Fibrinolysis refers to a process in which fibrin is proteolytically cleaved into soluble products. The natural fibrinolytic system which removes fibrin clots from the blood vessels consists of plasminogen and PA. Plasminogen is a zymogen which can be converted to the active enzyme plasmin following cleavage of the Arg_{560}–Val_{561} peptide bond by PA (Robins *et al.*, 1967). Plasmin is a serine protease with low specificity. Its primary function is to degrade the fibrin network in the thrombus, but it also degrades several other proteins that are essential for thrombosis including fibrinogen, factor V, and factor VIII (Collen, 1987). Plasma contains a plasmin inhibitor α_2-antiplasmin that binds and inhibits free circulating plasmin. However, when plasmin is produced abundantly, following an extensive systemic activation of plasminogen, α_2-antiplasmin may be depleted resulting in a systemic fibrinolytic state.

The endogenous level of t-PA in human blood is between 1 and 5 ng/ml (Bergdorf *et al.*, 1983). Under normal hemodynamic conditions t-PA is relatively inactive, but its ability to activate plasminogen is markedly enhanced in the presence of fibrin (Camiolo *et al.*, 1971). Kinetic studies indicate that the potentiation of t-PA by fibrin is the result of a marked decrease in the Michaelis constant (K_m) of the reaction whereas the catalytic rate constant (k_{cat}) remains unchanged. Hoylaerts *et al.* (1982) suggested that t-PA first binds to fibrin forming a bimolecular complex with high affinity for plasminogen. Plasminogen then adsorbs to the t-PA–fibrin complex and the conversion of plasminogen to plasmin occurs primarily on the surface of the fibrin clot with negligible systemic activation. Plasmin generated on the thrombus can efficiently exert its proteolytic activity. Its inaccessibility to circulating inhibitors results in sustained thrombolysis despite the short half-life of t-PA in the circulation (Eisenberg *et al.*, 1987). This rationale is currently the basis of clot-specific thrombolysis therapy.

1.2. Tissue Plasminogen Activator

t-PA has been identified and isolated from many tissues including pig heart, hog ovary, blood and blood vessel perfusate (Astrup and Permin, 1947; Cole and Bachman, 1977; Kok and Astrup, 1969; Rijken *et al.*, 1980). Human endogenous t-PA was first purified from uterine tissue (Rijken *et al.*, 1979a, b) and shown to be immunologically identical to vascular t-PA and t-PA derived from blood (Rijken *et al.*, 1980). t-PA was also found in certain cells in culture (Reich, 1975). The Bowes melanoma cell line (Rijken and

Collen, 1981; Collen *et al.*, 1982) produced t-PA in sufficient amounts for in-depth studies of its biological and biochemical properties (Collen *et al.*, 1982; Kruithof *et al.*, 1985; Wallen *et al.*, 1983). The melanoma-derived t-PA was subsequently used in two limited clinical trials and shown to be a promising agent for thrombolytic therapy (Weimar *et al.*, 1981; Van de Werf *et al.*, 1983).

rt-PA was initially cloned and expressed in *E. coli* which produced an active but nonglycosylated version of the protein (Pennica *et al.*, 1983). Subsequent expression in mammalian cell lines has yielded glycosylated rt-PA and alteplase is a t-PA produced by expressing natural human t-PA in CHO cells.

1.3. Pharmacology and Clinical Use of Alteplase

Pharmacokinetics and Metabolism

In normal male volunteers the disposition of alteplase following a 10-min constant-rate intravenous infusion of 0.25 mg/kg can be described by a two-compartment model. The disappearance of alteplase from the central compartment is dominated by the initial α phase with a half-life ($t_{1/2}$) of 3 to 5 min; the half-life of the β phase is approximately 30 min (Baughman, 1987). As a result, alteplase is cleared rapidly from the systemic circulation at a rate of approximately 650 to 700 ml/min. From the same study, it was concluded that normal, younger volunteers have a slightly increased clearance when compared with an older population.

The same dosage regimen was used to characterize the pharmacokinetic parameters in eight patients with thromboocclusive disease (Baughman, 1987). The results are consistent with those obtained from normal volunteers. A slightly lower clearance (549 ml/min, range: 269 to 926 ml/min) was reported in this study. Additional pharmacokinetic parameters are shown in Table I.

In a different trial, alteplase was studied in patients with myocardial infarction at infusion rates of 4.0, 5.0, 5.5, 7.7 μg/kg per min with 10% of the dose administered initially as a bolus (Garabedian *et al.*, 1987). A fifth group of patients received 8.3 μg/kg per min over 60 min without a loading dose. The five dosing regimens resulted in plateau plasma concentrations of 0.52 μg/ml to 1.4 μg/ml and over the dose range studied, linear pharmacokinetics were observed (Seifried *et al.*, 1986; Baughman, 1987).

Rabbits and rats that were given an intravenous injection of ^{125}I-labeled rt-PA showed an initial accumulation in the liver (40–60% of administered

Table I
Pharmacokinetic Parameter Estimates following a 0.25 mg/kg rt-PA Infusion over 10 min in Eight Patients with Thromboocclusive Disease

	Mean	±SD	Range
Two-compartment model			
$t_{1/2}$ (min), α	4.36	0.94	3.35–5.77
$t_{1/2}$ (min), β	26.5	11.0	15.0–47.5
V_1 (liters)	3.82	1.40	1.69–6.11
AUC (ng/min per liter)	36.3	11.9	21.0–55.8
CL (ml/min)	549	180	269–926
Compartment-independent			
AUC (ng/min per liter)	34.3	8.98	23.2–49.3
VD_{ss} (liters)	10.3	3.05	5.336–15.1
CL (ml/min)	554	175	304–861

dose) which rapidly decreased to around 5–10% of the total dose after 60 min (Bakhit *et al.*, 1987; Nilsson *et al.*, 1985). The disappearance of labeled alteplase from the liver was accompanied by an increase in the trichloroacetic acid (TCA)-soluble fragment in the blood, kidney, and urine which suggests that t-PA is metabolized in the liver, secreted into the blood, and then excreted in the urine.

Recently, Activase was approved for thrombolytic therapy in acute pulmonary embolism in adult. The recommended dose is 100 mg given as a slow intravenous infusion over 2 hr. Clinical trials are being conducted to assess the possibility of using alteplase in the management of stroke and peripheral arterial occlusion.

2. STRUCTURE AND PROPERTIES OF ALTEPLASE

2.1. Primary Structure

Alteplase is a single polypeptide chain composed of 527 amino acids held together by 17 disulfide bridges. The primary sequence of alteplase and its amino acid composition are illustrated in Fig. 1. Its molecular mass is 59,042 Da based on amino acid composition; however, the molecule migrates as a 63-kDa protein on SDS-PAGE (Section 3.2.1). This apparent increase in size is attributed to the presence of the *N*-linked carbohydrate chains. Results obtained from sedimentation velocity experiments indicate that alteplase exists primarily as a monomer which migrates at 4.5 S (Shire

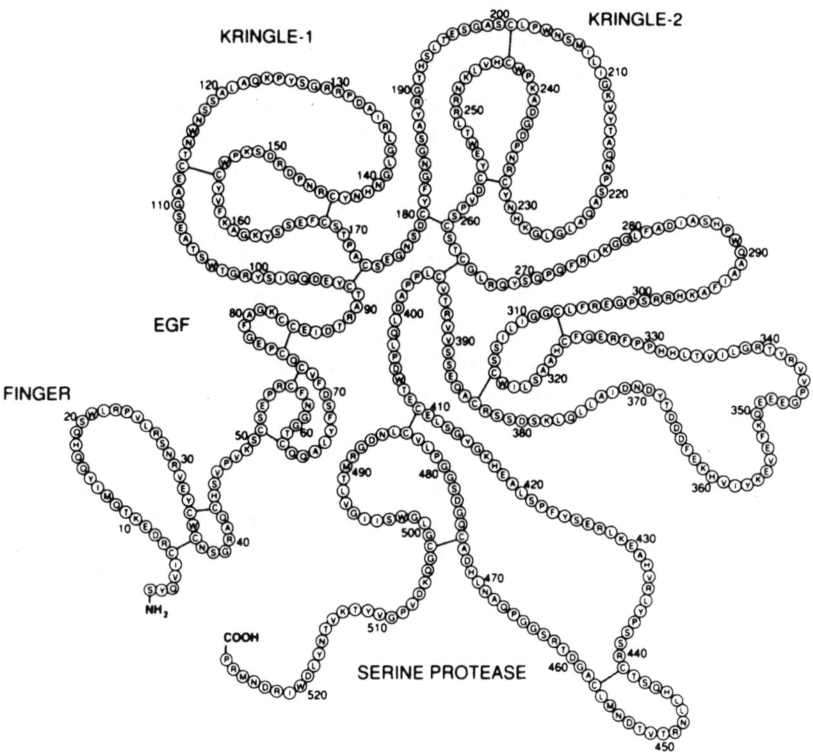

Figure 1. Primary structure of alteplase.

and Nguyen, unpublished data) in arginine phosphate buffer solution at neutral pH.

By homology with other proteins, alteplase is divided into five domains. A domain homologous to fibronectin type I finger extends from residues 1 to 43 (Banyai *et al.*, 1983); residues 44 to 91 are homologous to human epidermal growth factor; the kringle 1 (residues 92 to 173) and kringle 2 (residues 180 to 261) regions are homologous to the kringle regions found in plasminogen and prothrombin; the COOH-terminal region of alteplase, comprising residues 276 to 527 is homologous to the trypsin family of serine proteases and contains a catalytic active site formed by His_{322}, Asp_{371}, and Ser_{478} (Pennica *et al.*, 1983).

Alteplase interacts with the fibrin clot at several sites. Studies with deletion mutants suggested that the finger region, the epidermal growth factor domain, and kringle 2 are all involved in the formation of t-PA–fibrin complex (Van Zonnenfeld *et al.*, 1986a,b; Kalyan *et al.*, 1988; Larsen *et al.*,

1988). The stoichiometry of binding is about 1 mole of t-PA to 1 mole of fibrin monomer (Higgins and Vehar, 1987) with a dissociation constant (K_D) of approximately 0.6 μM (Liu and Wallen, 1984). The kringle 2 region also contains a binding site for lysine with a K_D of approximately 100 μM. Equilibrium dialysis experiments indicated that a charge interaction with the ϵ-amino group is essential, as evident from 2000- and 200-fold decreases in binding affinity to *N*-acetyl-L-lysine and L-arginine, respectively. Methylation of the ligand α-carboxylate moiety or peptide bond formation minimally affects the binding constant (Cleary *et al.*, 1989).

The peptide bond between Arg_{275} and Ile_{276} is particularly susceptible to cleavage. The reaction produces two fragments: a heavy chain consisting of the first 276 amino acids and the COOH-terminal region constituting the light chain. The two chains are linked together by a single disulfide bridge, formed by Cys_{264} and Cys_{395} (Rijken and Collen, 1981). The cleaved molecule is commonly referred to as "two-chain t-PA" as opposed to "one-chain" or the intact form. One-chain t-PA is catalytically less active toward low-molecular-weight substrates than the two-chain form (Wallen *et al.*, 1982) but their activities toward plasminogen are comparable (Rijken *et al.*, 1982).

2.2. Carbohydrate Composition

Alteplase contains four potential sites for *N*-linked glycosylation identified by the sequence Asn-X-Ser/Thr. These are Asn_{117}, Asn_{184}, Asn_{218}, and Asn_{448}. However, not all sites are occupied. Type II alteplase has a high-mannose oligosaccharide linked to Asn_{117} and a complex oligosaccharide at Asn_{448}. Type I alteplase has an additional complex oligosaccharide at Asn_{184} (Table II). The current product is a 50 : 50 mixture of type I and type II alteplase. The oligosaccharide chains are sialated via the $\alpha_{2,3}$ linkages to

Table II
Summary of Various Subspecies of Alteplase

Characteristics	Designation
Glycosylation at Asn_{117}, Asn_{184}, Asn_{448} Arg_{275}–Ile_{276} intact	Type I, one-chain alteplase
Glycosylation at Asn_{117}, Asn_{184}, Asn_{448} Arg_{275}–Ile_{276} cleaved	Type I, two-chain alteplase
Glycosylation at Asn_{117}, Asn_{448} Arg_{275}–Ile_{276} intact	Type II, one-chain alteplase
Glycosylation at Asn_{117}, Asn_{448} Arg_{275}–Ile_{276} cleaved	Type II, two-chain alteplase

galactose. Asn_{218} is not glycosylated. Carbohydrates account for approximately 5–7% of the molecular weight of the molecule (Spellman *et al.*, 1989).

Although the glycosylation pattern of t-PA derived from melanoma cells is identical to that of alteplase, the specific carbohydrate structures identified on alteplase from Bowes melanoma cells are somewhat different (Spellman *et al.*, 1989). There are no published data on the glycosylation pattern of endogenous t-PA produced by human endothelial cells.

Although nonglycosylated rt-PA expressed in *E. coli* and *N*-deglycosylated alteplase are active (Pennica *et al.*, 1983; Little *et al.*, 1984), the number of glycosylated sites and carbohydrate composition can influence the *in vitro* specific activity of alteplase as well as its *in vivo* clearance rate. Treatment of t-PA with neuraminidase and β-galactosidase resulted in a decrease in *in vitro* clot lysis activity; α-mannosidase caused an increase (Opdenakker *et al.*, 1986). Cell culture experiments showed that alteplase binds to the mannose receptor in liver endothelial cells and the galactose receptor in parenchymal cells. The protein is then endocytosed, metabolized by the hepatocytes (Ashwell and Hartford, 1982; Einarsson *et al.*, 1988; Smedsrod *et al.*, 1988). Oxidation of the carbohydrate residues or removal of the high-mannose oligosaccharides with endo-β-*N*-acetyl-glucosaminidase resulted in a two- to threefold decrease in the rate of clearance of alteplase in the rabbit. Conversely, when rt-PA was glycosylated only with high-mannose oligosaccharides, the clearance rate was increased by a factor of 2 (Hotchkiss *et al.*, 1988).

2.3. Solubility Behavior

Altephase is a heterogeneous mixture of approximately 50% type I and 50% type II rt-PA. It consists of 65 to 80% of the one-chain form and 20 to 35% of the two-chain form. While there is no evidence that one- and two-chain alteplase differ in their *in vivo* thrombolytic activity, carbohydrate content has a marked influence on the solubility of the protein. For example, the solubilities of type I and type II alteplase in 13 mM citrate pH 6.0 (ionic strength: 60 mM) are 0.12 and 0.02 mg/ml, respectively (Hwang-Felgner and Nguyen, unpublished data).

As with most proteins, the solubility of alteplase is pH dependent as illustrated in Fig. 2. The pH–solubility profile exhibits a typical V shape with a minimum around pH 6.0. At pH 6.0 and buffer concentration of 20 mM, the rank order observed was imidazole < phosphate < histidine < succinate < EDTA < citrate with an apparent solubility of alteplase in citrate buffer

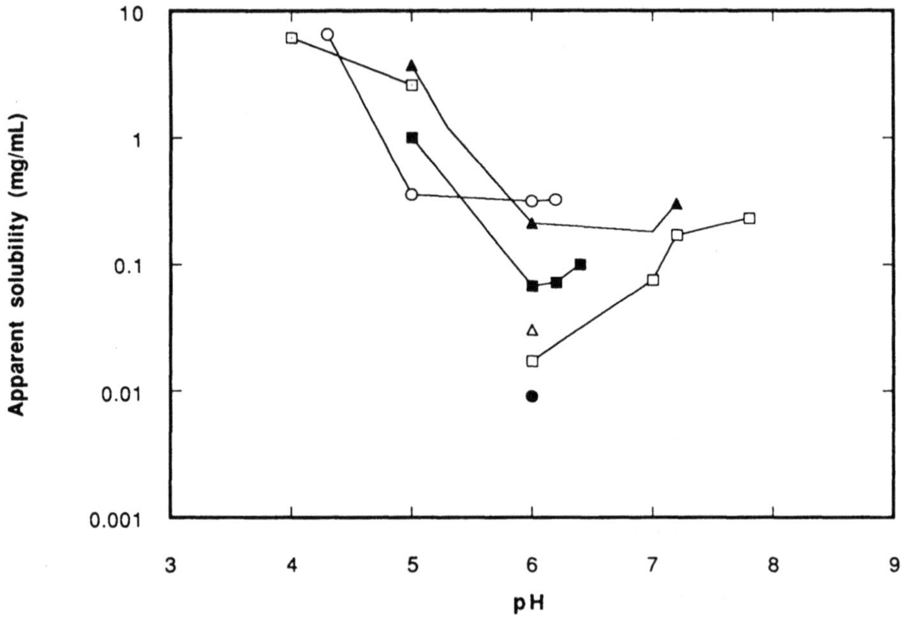

Figure 2. Effect of solution pH and buffer species on alteplase solubility at 25°C. Buffer concentration was maintained at 20 mM. ○, citrate; ▲, EDTA; □, phosphate; ■, succinate; △, histidine; ⊡, acetate; ●, imidazole.

being almost 50-fold greater than that observed in imidazole. Between pH 6.0 and 8.0, the dissolved alteplase was almost exclusively of type I. Type II alteplase was only significantly soluble at pH values below 5.0.

The solubility of alteplase in the neutral pH range was primarily driven by the solution ionic strength and buffer species. Figure 3 illustrates the dependence of alteplase solubility at pH 6.0 on ionic strength. A direct relationship between alteplase solubility and the amount of NaCl added to a 20 mM sodium succinate buffer at pH 6.0 was also observed. The solubilities of alteplase in this buffer were 0.06, 0.40, and 0.95 mg/ml at 0.0, 0.15, and 0.3 M NaCl, respectively (Ward and Gatlin, unpublished data).

Lysine, derivatives of lysine, ϵ-amino caproic acid, glycylglycine, arginine (Cleary *et al.*, 1989), and ornithine (Hasegawa *et al.*, 1986) bind to the alteplase kringle 2 region to various degrees. These ligands also solubilize alteplase when added to the protein solution. Figure 4 illustrates the relationship of the solubility of alteplase as a function of arginine concentration at pH 7.2. At concentrations below 75 mM there is no clear enhancement of solubility of either type I or type II alteplase. As the arginine concentration approaches 100 mM, the solubility of type I increases markedly to approxi-

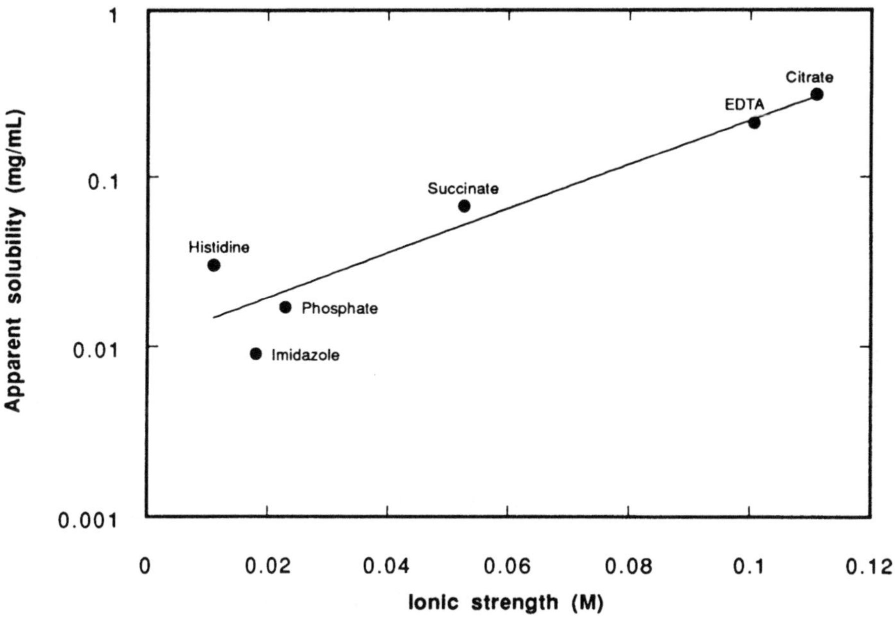

Figure 3. Effect of ionic strength on the apparent solubility of alteplase at pH 6.0. Buffer concentration was maintained at 20 mM with no other salt added. Ionic strength was calculated based on the pK_a of the acidic buffer species.

mately 30 mg/ml. At 175 mM arginine, Type I alteplase dissolves to more than 90 mg/ml. Its solubility at 200 mM arginine could not be determined; more than 150 mg of type I alteplase could be dissolved in 1 ml and the solution became gelatinous. On the other hand, type II solubility increases almost linearly with arginine concentration up to 175 mM at which point almost 5 mg/ml can be reached. At 200 mM arginine, the solubility of type II alteplase is approximately 18 mg/ml.

3. ANALYTICAL CHARACTERIZATION

Each lot of alteplase is subjected to standard identity and purity tests. The amino acid composition of the molecule is determined by total hydrolysis of the molecule followed by derivatization and reversed-phase liquid chromatographic (RP-HPLC) analysis. The identity of the protein is also ascertained by comparing the chromatogram of a tryptic digest of alteplase with a reference tryptic map (Chloupek *et al.*, 1989) as well as by a limited NH_2-ter-

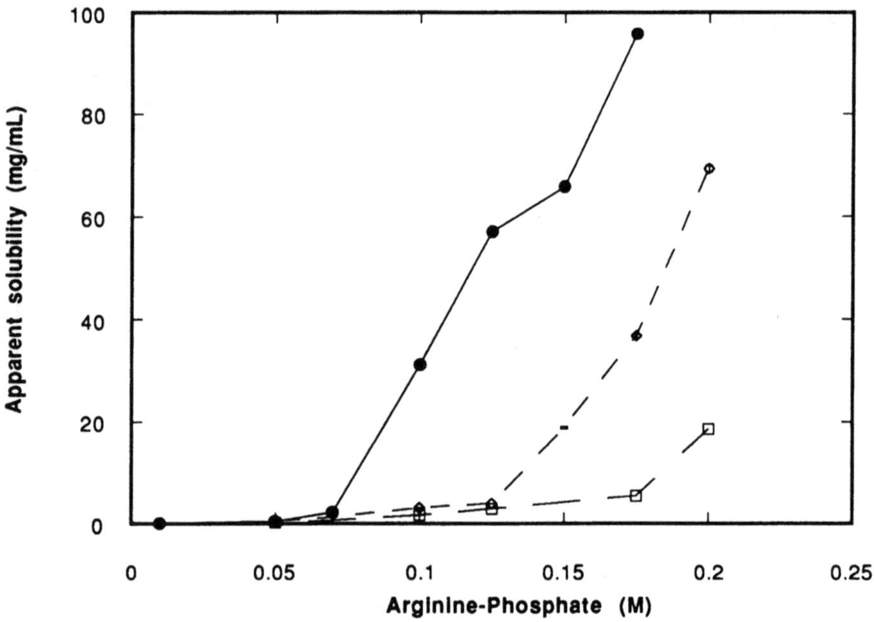

Figure 4. Effect of arginine on type I and type II alteplase solubility at pH 7.2 and 25°C. ●, type I alteplase; □, type II alteplase; ◇, 50 : 50 mixture of type I and type II alteplase.

minal sequencing through at least ten cycles. NH_2-terminal amino acid analysis and SDS-PAGE are employed to detect contaminant proteins. The concentration of purified alteplase in solution can be conveniently determined by UV spectrophotometry. An enzyme-linked immunosorbent assay (ELISA) using polyclonal goat antibody is available for the determination of the protein concentration in complex samples such as plasma or cell culture fluid (Mohler *et al.*, 1986; Amiral *et al.*, 1988; Ranby *et al.*, 1989).

Additionally, assays that are specific to alteplase were developed to monitor the bioactivity and the physical integrity of the molecule upon long-term storage of the formulation. These assays include size-exclusion (SE) HPLC, RP-HPLC, chromogenic enzymatic assay, and a clot lysis assay.

In the following sections, assays that are particularly useful for characterization of the identity, purity, and stability of alteplase are described.

3.1. Spectroscopy

Spectroscopic methods are used for the measurement of alteplase protein concentration and for qualitative evaluation of changes in secondary, tertiary, and quaternary structure in solution. Alteplase as manufactured by recombinant DNA technology consists of a mixture of four proteins, type I one-chain, type I two-chain, type II one-chain, and type II two-chain. This mixture is constant from lot to lot. Spectrophotometric studies were performed on the mixture to characterize the drug substance and to establish a specific absorptivity for use in concentration measurements. The individual species were also separated and their spectral characteristics compared for indications of structural differences.

3.1.1. UV Spectroscopy

The absorption of light by a protein in the UV range (175–350 nm) is a complex phenomenon and the resultant spectrum is a net summation of the absorption of light by the carbonyl group of the amide bond (190–210 nm), the aromatic amino acid side chain residues such as phenylalanine, tyrosine, and tryptophan (250–320 nm), and the disulfide bonds (250–300 nm).

Absorption of light by the individual chromophores can change due to changes in the environment of the individual chromophores that may occur

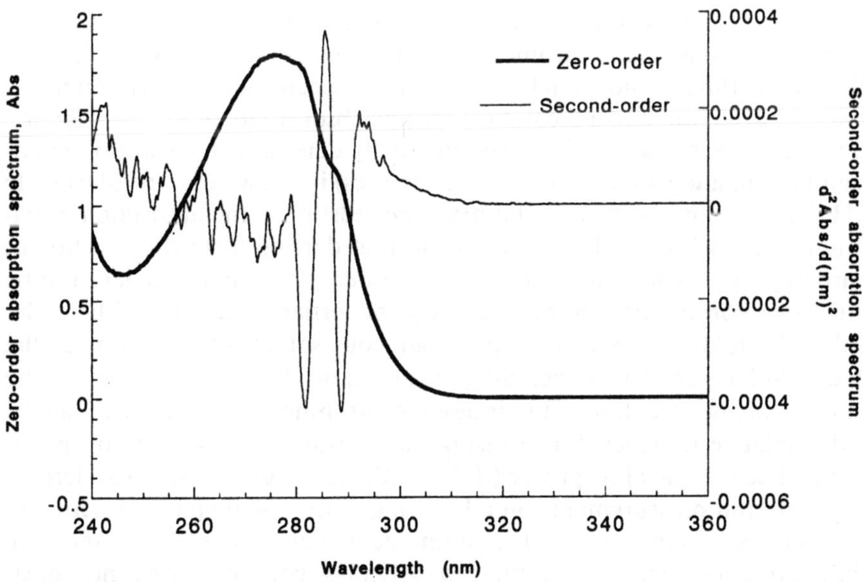

Figure 5. Theoretical UV absorption spectrum of alteplase as derived from the UV absorption spectra of its constituent chromophores.

Table III
Wavelength Minima from the Second Derivative of the UV Absorption Spectra of Various Alteplase Species

Peak No.	Type I, 1-chain (nm)	Type I, 2-chain (nm)	Type II, 1-chain (nm)	Type II, 2-chain (nm)
1	300.6	298.9	298.8	299.6
2	290.8	291.0	290.0	291.0
3	284.4	284.6	284.4	284.6
4	279.3	279.3	279.3	279.5
5	274.0	276.1	273.9	279.5
5a		274.1		
6	268.4	269.8	268.5	268.5
6a		268.1		
7	264.5	264.7	264.6	264.6
8	259.8	261.1	258.4	261.5
8a	258.6	258.6		258.9
9	253.4	253.0	252.7	253.0

from perturbations of the primary, secondary, tertiary, and quaternary structure of the protein.

Alteplase contains 16 phenylalanine, 24 tyrosine, and 13 tryptophan residues, and its tertiary structure is held together by 17 disulfide bonds. The theoretical UV absorption spectrum of alteplase derived from the summation of the constituent chromophores (Bewley, personal communication; Wetlaufer, 1962) is shown in Fig. 5. From this spectrum an absorption maximum of 276.0 nm and a theoretical specific absorptivity of $A^{0.1\%} = 1.8\ \text{cm}^{-1}$ $(\text{mg/ml})^{-1}$ are predicted. The experimentally determined absorption maximum for an alteplase mixture (all four species) in 0.2 M arginine phosphate pH 7.3 solution occurs at 280.0 nm. The redshift of the absorption maximum of the whole molecule relative to that derived from the constituent chromophores is well documented (Wetlaufer, 1962) and is believed to be due to the contribution of the secondary and tertiary structures of the molecule. The folding of a protein in an aqueous solution tends to place the hydrophobic side chains, including the aromatic chromophores, within the interior of the structure. This interior environment has a high refractive index relative to water, thus causing the redshift in the absorption maximum. The specific absorptivity ($A^{0.1\%}$) of the alteplase mixture was derived from spectral measurements on a known quantity of alteplase protein and the value is $1.9\ \text{cm}^{-1}$ $(\text{mg/ml})^{-1}$. Amino acid analysis was used to measure the alteplase concentration in this case. Alteplase concentration is now measured routinely by UV spectroscopy using this experimentally determined specific absorptivity value.

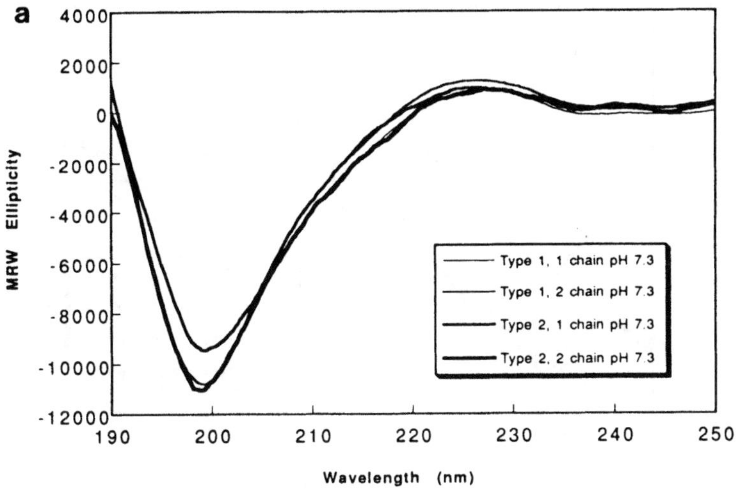

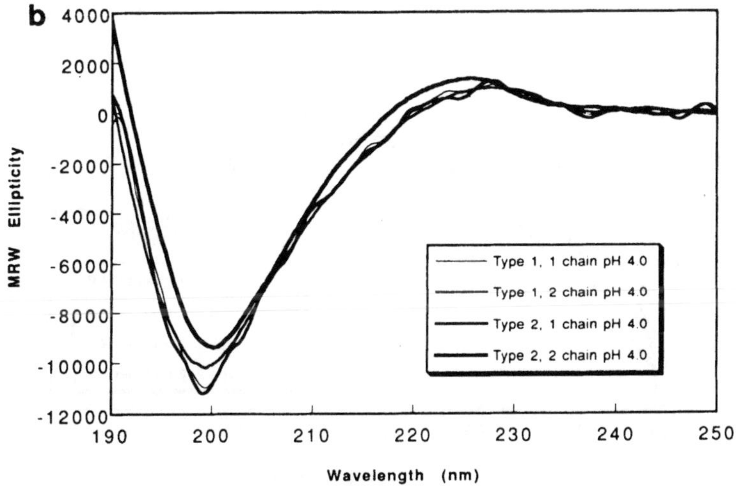

Figure 6. Far-UV CD spectra of various alteplase species in (a) 0.2 M sodium phosphate pH 7.3 and (b) 0.2 M sodium acetate pH 4.0.

The second-derivative spectrum (Fig. 5) is a useful tool for comparison of spectra in that spectral differences are amplified, and by implication, possible differences in chromophore environments are easier to observe. Comparison of the second-derivative minima of the four alteplase species measured in 0.2 M sodium phosphate buffer pH 7.0 is shown in Table III. The molecules exhibit slightly different spectral characteristics, especially in the region of phenylalanine absorption (274–253 nm), suggesting that some small conformational differences might exist among the four species.

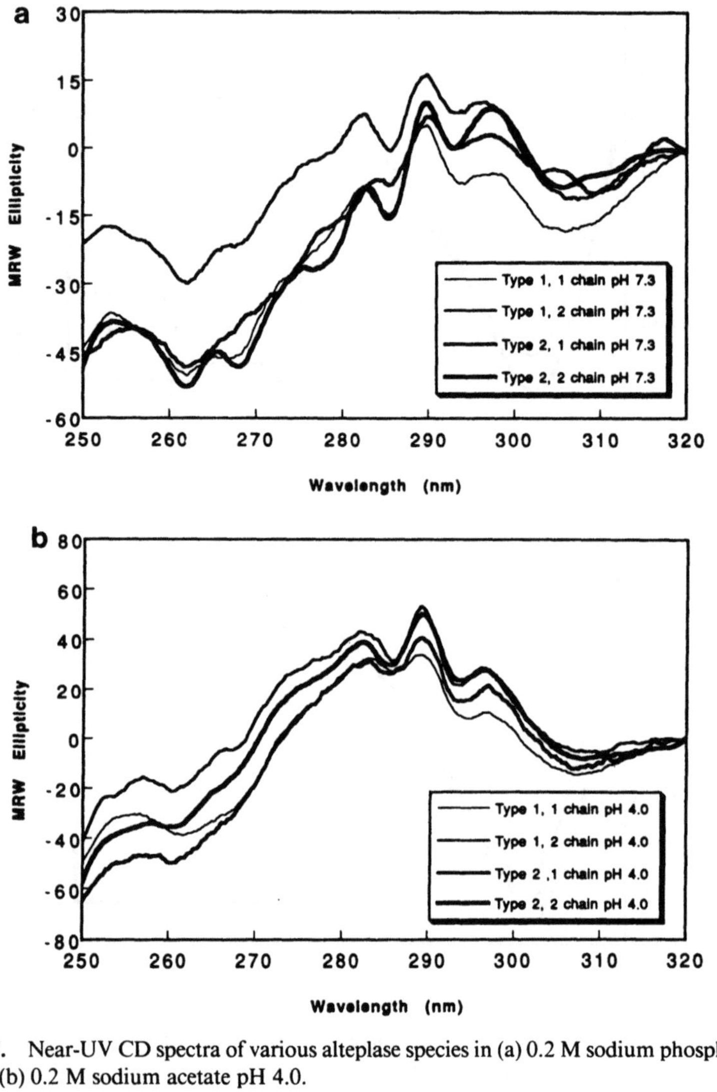

Figure 7. Near-UV CD spectra of various alteplase species in (a) 0.2 M sodium phosphate pH 7.3 and (b) 0.2 M sodium acetate pH 4.0.

3.1.2. CD Spectroscopy

CD spectrophotometry measures the amount of right and left circularly polarized light absorbed by a protein. Chromophores that are optically active absorb different amounts of the right and left polarized light and this difference results in either a positive or negative absorption band (conventionally, the right polarized spectrum is subtracted from the left polarized spectrum). Optical activity of a chromophore increases as its rotational movement in its

Table IV
Far-UV CD Spectral Analysis of Alteplase

Alteplase species	pH	% β-sheet and β-turn Y/P[a]	% random coil Y/P
Type I			
1-chain	4.0	60/72	40/28
2-chain	4.0	62/73	40/27
Type II			
1-chain	4.0	62/73	38/27
2-chain	4.0	59/73	41/27
Type I			
1-chain	7.3	60/73	40/27
2-chain	7.3	60/73	40/27
Type II			
1-chain	7.3	59/73	41/27
2-chain	7.3	64/73	36/27

[a] Value obtained based on the Yang model/value obtained based on the Provencher model.

environment is reduced. Therefore, the CD spectrum is a way to evaluate the protein structure as a reflection of the environment of the chromophores. Changes in the chromophore environment may signal changes in the conformation of the protein (Strickland, 1974; Bewley, 1979; Manning, 1989).

Analysis of the CD spectrum of a protein in the region from 190 to 250 nm (far-UV CD) provides information on the environment of the carbonyl group of the amide bond and consequently the secondary structure of the protein. Several researchers have provided a means to analyze the α-helical, β-sheet, β-turn, and random-coil contributions to the far-UV spectrum (Yang *et al.*, 1978; Provencher and Glockner, 1981). Analysis of the CD spectrum in the region of 250–350 nm (near-UV CD) provides information on the environment of the aromatic chromophores (phenylalanine, tyrosine, and tryptophan) as well as the disulfide bonds.

Due to the heterogeneous nature of alteplase, CD spectroscopy was only used as a tool to monitor gross conformational changes during formulation development of Activase. However, some exploratory structural studies were also performed on alteplase that had been separated into the four individual species. Figure 6 shows the CD spectra of the four alteplase species in the far-UV range (190–250 nm). Near-UV data (250–360 nm) are presented in Fig. 7a and b (pH 7.3 and 4.0, respectively).

The far-UV spectra were analyzed for secondary structural components by the methods of Yang *et al.* (1978) and Provencher *et al.* (1981). Secondary structure estimates of the various alteplase species are presented in Table IV.

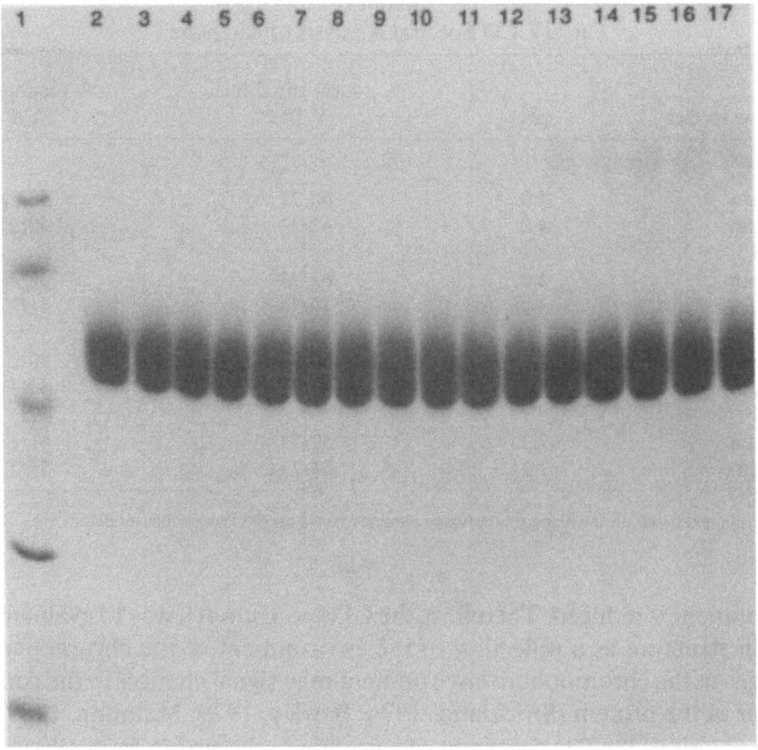

Figure 8. (a) Nonreduced SDS-PAGE of lyophilized alteplase, Activase. Lane 1: molecular weight standard; lane 2: reference standard; lanes 3–7: samples of five different lots stored at 5°C for 1 year; lanes 8–12: samples of five different lots stored at 30°C for 1 year; lanes 13–17: samples of five different lots stored at 50°C for 1 year. (b) Reduced SDS-PAGE of lyophilized alteplase, Activase. Lane 1: molecular weight standard; lane 2: reference standard; lanes 3–7: samples of five different lots stored at 5°C for 1 year; lanes 8–12: samples of five different lots stored at 30°C for 1 year; lanes 13–17: samples of five different lots stored at 50°C for 1 year.

Both methods of analysis indicated that the protein is composed of 59–73% β-sheet and β-turn with the remaining 27–41% as random coil. All four species of the protein exhibited similar far-UV CD spectra. Between 7.5 and 4.0, the pH of the solution did not appear to affect the secondary structure of the protein.

In contrast, the near-UV spectra of the four species, although homologous in shape, were not superimposable, suggesting that the environment around the chromophores may be slightly different in each species. The CD spectra at pH 4.0 are quite different from those at pH 7.3 in that the mean residue weight ellipticity is shifted toward a more positive region at lower

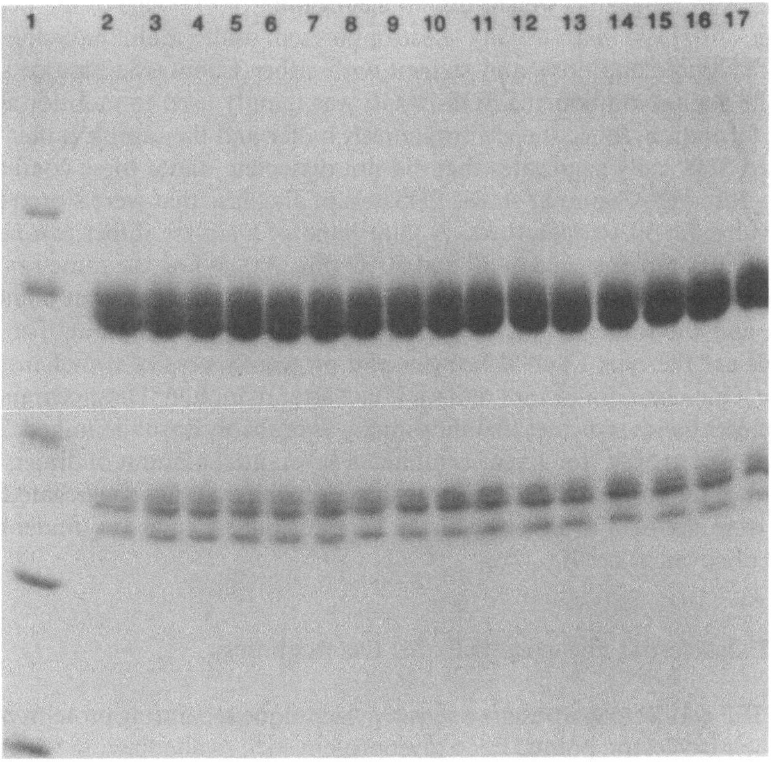

Figure 8. (*Continued*)

pH. The data suggest that the tertiary structure of alteplase changes as the pH decreases from 7.3 to 4.0. If this difference in spectra signals an unfolding of the molecule in this pH range, it may explain the decreased bioactivity and dramatically increased solubility that was seen when the pH was decreased from 7.0 to 4.0 (Fig. 2).

3.2. Electrophoresis

3.2.1. SDS-PAGE

SDS-PAGE was used for determining the purity and homogeneity of alteplase during the manufacturing process, and for qualitative estimation of changes in molecular weight due to aggregation or cleavage of the peptide chain. Alteplase was usually electrophoresed under both reducing and nonreducing conditions and stained with either Coomassie blue or silver nitrate stain. Nonreducing SDS-PAGE was mainly used to monitor aggregate formation. Since the electrophoresis buffer and the sample buffer contained SDS, only aggregates that did not dissociate under these conditions were observed. Figure 8t shows SDS gels of alteplase that were stored for 1 year at different temperatures. A faint band of alteplase dimer can be observed in samples stored at 30 and 50°C (Fig. 8a). When the same samples are reduced before electrophoresis, several new bands appear at about 30 kDa and the dimer bands in 30°C samples disappear (Fig. 8b). The new bands are the type I and II kringles and protease pieces of two-chain alteplase. One-chain molecules remain intact after reduction. Disappearance of the dimer bands indicates that the dimer was probably disulfide-linked. Samples stored at 50°C for 1 year exhibited a substantial amount of dimers that were not reducible by DTT treatment, suggesting that these aggregated species were not held together by disulfide bonds but by an as yet unidentified type of covalent bond.

3.2.2. Isoelectric Focusing (IEF) Gel Electrophoresis

IEF gel electrophoresis is a sensitive technique separating proteins based on their isoelectric points. For a glycoprotein such as alteplase, in which the carbohydrate moieties are capped by sialic acid, IEF electrophoresis usually reveals the microheterogeneity of highly purified samples due to varying sialic acid content or distribution. IEF electrophoresis of alteplase was carried out in the presence of urea to improve its solubility. The gel consisted of 4% acrylamide, 2% of a 2 : 1 mixture of FMC Ampholines (pH 5–8) and Bio-Rad ampholytes (pH 3–10), 1% Triton X-100, and 8 M urea. Unreduced alteplase appeared as a series of bands in the pl range between 6.5 and 8.5. All protein bands were active alteplase since they were immunoreactive on Western blot and, when overlaid on polyacrylamide gel containing fibrin, all bands exhibited fibrin lysis activity (Bridenbaugh and Mulholland, unpublished data). The two-chain form of alteplase, upon reduction by β-mercaptoethanol, exhibited two groups of bands attributable to the protease region which centers around pl 6.5 (consistent with its theoretical pl), and the kringle portion in the basic region of the gel (pl $\approx$ 9.0). When alteplase was treated with the enzyme neuraminidase, which cleaved sialic acid from the carbohydrate chains, the pattern simplified significantly, but did not collapse into a single band as would be expected if the microheterogeneity of the sample was due solely to the presence of sialic acid. Figure 9 illustrates

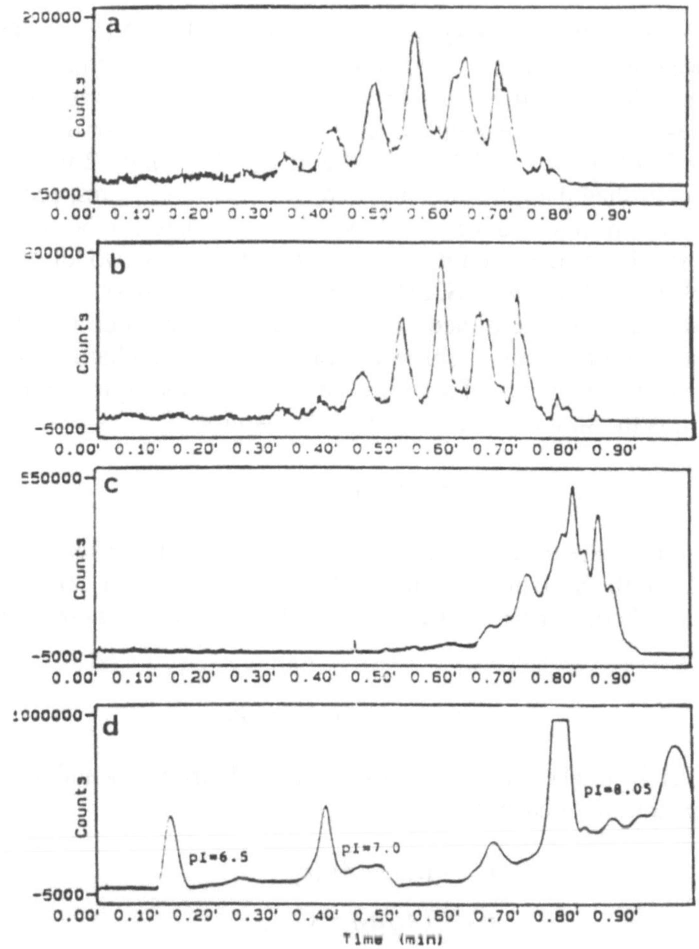

Figure 9. Densitometer tracings of a Coomassie-stained isoelectric focusing gel. *y*-axis units are normalized area units. *x*-axis units are time in minutes at scan speed of 40 mm/min across the gel. (a) Alteplase; (b) alteplase treated with plasmin; (c) neuraminidase-treated alteplase; (d) pl marker.

densitometer tracings of Coomassie-stained IEF gels of intact alteplase and neuraminidase-treated alteplase (Bridenbaugh and Mulholland, unpublished data).

3.3. Chromogenic Enzymatic Assay

The integrity of the protease active site was assessed by measuring the peptidolytic (and by inference the proteolytic) activity of alteplase on a synthetic substrate D-Ile-Pro-Arg-*p*-nitroanilide (S2288, Kabi Vitrum). The rate of cleavage was monitored by following the formation of *p*-nitroaniline spectrophotometrically at 405 nm (chromophore absorbance was 9520 cm^{-1} M^{-1}). One- and two-chain alteplase have different affinities for the substrate, with K_m of 14.6×10^{-4} and 4.9×10^{-4} and k_{cat} of 5.7 and 12 sec^{-1} (Ward, 1986, unpublished data) for the one- and two-chain forms, respectively.

The concentrations of one- and two-chain alteplase can be determined based on the difference in amidolytic activity between the two forms toward S2288 substrate. The rate of S2288 cleavage by a mixture of one- and two-chain rt-PA is first determined. The assay is then repeated on the same sample after all one-chain alteplase is cleaved to the two-chain form by the addition of a trace amount of plasmin. The total chromogenic activity of the sample, A, is the sum of the activities of the two forms:

$$A = C_1 a_1 + C_2 a_2$$

where C_1 and C_2 are the concentrations of the one- and two-chain forms and a_1 and a_2 are the specific activities of the two forms. Upon conversion by plasmin, all alteplase present reacts as two-chain, hence the new total activity, A_p, is:

$$A_p = C_T a_2 = (C_1 + C_2) a_2$$

where C_T is the total concentration of alteplase. Rearranging and solving for the concentration of each species:

$$C_1 = (A_p - A)/(a_2 - a_1)$$
$$C_2 = (A_p/a_2) - C$$
$$C_T = A_p/a_2 = C_1 + C_2$$

The specific activities of the two forms of alteplase are expressed in terms of moles of *p*-nitroaniline generated per minute by 1 mg of alteplase at 37°C. The specific activities a_1 and a_2 are experimentally determined to be 2.46 $\times 10^{-5}$ and 1.74×10^{-4} mole sec^{-1} mg^{-1}, respectively (T. Nguyen, unpublished data).

Ranby (1982) described a physiologically relevant assay in which the rate of activation of plasminogen by t-PA was followed. Plasminogen, alteplase, and the plasmin-specific synthetic substrate D-Val-Leu-Lys-*p*-nitroanilide (S2251, Kabi) were mixed in a reaction vessel. Alteplase cleaved plasminogen to plasmin which catalyzed the hydrolysis of the substrate and generated *p*-nitroaniline. The rate of formation of *p*-nitroaniline was monitored spectrophotometrically at 405 nm.

The reaction scheme can be summarized as follows:

$$\text{plasminogen} + H_2O \xrightarrow[\text{[alteplase]}]{k_1} \text{plasmin}$$

$$\text{D-Val-Leu-Lys-}p\text{-anilide} + H_2O \xrightarrow[\text{[plasmin]}]{k_2} p\text{-nitroaniline}$$

The initial rates of formation of the various products can be approximated by first-order kinetics:

$$d[\text{plasmin}]/dt = k_1[\text{alteplase}]$$
$$d[p - \text{NA}]/dt = k_2[\text{plasmin}]$$

k_1 and k_2 are rate constants for the conversion of plasminogen to plasmin and cleavage of S2251 to *p*-nitroaniline, respectively. The rate of formation of *p*-nitroaniline was followed by monitoring the absorbance of the reaction vessel at 405 nm as a function of time and by fitting the data to the following quadratic equation:

$$A = A_0 + \Delta\epsilon_{405\text{ nm}}\, k_2[\text{plasmin}]_0 t + 0.5\Delta\epsilon_{405\text{ nm}}\, k_1 k_2[\text{alteplase}]t^2$$

A_0 and $[\text{plasmin}]_0$ are the absorbance and plasmin concentration at initial time, respectively. The coefficient of the time squared term is equal to 0.5 $\Delta\epsilon_{405\text{ nm}}\, k_1 k_2[\text{alteplase}]$. Under the experimental conditions $\Delta\epsilon_{405\text{ nm}}$ is 9520 $cm^{-1}M^{-1}$; the alteplase concentration was known and did not change during the reaction. At fixed concentrations of substrate and alteplase, k_2 could be determined independently, thus allowing the solution for k_1. The reaction was also carried out in the presence of fibrin to quantitate the potentiation of alteplase by fibrin.

3.4. Clot Lysis Assay

The potency of alteplase was assessed by an *in vitro* clot lysis assay. This assay was based on measuring the time taken for a fixed amount of alteplase to dissolve a fibrin clot. Typically, a fibrin clot was produced by combining fibrinogen and thrombin in the presence of plasminogen. Alteplase was then introduced to initiate the lytic reaction. If alteplase is present in a limiting amount, the time until clot lysis is directly related to its concentration. The accuracy of the assay depended on the ability to determine reproducibly the reaction end point. Several methods for end point determination were proposed. These included the release of entrapped air bubbles from the clot (bubble release method) or the dropping of glass beads through the fibrin clot (Gaffney and Curtis, 1985; Philo and Gaffney, 1981a,b; Wilkens and Back, 1976; Collen *et al.,* 1968). Carlson *et al.* (1988) described an assay in which end point detection is based on turbidimetric measurement using a commercially available microcentrifugal analyzer. The instrument provides automation in sample preparation, data collection, and final result calculation and analysis. A typical tracing of the turbidity of the fibrin clot as monitored spectrophotometrically at 340 nm versus time is reproduced in Fig. 10. The formation of a stable fibrin clot occurred rapidly following the mixing of thrombin and fibrinogen which was reflected in a rapid increase in absorbance that reached a plateau within approximately 30 sec. Lysis of the fibrin clot was evident at around 120 sec, accompanied by a decrease in absorbance at 340 nm. Complete dissolution of the clot occurred around 300 sec when the absorbance returned to baseline. The method has an accuracy of 99.5% with a precision of 5% and a useful assay range of between 40 and 1000 ng/ml of alteplase. It is reliable, reproducible, and requires minimum sample

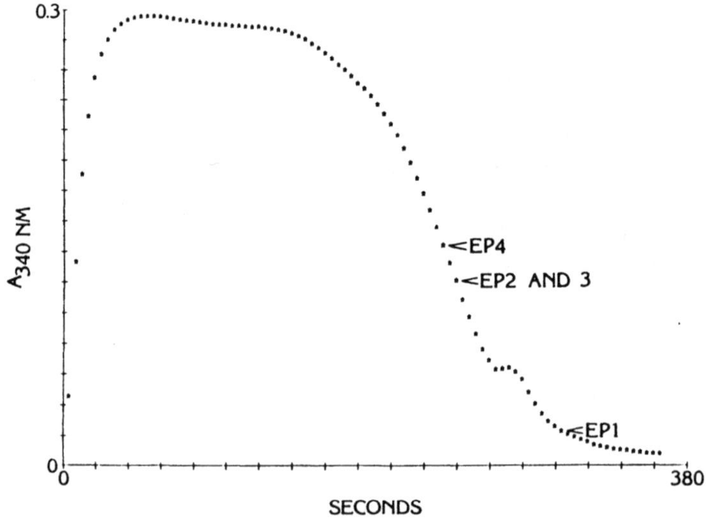

Figure 10. Absorbance of fibrin clot at 340 nm following addition of 950 ng/ml alteplase. (Adapted from *Anal. Biochem.* **168:**428–435.)

handling. This assay is currently one of the stability-indicating assays used to determine the shelf life of the drug product. Recently, this assay was also adapted to the microtiter plate format (Jones and Meunier, 1990).

3.5. HPLC

Several HPLC methods have been developed for alteplase. These include methods for characterization of the protein, as well as methods to assess the stability of the protein with time. The latter includes size-exclusion methods for determination of the relative amounts of monomer, aggregate, and one- and two-chain forms of the protein, as well as a reversed-phase method to monitor the hydrolytic cleavage of alteplase.

3.5.1. Size-Exclusion Chromatography (SEC)

Alteplase is hydrophobic and binds tightly to the silica solid-phase column matrix often used for SEC. Complete elution of alteplase from the column requires the addition of an anionic surfactant such as SDS to the mobile phase. SDS dissociates aggregates that are not covalently bound, resulting in a possible underestimation of the amount of aggregate in the sample. Nonetheless, the method has been extremely useful as a measurement of protein concentration, as an indicator of aggregation and, with slight modification, is the method of choice for determining the relative amounts of one- and two-chain alteplase.

3.5.1a. SEC for Determination of Protein Concentration and Nondissociable Aggregates. An isocratic chromatographic profile of an Activase sample containing alteplase on a TSK3000SWXL is shown in Fig. 11a (mobile phase = 0.2 M sodium phosphate, pH 6.8, and 0.1% SDS). Alteplase eluted at a retention time corresponding to a molecular mass of approximately 63 kDa (peak 1) relative to molecular mass standards run on the same column. The later eluting peak 2 was the excipient arginine in Activase. Immediately preceding the alteplase peak were smaller peaks that were identified as non-SDS-dissociable oligomers of alteplase; polysorbate 80 appeared as a small peak at approximately 17 min. Quantification of these peaks was used to monitor aggregation of alteplase during stability studies.

3.5.1b. SEC for Determination of the Relative Amounts of One- and Two-Chain Alteplase. The SEC assay was also used to determine the relative amounts of one- and two-chain alteplase. Two-chain alteplase has a clip

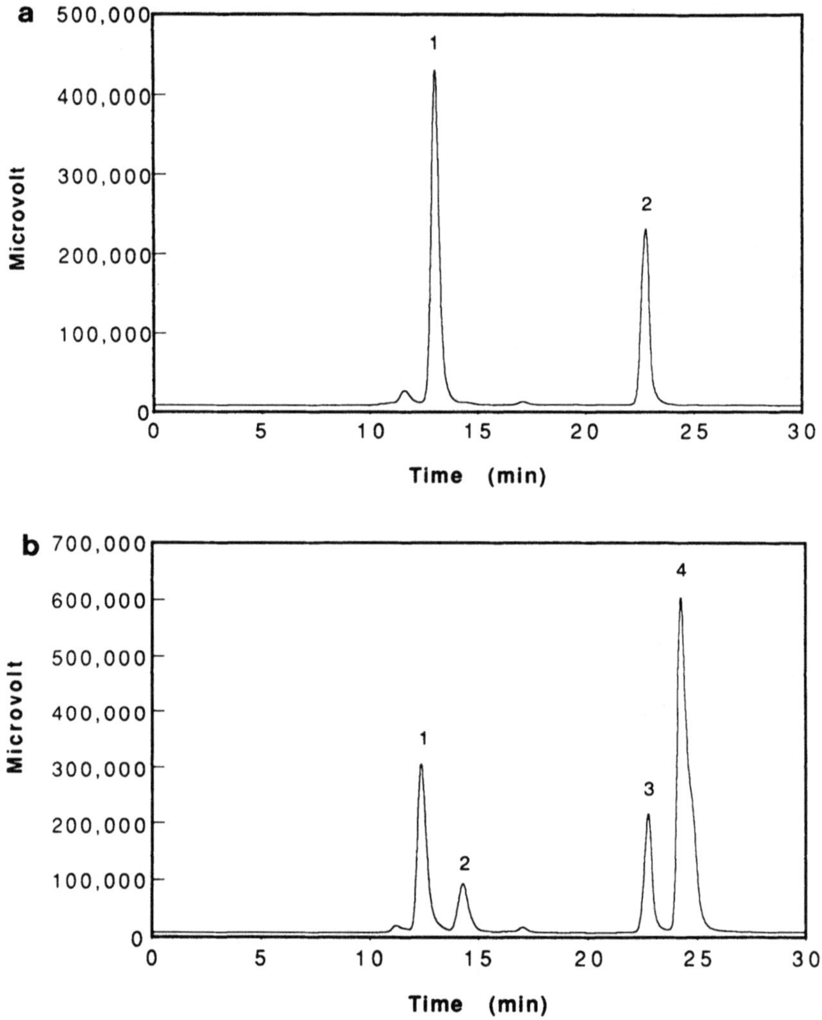

Figure 11. Chromatograms of Activase stored at 50°C for 12 months, analyzed by size-exclusion chromatography. (a) Chromatographed under nonreducing conditions; (b) sample reduced by dithiothreitol prior to analysis.

between Arg_{275} and Ile_{276}. When two-chain alteplase was reduced with either β-mercaptoethanol or dithiothreitol, the molecule dissociated into two fragments following the reduction of the disulfide bond between Cys_{264} and Cys_{395} that holds the two-chain molecule together. One fragment contains the kringle region and the other contains the protease region. Each of these fragments was about 30 kDa in size. Chromatography of the reduced protein sample gave rise to two peaks, peak 1 corresponding to the intact one-chain molecule at ~63 kDa and peak 2 eluting at ~30 kDa containing the kringle and protease fragments derived from the two-chain molecule (Fig. 11b). The peak area of each species was proportional to its relative amount in the original sample. Polysorbate 80 eluted at 17 min and peaks 3 and 4 were arginine and DTT, respectively.

3.5.2. RP-HPLC

RP-HPLC was used to detect fragments of the molecule which result from the hydrolytic cleavage of the peptide backbone. There are three sites on alteplase that are particularly susceptible to hydrolytic cleavage. In addition to the clip between Arg_{275} and Ile_{276} which converts one-chain alteplase to the two-chain form, cleavage between Arg_7 and Asp_8 and Arg_{27} and Ser_{28} occurs quite readily. Indeed, the 1-27 peptide fragment was usually detected first when alteplase was subjected to digestion with low levels of trypsin. This and other fragments can be readily detected by RP-HPLC. Alteplase was first reduced by incubation in 20 mM DTT at 37°C for 30 min in the presence of 8 M urea. The reduced sample was then analyzed on a Vydac C4 column at

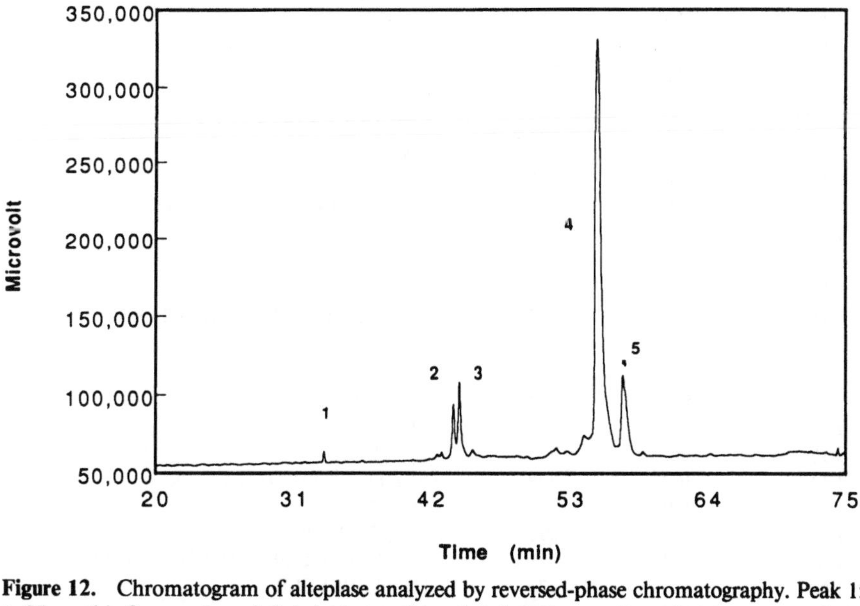

Figure 12. Chromatogram of alteplase analyzed by reversed-phase chromatography. Peak 1: 1–27 peptide fragment; peak 2: kringle type I; peak 3: kringle type II; peak 4: intact one-chain alteplase; peak 5: serine protease fragment.

Table V
Theoretical Tryptic Peptides of rt-PA

Tryptic peptide	Residue No. Start	End	Sequence[a]
T1	1	7	SYQVICR
T2	8	10	DEK
T3	11	27	TQMIYQQHQSWLRPVLR
T4	28	30	SNR
T5	31	40	VEYCWCNSGR
T6	41	49	AQCHSVPVK
T7	50	55	SESEPR
T8	56	82	CFNGGTCQQALYFSDFVCQCPEGFAGK
T9	83	89	CCEIDTR
T10	90	101	ATCYEDQGISYR
T11	102	129	GTWSTAESGAECTNWNSSALAQKPYSGR
T12	130	135	RPDAIR
T13	136	145	LGLGNHNYCR
T14	146	149	NPDR
T15	150	159	DSKPWCYVFK
T16	160	162	AGK
T17	163	189	YSSEFCSTPACSEGNDCYFGNSAYR
T18	190	212	GTHSLTESGASCLPWNSMILIGK
T19	213	228	VYTAQNPSAQALGLK
T20	229	233	HNYCR
T21	234	247	NPDGDAKPWCHVLK
T22	248	249	NR
T23	250	250	R
T24	251	267	LTWEYCKVPSCSTCGLR
T25	268	275	QYSQPQFR
T26	276	277	IK
T27	278	296	GGLFADIASHPWQAAIFAK
T28	297	298	HR
T29	299	299	R
T30	300	304	SPGER
T31	305	327	FLCGGILISSCWILSAAHCFQER
T32	328	339	FPPHHLTVILGR
T33	340	342	TYR
T34	343	351	VVPGEEEQK
T35	352	356	FEVEK
T36	357	361	YIVHK
T37	362	378	EFDDDTYDNDIALLQLK
T38	379	383	SDSSR
T39	384	392	CAQESSVVR
T40	393	416	TVCLPPADLQLPDWTECELSGYGK
T41	417	427	HEALSPFYSER
T42	428	429	LK
T43	430	434	EAHVR

(continued)

Table V. (*Continued*)

Tryptic peptide	Residue No. Start	End	Sequence[a]
T44	435	440	LYPSSR
T45	441	449	CTSQHLLNR
T46	450	462	TVTDNMLCAGDTR
T47	463	489	SGGPQANLHDACQGDSGGPLVCLNDGR
T48	490	505	MTLVGIISWGLGCGQK
T49	506	513	DVPGVYTK
T50	514	522	VTNYLDWIR
T51	523	527	DNMRP

[a] Single-letter symbols for amino acids: A, Ala; C, Cys; D, Asp; E, Glu; F, Phe; G, Gly; H, His; I, Ile; K, Lys; L, Leu; M, Met; N, Asn; P, Pro; Q, Gln; R, Arg; S, Ser; T, Thr; V, Val; W, Trp; Y, Tyr.

40°C using a gradient of 10–40% acetonitrile water–0.1% TFA over 55 min. A typical chromatogram of alteplase is shown in Fig. 12. The kringle regions of type I and II alteplase, the protease domain (resulting from the reduction of two-chain alteplase), and the intact one-chain molecule were resolved by this method. Peak 1 is the NH_2-terminus fragment containing the first 27 amino acids. Peaks 2 and 3 are the type I and type II alteplase kringles, respectively, peak 4 is the one-chain species from either type I or II, and peak 5 is the two-chain protease fragment from either type. The shoulder in front of peak 4 has not been identified.

3.5.3. Peptide Mapping

Tryptic digestion of alteplase, coupled with RP-HPLC separation of the resulting peptides, has been successfully applied to elucidate the primary structure and to identify chemical alterations in several recombinant proteins such as methionyl-human growth hormone (Kohr *et al.,* 1982), bovine growth hormone (Hartman *et al.,* 1986), and interleukin (Kunitani *et al.,* 1986).

On a more routine basis, tryptic mapping is used as an identity assay for alteplase to demonstrate the reproducibility of the manufacturing process. In this assay the tryptic map of a new batch of alteplase is compared with that of a reference standard. It is anticipated that tryptic digestion could produce as many as 51 peptides based on the presence of arginine and lysine (Table V). To develop the tryptic map, alteplase was digested with trypsin, and the peptide fragments were separated by reversed-phase chromatography

(Chloupek *et al.*, 1989). Each peak was recovered from the original chromatographic run and identified using a combination of amino acid analysis and NH_2-terminal sequencing. A representative tryptic map of alteplase is shown in Fig. 13.

4. STABILITY OF ALTEPLASE

4.1. Differential Scanning Calorimetry (DSC) Studies

The thermal stability of alteplase was studied by DSC and reported by Radek and Castellino (1988). In 20 mM Hepes buffer, pH 7.4, with 250 mM arginine, the thermogram obtained for alteplase exhibited a melting temperature (T_m) of 72.8°C when the temperature scan rate was 60°C/hr. The total heat liberated in the transition (ΔQ_t) was calculated to be 286 ± 30 kcal/mole. The endotherm can be deconvoluted into two two-state regions undergoing independent melting with transition temperatures at 70.2 and 72.6°C,

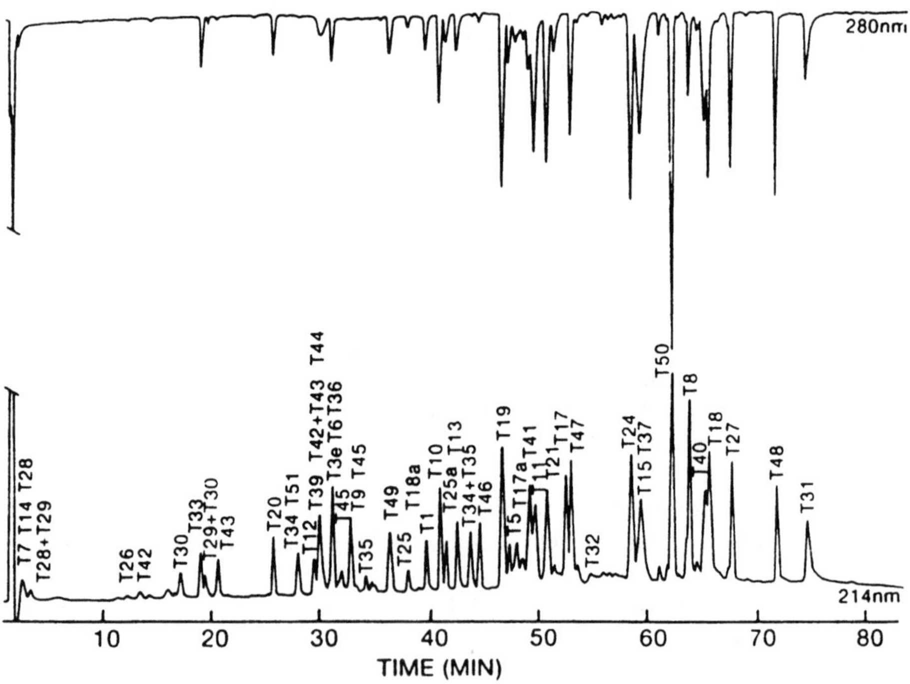

Figure 13. Tryptic map of alteplase. (Adapted from *J. Chromatogr.* **463**:375–396.)

respectively. The ratio of the heat of transition to the van't Hoff enthalpy (ΔH_{vh}) was two, in good agreement with the mathematical model. However, the thermodynamic significance of these data must be interpreted with caution since alteplase appears to be irreversibly denatured in these experiments. Studies at scan rates between 10 and 60°C/hr showed that there is a decrease in T_m of approximately 1°C for each 10°C/hr drop in scan rate; ΔQ_t and $\Delta Q_t/\Delta H_{vh}$ remained essentially unchanged in these experiments. These data suggest that the transition is kinetically controlled and that the activation energy of the process was estimated to be around 81.5 kcal/mole (Radek and Castellino, 1988).

The influence of pH on the thermal stability of alteplase was examined in the range between pH 3.0 and 7.4. There were no significant differences in DSC properties of alteplase at pH 7.4 and 6.0. However, T_m's at pH 5.0 and 4.0 were 69.5 and 59.4°C, respectively, and ΔQ_t values at the corresponding pH's were 220 and 105 kcal/mole. While the thermal transitions at pH's above 5.0 are irreversible, the transition obtained at pH 5.0 shows partial reversibility upon heating and cooling, and, at pH 4.0, the process is completely reversible. These results suggest that the conformation of alteplase at pH 4.0 is significantly different from the native conformation at neutral pH and that a secondary reaction, possibly aggregation, occurs following unfolding at higher pH's and is inhibited in the acid pH range. No observable endothermic transition was recorded at pH 3.0, indicating that extensive denaturation has taken place under these conditions. The thermal behavior of alteplase and a preparation of all two-chain alteplase are almost identical.

Separately, Northey *et al.* (1987) presented similar results. In their study, alteplase was dissolved in 500 mM arginine phosphate pH 7.2 and the scan rate was 60°C/hr. The endothermic tracing of the transition shown in Fig. 14 exhibits a maximum at 71.3°C and ΔQ_t of 332.5 ± 14 kcal/mole. The ratio of $\Delta Q_t/\Delta H_{vh}$ was 3.4, suggesting that the experimental results may be composed of more than two independent regions as suggested by Radek and Castellino. In fact, the thermogram of alteplase can be deconvoluted into three two-state independent regions described by three Gaussian curves having maxima at 65.7, 71.2, and 73.2°C. It is hypothesized that the three deconvoluted transition peaks represent the three largest domains of alteplase, namely, the kringle 1, the kringle 2, and the serine protease domains. The smaller epidermal growth factor and the finger regions of alteplase may unfold as part of the other domains or, if they are separate domains, their melting enthalpies are negligible. DSC studies on purified kringle 2, and on the COOH-terminal serine protease domains indicate that the lowest melting and the highest melting domains correspond to the serine protease and the kringle 2, respectively. Additionally, two deletion mutants in which either the kringle 1 (ΔK_1) or the kringle 2 (ΔK_2) were removed were also

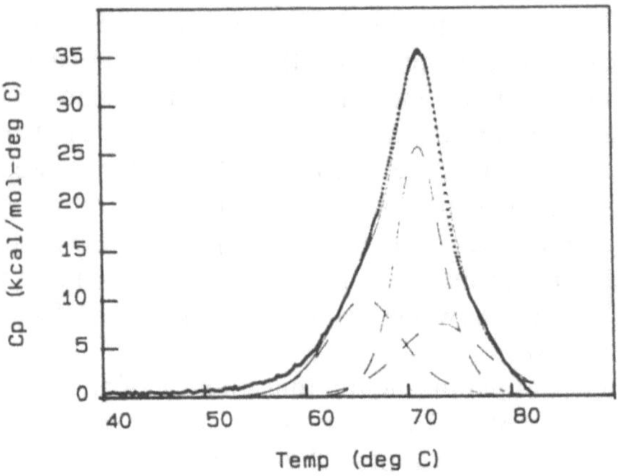

Figure 14. Differential calorimetry scan of alteplase. Scan rate 60°C/hr. Data deconvoluted to three independent two-state domains. · · · · ·, actual data; ——, fitted curve; ---, deconvoluted curves.

examined by DSC. The thermogram of the ΔK_1 mutant is composed of two unfolding domains with T_m values matching those of the serine protease and the kringle 2. On the other hand, the thermogram of the ΔK_2 mutant can be deconvoluted into two domains corresponding to those of the serine protease and kringle 1 on alteplase (Northey *et al.*, 1987).

Arginine and ε-amino caproic acid are known to bind to the lysine-binding site on kringle 2 (Cleary *et al.*, 1989). The association of these ligands shifted the isolated kringle 2 T_m upward by 10–13°C and the T_m of kringle 2 in alteplase upward by 5–7°C. These amino acids also bind to the protease region, raising the T_m of the isolated protease by approximately 10°C. Arginine also raised the melting temperature of the COOH-terminal serine protease region on alteplase, albeit to a smaller extent (Northey *et al.*, 1987).

4.2. Stability in Solution

4.2.1. Influence of pH

The kinetics and mechanism of alteplase degradation in arginine–phosphate buffers were studied in the pH range 3.0–10.0. The concentration of alteplase was 1.0 mg/ml and the buffer concentration was maintained at 0.2 M. As with most proteins, the selection of sensitive and relevant stability-indicating assays is crucial. For this study, the conversion of one-chain alteplase to the two-chain form was monitored by SEC using the TSK3000SW column. The integrity of the serine protease active site was assessed by a chromogenic enzymatic assay using Kabi S2288 substrate in the presence of plasmin. The bioactivity of the molecule was followed using *in vitro* clot lysis assay. RP-HPLC was employed in an effort to elucidate structural changes of alteplase following long-term storage in solution.

At 35°C, the degradation rate in solution can be described satisfactorily using pseudo-first-order kinetics. The slope of the semilog plots of alteplase activity or protein concentration versus time were used to construct the pH–rate profile shown in Fig. 15. The stability of alteplase was pH dependent where the rate of loss of clot lysis activity at pH 5.5 was slowest, and was significantly accelerated at higher and lower pH's. The rate constant at pH 3.0 was approximately one order of magnitude higher than at pH 5.5, and at 9.0, the degradation rate was 7-fold higher. The overall pH–rate profile is V-shaped, suggesting that the process is acid–base catalyzed. The low point of the V corresponds to the pH of maximum stability.

Similarly, the rate of loss in peptidolytic activity as measured by the S2288 enzymatic assay exhibited a V-shaped pH–degradation rate profile. However, at all pH's studied, the rate of loss in peptidolytic activity was 2 to

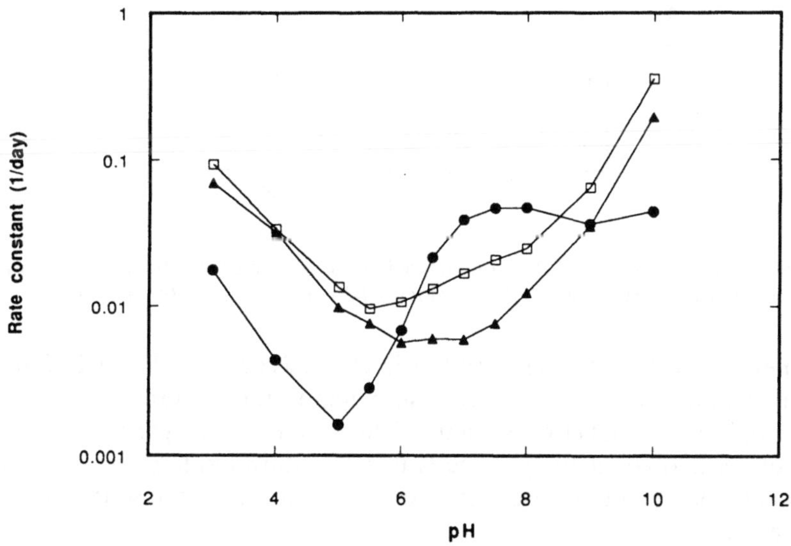

Figure 15. pH–degradation rate profile of alteplase in 0.2 M arginine phosphate at 35°C. ●, rate of cleavage between Arg_{275} and Ile_{276} converting one-chain alteplase to the two-chain form monitored by SEC-HPLC; □, rate of loss in clot lysis activity; ▲, rate of loss in peptidolytic activity based on S2288 assay.

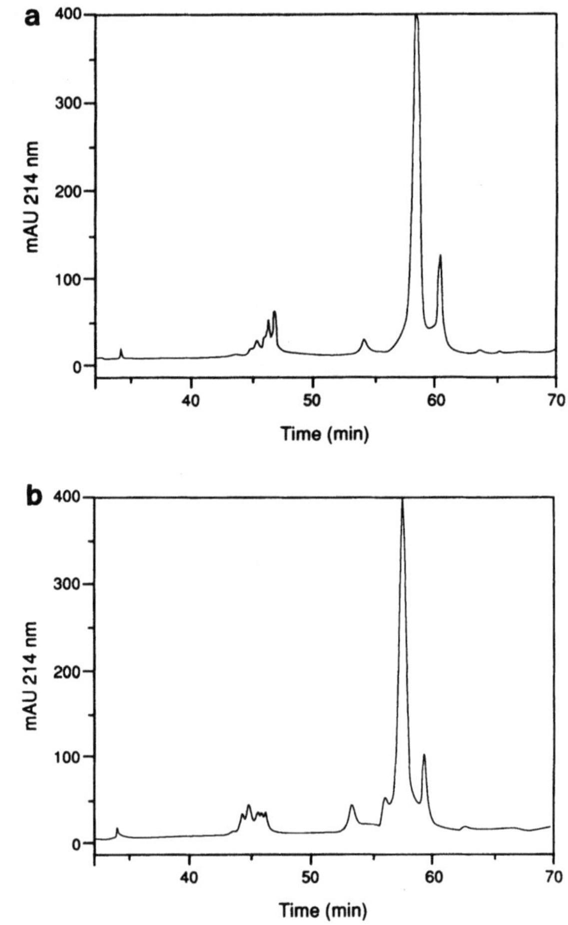

Figure 16. Chromatogram of alteplase analyzed by reversed-phase chromatography on Vydac C4 column. Samples stored at 5°C for 18 months. (a) pH 5.5; (b) pH 4.0; (c) pH 7.5.

3 times slower than the observed loss in clot lysis activity. The pH of maximum stability was also wider, corresponding to the pH range between 5.5 and 7.0. The fact that clot lysis activity deteriorated at a higher rate than the rate of peptidolytic activity suggests that the fibrin binding regions on the molecule may be more susceptible to degradation than the serine protease active site.

Cleavage of the peptide bond between Arg_{275} and Ile_{276} converts the one-chain form of alteplase to its two-chain form. This reaction was often observed during purification of t-PA. In the pH range between 5.0 and 9.0, the pH–rate profile exhibits a partial bell shape with a maximum around pH

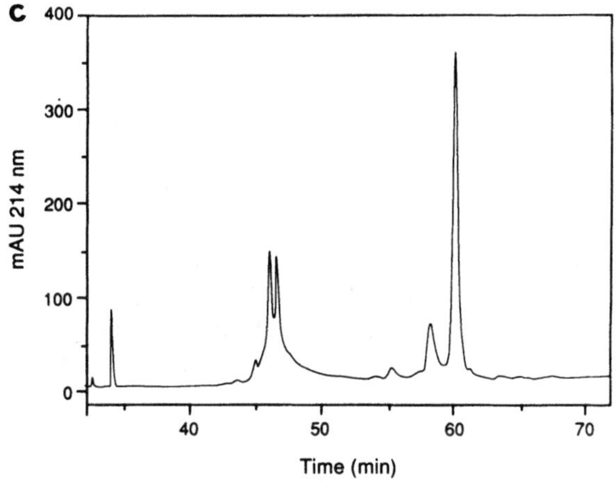

Figure 16. (*Continued*)

7.5 to 8.0. At pH 5.0, the rate is reduced 25-fold relative to the rate observed at pH 7.5. In a separate study, the peptidolytic activity of alteplase on S2288 substrate was assessed as a function of pH. The resulting pH–activity curve exhibits a similar bell shape profile (Nguyen, unpublished data). Thus, the data suggest that in the pH range between 5.0 and 9.0, the cleavage of one-chain alteplase to the two-chain form was autocatalyzed. At a pH above 10, the rate of loss of one-chain alteplase increased rapidly with the appearance of insoluble protein precipitates. In the region below pH 5.0, the rate of disappearance of one-chain alteplase was inversely proportional to the pH of the solution. However, loss of one-chain alteplase was not accompanied by the appearance of the two-chain form as in the more neutral pH range. A broad peak appearing in the low-molecular-weight range of the chromatogram suggested that in an acid medium, the molecule was cleaved extensively at several points when stored at 35°C.

In summary, kinetic studies indicate that loss in clot lysis activity is stability limiting below pH 6.0, and that the clot lysis assay is the most appropriate method for shelf life determination. In the pH range between 6.5 and 9.0, conversion of one-chain alteplase to the two-chain form occurs more rapidly, and monitoring this reaction by SEC-HPLC and/or chromogenic assay would be more relevant in determining the shelf life of the product.

Representative chromatograms of alteplase stored at 5°C for 18 months obtained by RP-HPLC are shown in Fig. 16a (pH 5.5), 16b (pH 4.0), and 16c (pH 7.5). In the neutral to slightly basic pH range, cleavage of the Arg_{275}–

Ile_{276} bond is predominant as evident from the increase in peak area corresponding to the kringle and protease regions, and the concomitant decrease in peak area due to the intact molecule. Clipping also occurred at the NH_2-terminus producing 1–11 and 1–27 peptide fragments. In the acid pH range (i.e., pH 4.0), the one- to two-chain conversion reaction was inhibited, but fragmentation of the molecule occurred at several other sites. The chromatogram of the degraded sample of alteplase shows a new peak at approximately 54 min, tentatively identified as the 1–462 fragment (Lee *et al.,* 1989). Two other peaks also appeared adjacent to the type I and II kringles, tentatively attributed to the altered form of type I and II kringles. The site of degradation has yet to be determined. The molecule again appeared to be least affected in the pH range between 5.0 and 5.5.

4.2.2. Influence of Arginine Concentration

In the neutral pH range, the stability-limiting reaction is cleavage of the Arg_{275}–Ile_{276} peptide bond converting one-chain alteplase to its two-chain form. The effect of alteplase and arginine concentrations on the kinetics of the reaction at pH 7.2 is illustrated in Fig. 17. At fixed arginine concentration, the rate of cleavage was directly proportional to the alteplase concentra-

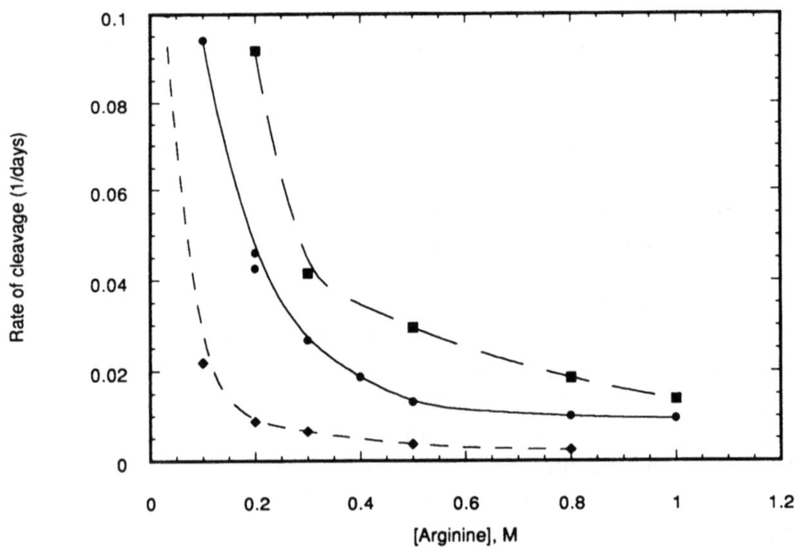

Figure 17. Influence of arginine concentration on the rate of cleavage between Arg_{275} and Ile_{276} converting one-chain alteplase to the two-chain form. ♦, 0.5 mg/ml alteplase; ●, 1.8 mg/ml alteplase; ■, 5.0 mg/ml alteplase.

tion as evident from the approximately tenfold increase in cleavage rate of 0.5 mg/ml to 5.0 mg/ml alteplase. On the other hand, the rate of one- to two-chain conversion of 1.8 mg/ml alteplase was reduced from 1.09×10^{-6} sec^{-1} to 1.16×10^{-7} sec^{-1} when the arginine concentration was increased from 0.1 M to 0.8 M. The enzyme–inhibitor dissociation constant K_i as determined by plotting the inverse of the reaction rate versus arginine concentration (Dixon plot) was approximately 0.04 M (Nguyen and Adair, unpublished data).

4.2.3. Influence of Storage Temperature

The influence of temperature between 5 and 50°C on the degradation rate of alteplase in solution was studied. Figure 18 is a typical Arrhenius plot of the various reactions at pH 6.0. The rates of loss in peptidolytic and clot lysis activities are directly related to the storage temperature with apparent activation energies (E_a) of 34.2 ± 3.6 and 27.2 ± 3.4 kcal/mole, respectively. However, the most rapid reaction observed involves the conversion of one-chain alteplase to the two-chain form. This rate exhibits Arrhenius behavior

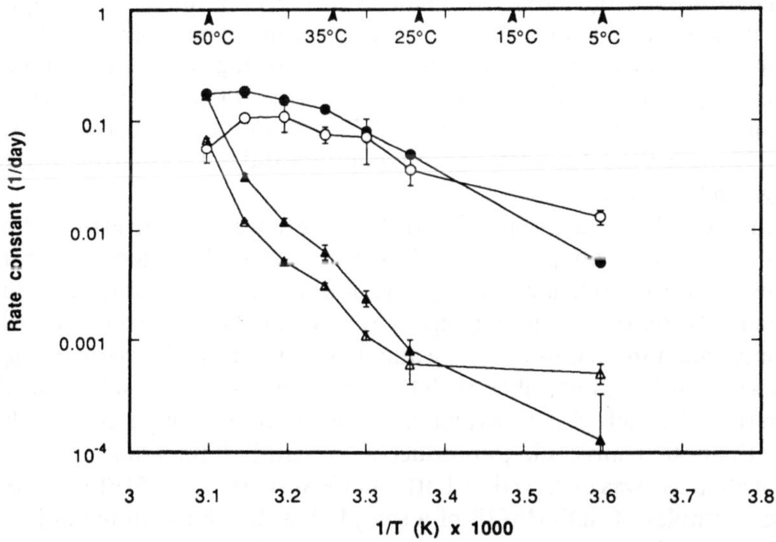

Figure 18. Arrhenius plot of the rate degradation of alteplase in 0.2 M arginine phosphate pH 6.0. ▲, rate of loss in clot lysis activity; △, rate of loss in peptidolytic activity based on S2288 assay; ●, rate of cleavage between Arg_{275} and Ile_{276} converting one-chain alteplase to the two-chain form monitored by SEC-HPLC; ○, rate of cleavage between Arg_{275} and Ile_{276} converting one-chain alteplase to the two-chain form monitored by S2288 chromogenic assay.

between 5 and 35°C with a calculated E_a of 21.1 kcal/mole. The break in linearity at temperatures above 35°C supports the hypothesis that cleavage of the Arg_{275}–Ile_{276} peptide bond is catalyzed by a protease, purportedly alteplase itself, with maximum activity between 35 and 40°C.

4.3. Stability of Alteplase in the Lyophilized State (Activase)

Activase is a lyophilized preparation containing alteplase in arginine phosphate buffer with approximately 0.01% polysorbate 80. It is supplied as a sterile, lyophilized powder in 20- and 50-mg vial sizes.

Assays used to monitor the stability of Activase included the clot lysis and S2288 chromogenic assays for bioactivity, SEC-HPLC for monitoring aggregate formation and determination of relative amounts of one- and two-chain protein, UV spectroscopy and Bradford analysis for determination of protein content, SDS-PAGE for determination of changes in molecular shape and size, and CD spectroscopy for monitoring changes in the tertiary structure of the molecule.

Alteplase, when formulated as Activase lyophilized powder, showed no significant loss of bioactivity or changes in the molecule after storage for more than 4 years at controlled room temperature. Therefore, only limited degradation data are available under real-time storage conditions. Stability studies suggested that the degradation rate does not follow the Arrhenius relationship, and there is a break in the $1/T$ versus $\log k$ plots between 30 and 40°C. However, the degradation product profile did not seem to change with storage temperature.

In marked contrast to the liquid state, there was no conversion of one-chain to two-chain alteplase in the lyophilized state. Elevated temperature studies (40–60°C) indicated that aggregation was the main route of degradation in the lyophilized state. Aggregates were formed through either covalent or noncovalent interactions among alteplase molecules. Noncovalent aggregates were usually dissociable by SDS. Alteplase has a single free sulfhydryl group (Cys_{83}) which may be available for interaction through disulfide linkage with another molecule to produce a covalently linked dimer. Aggregation of alteplase was observed by both SDS-PAGE and SEC-HPLC. Figure 8 shows examples of SDS-PAGE of a sample that had been stored at several temperatures for 1 year and then electrophoresed under both nonreducing and reducing conditions, respectively. There was no significant difference between the sample and the reference standard at 5 and 30°C. At 40 and 50°C, evidence of dimerization was seen with the appearance of a band at approximately 120 kDa. The intensity of this band diminished when the sample was treated with a reducing agent before SDS-PAGE analysis, indicating that a significant fraction of the aggregates were disulfide linked (Fig. 8b).

Figure 11a shows an SEC-HPLC chromatogram of an Activase sample that was stored at 50°C for up to 12 months. A significant amount of aggregate was observed as evident from the appearance of a peak eluting in front of one-chain alteplase. Alteplase stored at 5°C for 12 months was indistinguishable from the initial sample (data not shown). Reduction of alteplase with DTT prior to analysis results in a lower level of aggregate, again indicating that part of the aggregation observed was due to intermolecular disulfide formation (Fig. 11b). As the storage temperature and time increase, the amount of nonreducible aggregates increases. The nature of this high-molecular-weight species has not been elucidated. SEC on a TSK3000 column did not offer an accurate quantification of the amount of aggregate since the mobile phase used in this method contained SDS, which dissociates noncovalently bound high-molecular-weight species.

S-300 Sepharose open column liquid chromatography in an arginine phosphate–polysorbate 80 buffer at pH 7.3 was used to isolate the aggregate from stability samples under nondissociating conditions. A chromatographic profile of a sample that was stored for 11 months at 45°C is shown in Fig. 19. The bioactivity of each fraction was determined using the clot lysis

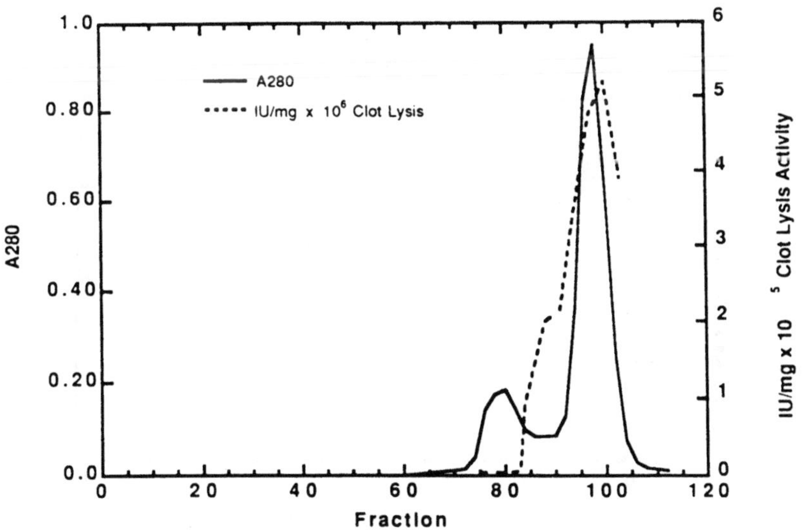

Figure 19. Separation and characterization of alteplase aggregate by native size-exclusion chromatography on S-300 Sepharose.

assay. Protein concentration was measured by UV spectrophotometry at 280 nm and Bradford protein assay. Collected fractions 90 to 110 contained the monomeric form of alteplase. Fractions 70 to 90 contained aggregated alteplase of various forms. Dimers of alteplase resided primarily in fractions 70 to 84. Fractions 84 to 90 contained a mixture of monomer, dimer, and aggregate having a molecular size equivalent to 1.5-fold the molecular weight of alteplase. It is evident from the bioactivity profile that alteplase dimer was inactive. Figure 20 illustrates a direct relationship between the specific activity of Activase stored for various lengths of time at temperatures up to 50°C and the amount of non-SDS-dissociable aggregates in the samples, as determined by SEC-HPLC.

5. CONCLUSIONS

In this chapter we briefly discussed the isolation of t-PA, summarized its physical and pharmacological properties, and presented developmental work leading to the drug product, Activase.

The existence of an endogenous enzyme responsible for dissolving unwanted blood clots was demonstrated more than 40 years ago. However, it was only with the advent of biotechnology that sufficient amounts could be produced for detailed studies of its molecular structure and its pharmacological effect at doses much higher than the endogenous levels. The capability to

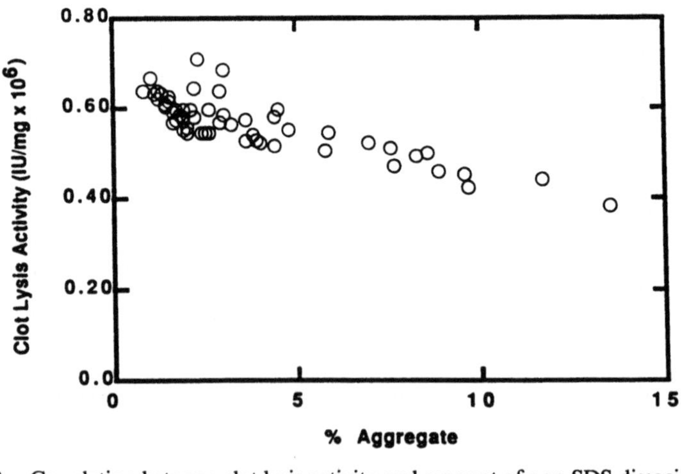

Figure 20. Correlation between clot lysis activity and amount of non-SDS-dissociable aggregate. (The specific activity of Activase reference standard is 0.68×10^6 IU/mg.)

mass produce alteplase at an affordable cost has made it a successful pharmaceutical product that has been administered to hundreds of thousands of patients. Its specificity for the blood clot and its short biological half-life ensure a safe and efficacious thrombolysis therapy.

As with most serine proteases, alteplase degraded in solution via hydrolytic cleavage at several sites. Its specific interaction with lysine, arginine, and molecules of similar structure was used to overcome its limited aqueous solubility and minimize its degradation in aqueous solution. It is possible that alteplase self-associated via the lysine binding site, resulting in low intrinsic aqueous solubility. Being a serine protease, alteplase could undergo autocatalytic degradation. Aggregation and precipitation of alteplase enhanced the rate of self-cleavage, possibly due to a proximity effect. Arginine introduced into the formulation at fairly high concentration competed for the binding site and thereby prevented self-association with increased solubility of alteplase. DSC studies indicated that arginine also stabilized the molecule to heat denaturation. Finally, arginine inhibited the cleavage reaction of one-chain alteplase to its two-chain form. This, combined with a careful selection of the formulation pH, laid the foundation for the design of a parenteral formulation of alteplase that has an acceptable shelf life.

In the lyophilized state, alteplase was stabilized by both excipients and pH. This resulted in a stable product with a shelf life of 4 years at controlled room temperature. Degradation only occurred following extended storage at high temperatures, where the main degradation pathway in the solid state was the formation of higher-molecular-weight species.

ACKNOWLEDGMENTS

The authors thank Ms. Jessica Burdman for her help in editing and preparing this manuscript, and Dr. Mike Powell for his critical review of the chapter.

REFERENCES

Amiral, J., Plassard, V., Grosley, M., Mimilla, F., and Constant, G., 1988, Measurement of t-PA and t-PA–PAI-1 complexes by ELISA using monoclonal antibodies: Clinical relevance, *Thromb. Res. Suppl.* **8**:99–113.

Ashwell, G., and Hartford, J., 1982, Carbohydrate-specific receptors of the liver, *Annu. Rev. Biochem.* **51**:532–554.

Astrup, T., and Permin, P., 1947, Fibrinolysis in the animal organism, *Nature* **159**:681–682.

Bakhit, C., Lewis, D., Billings, R., and Malfroy, B., 1987, Cellular catabolism of recombinant tissue-type plasminogen activator, *J. Biol. Chem.* **262**:8716–8720.

Banyai, L., Varadi, A., and Patthy, L., 1983, Common evolutionary origin of plasminogen activator, *FEBS Lett.* **163**:37–41.

Baughman, R. A., 1987, Pharmacokinetics of t-PA in tissue plasminogen activator thrombolytic therapy, in: *Tissue Plasminogen Activation in Thrombolytic Therapy* (B. E. Sobel, D. Collen, and E. Grossbard, eds.), Dekker, New York, pp. 41–53.

Bergdorf, N., Nilsson, T., and Wallen, P., 1983, An enzyme linked immunosorbent assay for determination of tissue plasminogen activator applied to patients with thromboembolic disease, *Thromb. Haemostasis* **50**(3): 740–744.

Bewley, T. A., 1979, Circular dichroism of pituitary hormones, *Recent Prog. Horm. Res.* **35**:155–212.

Camiolo, S. M., Thorsen, S., and Asrup, T., 1971, Fibrinogenolysis and fibrinolysis with tissue plasminogen activator, urokinase, streptokinase-activated human globulin and plasmin, *Proc. Soc. Exp. Biol. Med.* **138**:277–280.

Carlson, R. H., Garnick, R. L., Jones, A. J. S., and Meunier, A. N., 1988, The determination of recombinant human tissue-type plasminogen activator by turbidimetry using a microcentrifugal analyzer, *Anal. Biochem.* **168**:428–435.

Chloupek, R. C., Harris, R. J., Leonard, C. K., Keck, R. G., Keyt, B. A., Spellman, M. W., Jones, A. J. S., and Hancock, W. S., 1989, Study of the primary structure of recombinant tissue plasminogen activator by reversed-phase high performance liquid chromatographic tryptic mapping, *J. Chromatogr.* **463**:375–396.

Cleary, S., Mulkerrin, M., and Kelley, R., 1989, Purification and characterization of tissue plasminogen activator kringle-2 domain expressed in Escherichia coli, *Biochemistry* **28**:1884–1890.

Cole, E. R., and Bachman, F. W., 1977, Purification and properties of plasminogen activator from pig heart, *J. Biol. Chem.* **52**:3729–3737.

Collen, D., 1987, Molecular mechanism of fibrinolysis and their application to fibrin-specific thrombolytic therapy, *J. Cell. Biochem.* **33**:77–86.

Collen, D., Tytgat, G., and Verstraete, M., 1968, Semi-automatic method for routine evaluation of fibrinolytic components, *J. Clin. Pathol.* **21**:705–707.

Collen, D., Rijken, D. C., Van Damme, J., and Billiau, A., 1982, Purification of human tissue-type plasminogen activator in centigram quantities from human melanoma cell culture fluid and its conditioning for use in vivo, *Thromb. Haemostasis* **48**:294–296.

Einarsson, M., Smedsrod, B., and Pretoft, H., 1988, Uptake and degradation of tissue plasminogen activator in rat liver, *Thromb. Haemostasis* **59**:474–479.

Eisenberg, P., Sherman, L. A., Tiefenbrunn, A. J., Ludbrook, P. A., Burton, S. E., and Jaffe, A. S., 1987, Sustained fibrinolysis after administration of t-PA despite its short half-life in the circulation, *Thromb. Haemostasis* **57**:35–40.

Gaffney, P. J., and Curtis, A. D., 1985, A collaborative study of a proposed international standard for tissue plasminogen activator (t-PA), *Thromb. Haemostasis* **53**:134–136.

Garabedian, H. D., Gold, H. K., Leinbach, R. C., Johns, J. A., and Yasuda, T., 1987, Comparative properties of two clinical preparations of recombinant tissue-type plasminogen activator in patients with acute myocardial infarction, *J. Am. Coll. Cardiol.* **9**:599–607.

Hartman, P. A., Stodola, J. D., Harbour, G. C., and Hoogerheide, J. G., 1986, Reversed-phase high-performance liquid chromatography peptide mapping of bovine somatotropin, *J. Chromatogr.* **360**:385–395.

Hasegawa, A., Shukei, N., and Kondo, F., 1986, Aqueous solution of a tissue plasminogen activator dissolved at an increased concentration and a method, U.S. Patent 4,568,544.

Higgins, D. L., and Vehar, G. A., 1987, Interaction of one chain and two chain tissue plasminogen activator with intact and plasmin degraded fibrin, *Biochemistry* **26**:7786–7791.

Hotchkiss, A., Refino, C. J., Leonard, C. K., O'Connor, J. V., Crowley, C., McCabe, J., Tate, K., Nakamura, G., Powers, D., Levinson, A., Mohler, M., and Spellman, M. W., 1988, The influence of carbohydrate structure on the clearance of recombinant tissue type plasminogen activator, *Thromb. Haemostasis* **60**:255–261.

Hoylaerts, M., Rijken, D. C., Lijnen, H. R., and Collen, D., 1982, Kinetics of the activation of plasminogen by human tissue plasminogen activator: Role of fibrin, *J. Biol. Chem.* **257**:2912–2919.

Jones, A. J., and Meunier, A. M., 1990, A precise and rapid microtitre plate clot lysis assay: Methodology, kinetic modeling and measurement of catalytic constants for plasminogen activation during fibrinolysis, *Thromb Haemostasis* **64:** 455–463.

Kalyan, N. K., Lee, S. G., Wilhelm, J., Fu, K. P., and Hum, W. T., 1988, Structure function analysis with tissue-type plasminogen activator. Effect of deletion of NH_2-terminal domains on its biochemical and biological properties, *J. Biol. Chem.* **263**:3971–3978.

Kohr, W. J., Keck, R., and Harkins, R. N., 1982, Characterization of intact and trypsin-digested biosynthetic human growth hormone by high-pressure liquid chromatography, *Anal. Biochem.* **122**:348–359.

Kok, P., and Astrup, T., 1969, Isolation and purification of a tissue plasminogen activator and its comparison with urokinase, *Biochemistry* **8**:79–86.

Kruithof, E. K. O., Schleuning, W. D., and Bachman, F., 1985, Human tissue-type plasminogen activator: Production in continuous serum-free cell culture and rapid purification, *Biochem. J.* **226**:631–636.

Kunitani, M., Hirtzer, P., Johnson, D., Halenbeck, R., Boosman, A., and Koths, K., 1986, Reversed-phase chromatography of interleukin-2 muteins, *J. Chromatogr.* **359**:391–402.

Larsen, G. R., Henson, K., and Blui, Y., 1988, Fibrin binding, fibrinolytic and fibrinogenolysis characterization of generic variants lacking the fibrinolectin finger-like and/or epidermal growth factor domains, *J. Biol. Chem.* **263**:1023–1029.

Lee, J., Keyt, B. A., and Canova-Davis, E., 1989, Third Symposium of the Protein Society, Seattle.

Little, P. S., Bang, N. U., Harino, C. S., Marko, C. A., and Mattler, L. E., 1984,

Functional properties of carbohydrate-depleted tissue plasminogen activator, *Biochemistry* **23**:6191–6195.

Liu, C. Y., and Wallen, P., 1984, The binding of tissue plasminogen activator by fibrin, *Circulation* **70**:365.

Manning, M. C., 1989, Underlying assumptions in the estimation of secondary structure content in proteins by circular dichroism spectroscopy—A critical review, *J. Pharm. Biomed. Anal.* **7**:1103–1119.

Mohler, M. A., Refino, C. J., Chen, S. A., Chen, A. B., and Hotchkiss, A. J., 1986, D-Phe-Pro-Arg-chloromethylketone: Its potential use in inhibiting the formation of in vitro artifacts in blood collected during tissue-type plasminogen activator thrombolytic therapy, *Thomb. Haemostasis* **56**:160–164.

Nilsson, S., Einarsson, M., Ekvarn, S., Haggroth, L., and Mattsson, C. H., 1985, Turnover of tissue plasminogen activator in normal and hepatectomized rabbits, *Thromb. Res.* **39**:511–521.

Northey, R. P., Jr., Hsu, C. C., Nguyen, T. H., Gatlin, L., Kelley, R., Mulkerrin, M., and Cleary, S., 1987, Tissue-type plasminogen activator domain structure analysis by scanning microcalorimetry, *Pharm. Res. Suppl.* **5**:530.

Opdenakker, G., Van Damme, J., Bosman, F., Billiau, A., and DeSomer, P., 1986, Influence of carbohydrate side chains on activity of tissue-type plasminogen activator, *Proc. Soc. Exp. Biol. Med.* **182**:248–257.

Pennica, D., Holmes, W. E., Kohr, W. J., Harkins, R. N., Vehar, G. A., Ward, C. A., Bennett, W. F., Yelverton, E., Seeburg, P. H., Heyneker, H. L., Goeddel, D. V., and Collen, D., 1983, Cloning and expression of human tissue-type plasminogen activator cDNA in E. Coli, *Nature* **301**:214–221.

Philo, R. D., and Gaffney, P. J., 1981a, Assay methodology for urokinase: Its use in assessing the composition of mixtures of high and low molecular weight urokinase, *Thromb. Haemostasis* **21**:81–88.

Philo, R. D., and Gaffney, P. J., 1981b, Plasmin potency estimates: Influence of the substrate used in the assay, *Thromb. Haemostasis* **45**:107–109.

Provencher, S. W., and Glockner, J., 1981, Estimation of globular protein secondary structure from circular dichroism, *Biochemistry* **20**:33–37.

Radek, J. T., and Castellino, F. J., 1988, A differential calorimetric investigation of the domains of recombinant tissue plasminogen activator, *Arch. Biochem. Biophys.* **267**:776–786.

Ranby, M., 1982, Studies on the kinetics of plasminogen activation by tissue plasminogen activator, *Biochim. Biophys. Acta* **704**:461–469.

Ranby, M., Nguyen, G., Scarabin, P. Y., and Samama, M., 1989, Immunoreactivity of tissue plasminogen activator and of its inhibitor complexes. Biochemical and multicenter validation of a two site immunosorbent assay, *Thromb. Haemostasis* **61**:409–414.

Reich, E., 1975, Plasminogen activators: Secretion of neoplastic cells and macrophages, in: *Proteases and Biological Control* (E. Reich, D. B. Rifkin, and E. Shaw, eds.), Cold Spring Harbor Laboratory, Cold Spring Harbor, N.Y., pp. 333–341.

Rijken, D. C., and Collen, D., 1981, Purification and characterization of the plasminogen activator secreted by human melanoma cells in culture, *J. Biol. Chem.* **256**:7035–7041.

Rijken, D. C., Wijngaards, G., Zaal de Jong, M., and Welbergen, J., 1979a, Purification and partial characterization of plasminogen activator from human uterine tissue, *Biochim. Biophys. Acta* **580**:140–153.

Rijken, D. C., Wijngaards, G., and Welbergen, J., 1979b, Biochemical and immunological characterization of plasminogen activator from human tissue, *J. Prog. Chem. Fibrinolysis Thrombolysis* **4**:349–354.

Rijken, D. C., Wijngaards, G., and Welbergen, J., 1980, Relationship between tissue plasminogen activator and the activators in blood and vascular wall, *Thromb. Res.* **18**:815–830.

Rijken, D. C., Hoylaert, M., and Collen, D., 1982, Fibrinolytic properties of one chain and two chain human extrinsic tissue type plasminogen activator, *J. Biol. Chem.* **257**:2920–2925.

Robins, K. C., Summaria, L., Hsieh, B., and Shah, R. J., 1967, The peptide chains of human plasmin, *J. Biol. Chem.* **242**:2333–2342.

Seifried, E., Tanswell, P., Su, C.A.P.F., Feurer, W., Pindur, G., and Heimpel, H., 1986, Recombinant tissue type plasminogen activator (rt-PA): Pharmacokinetics and effects on coagulation and fibrinolytic system in healthy volunteers, Abstract A242, 8th Int. Congress on Fibrinolysis, Vienna.

Smedrod, B., Einarsson, M., and Pertoft, H., 1988, Tissue plasminogen activator is endocytosed by mannose and galactose receptors of rat liver cells, *Thromb. Haemostasis* **59**:480–484.

Spellman, M., Basa, L., Leonard, C., Chakel, J., and O'Connor, J., 1989, Carbohydrate structures of human tissue plasminogen activator expressed in Chinese hamster ovary cells, *J. Biol. Chem.* **264**:14100–14111.

Strickland, E. H., 1974, Aromatic contributions to circular dichroim spectra of proteins, *CRC Crit. Rev. Biochem.* **1974**:113–175.

Van de Werf, F., Ludbrook, P. A., Bergman, S. R., Tiefenbrunn, A. J., Fox, K. A. A., de Geest, H., Verstraete, M. Collen, D., and Sobel, B. E., 1983, Coronary thrombosis with tissue-type plasminogen activator in patients with evolving myocardial infarction, *N. Engl. J. Med.* **310**:609–613.

Van Zonnenfeld, A. J., Veerman, H., and Pannekoek, H., 1986a, On the interaction of the finger and the kringle-2 domain of tissue-type plasminogen activator with fibrin, *J. Biol. Chem.* **261**:14214–14218.

Van Zonnenfeld, A. J., Veerman, H., and Pannekoek, H., 1986b, Autonomous functions of structural domains on human tissue-type plasminogen activator, *Proc. Natl. Acad. Sci. USA* **83**:4670–4674.

Wallen, P., Bergsdorf, N., and Ranby, M., 1982, Purification and identification of two structure variants of porcine tissue plasminogen activator by affinity adsorption on fibrin, *Biochim. Biophys. Acta* **719**:318–328.

Wallen, P., Pohl, G., Bergsdorf, N., Ranby, M., Ny, T., and Jornvall, H., 1983, Purification and characterization of a melanoma cell plasminogen activator, *Eur. J. Biochem.* **132**:681–686.

Weimar, W., Stibbe, J., Van Seyen, A. J., Billiau, A., De Somer, P., and Collen, D.,

1981, Specific lysis of an iliofemoral thrombus by administration of extrinsic tissue-type plasminogen activator, *Lancet* **2**:1018–1020.

Wetlaufer, D. B., 1962, Ultraviolet spectra of proteins and amino acids, *Adv. Prot. Chem.* **17**:303–390.

Wilkens, H. J., and Back, N., 1976, New instrument for automatic recording of clot lysis, *Am. J. Clin. Pathol.* **66**:124–131.

Yang, J. T., Chang, C. T., and Wu, C. C., 1978, Circular dichroism analysis of protein conformation: Inclusion of β-turns, *Anal. Biochem.* **91**:13–31.

Chapter 4

Orthoclone OKT3

Chemical Mechanisms and Functional Effects of Degradation of a Therapeutic Monoclonal Antibody

Patricia E. Rao and Daniel J. Kroon

1. BACKGROUND

1.1. History of Clinical Use of OKT3

Orthoclone OKT3 has been marketed since 1986 for reversal of human kidney graft rejections.* Widely used for reversal of rejections and for rescue of steroid-resistant rejections, some groups have experimented with the use

---

Patricia E. Rao and Daniel J. Kroon • The R. W. Johnson Pharmaceutical Research Institute, Raritan, New Jersey 08869. *Present address of P.E.R.:* Ortho Diagnostic Systems, Inc., Raritan, New Jersey 08869.

Stability and Characterization of Protein and Peptide Drugs: Case Histories, edited by Y. John Wang and Rodney Pearlman. Plenum Press, New York, 1993.

* This chapter summarizes some of the data collected over 6 years on the degradation of OKT3 and contains original information from our laboratories as well as the laboratory of Dr. An Tran. Creative ideas and technical contributions were made by Elizabeth Westberg, Alysia Baldwin-Ferro, Eric Meyer, Eileen Hannigan, Praful Lalan, and Mary Makowski. Helpful suggestions were received from Dr. Anneli DePaolis, Judy Schnell, Dr. Vincent Pigiet, Dr. Allan Williams, and Dr. Lewis Hanna.

of OKT3 in other antirejection applications (Bristow *et al.*, 1988; Cosimi *et al.*, 1990). OKT3 is a murine monoclonal antibody directed to a component of the antigen receptor present on all mature, human T cells. The OKT designation derives from its origin (*O*rtho Pharmaceutical Corporation, lab of Patrick *K*ung) and its specificity (*T* cells). It has been given the generic name, Muromonab CD3. When OKT3 is administered to a person experiencing rejection, all circulating peripheral T cells disappear from circulation within 30 min of the first injection (Chatenoud *et al.*, 1982; Norman *et al.*, 1987). Mature T cells remain absent throughout the first 7–10 days of treatment, although T cells bearing other T cell surface markers reappear earlier. However, these cells typically lack the T cell receptor for antigen and are unable to respond to foreign antigens, making graft rejection impossible. Patients receive a course of treatment for 10–14 days and clinical signs of rejection reversal are usually present by the third to fifth day of treatment (Ortho Multicenter Transplant Study Group, 1985; Thistlethwaite *et al.*, 1987). After the course of OKT3 therapy, the patient resumes his normal immunosuppressive regimen.

OKT3 was the first mouse monoclonal antibody approved by the FDA for human therapy, and so was the first monoclonal antibody produced on a large scale under GMP conditions for parenteral administration. Since its introduction onto the market, Orthoclone OKT3 has been formulated for intravenous use as a 1 mg/ml sterile solution in pH 7.0 phosphate-buffered saline containing 0.02% polysorbate 80.

1.2. Basic Biochemical Structure of OKT3

OKT3 is a mouse-derived antibody of the IgG2a class. Antibodies of this class have a molecular mass of approximately 150 kDa, of which 2–3% is carbohydrate. In mouse IgG2a antibodies, carbohydrates are generally *N*-linked to the Fc region (for a general review of antibody structure see Nisonoff *et al.*, 1975; Amzel and Poljak, 1979; Edelman, 1971). Each antibody molecule consists of two identical heavy polypeptide chains, and two identical light chains. These chains are linked, each light chain to a heavy chain, and the two heavy chains together, by disulfide bonds between cysteines. In OKT3, the heavy and light chains are composed of 449 and 213 amino acids, respectively (Fig. 1). A protein of this size can be expected to exhibit multiple decomposition pathways, complicating the analysis of product stability. Although antibody binding specificity is conferred by unique amino acid sequences, a major portion of the sequence is conserved among other antibodies of the same class; most of the variability is confined to the residues located

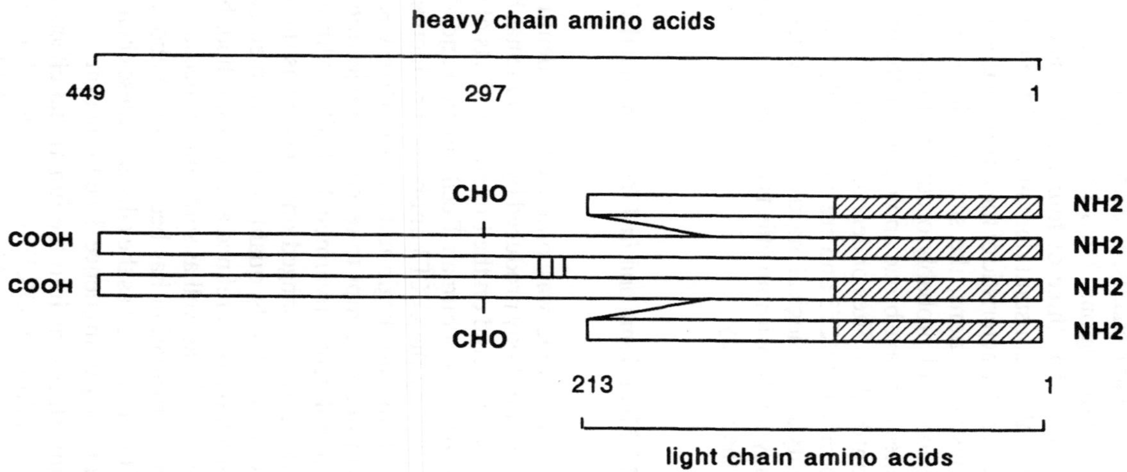

Figure 1. Schematic diagram of mouse IgG2a with amino acid numbering from OKT3. Schematic diagram of heavy- and light-chain structure of a mouse IgG2a antibody. Disulfide bonds linking the heavy chains together and joining heavy and light chains are shown as connecting lines. Hatched area indicates variable regions; CDRs are contained within this area. CHO indicates glycosylation linkage sites.

in patches near the NH_2-terminus of the four chains that participate in antigen binding (the hypervariable loops, or complementarity-determining regions, CDRs). The bulk of the antibody molecule, three-fourths of the heavy chains and one-half of the light chains, is composed of highly conserved sequences known as constant regions. Each light chain has one constant region and each heavy chain has three or four repetitions of the constant domain. The sequence of these constant domains and the number of times the sequence is repeated are characteristic of the antibody isotype. Because of the conservation of constant regions, the sites at which OKT3 degrades within the constant regions will probably occur in other antibodies; thus, identification of these sites will have broad application in the pharmaceutical development of other antibodies. Sites of degradation in the variable region are likely to be more specific to OKT3 and are more likely to result in loss of antigen binding capability and biological inactivity. However, damage to the Fc portion of OKT3 may not be without effects on the *in vivo* effector function or metabolism of the antibody.

1.3. Discovery of OKT3 Hybridoma and *in Vitro* Functional Effects

The OKT3 hybridoma clone was produced and identified in 1978 (Kung *et al.,* 1979) and the antibody produced by the clone has been studied since then. Original studies centered on the functional effects of the addition of this antibody to *in vitro* assays of human T cell function and identification of the T cell antigen bound by the OKT3 antibody. Production of the first material for human trials did not begin until 1980, when OKT3 was first administered to a person in an attempt to reverse the rejection of a kidney transplant after conventional steroid treatment had failed. Although the purification methods for OKT3 were changed only slightly as the method was scaled up to its final production levels, quality assurance methods for analysis of the finished product have been refined as newer methods have become available. Using these methods, as well as some more specialized analytical techniques, considerable information has been obtained about the mechanism of degradation of OKT3 as well as the effects of these chemical changes on the ability of this antibody to influence T cell function.

During the early investigations into the functional effects of OKT3, it was discovered that the addition of OKT3 to cultures of human T cells, in concentrations below that necessary to bind to all available antigen, would induce the cells to divide (VanWauwe *et al.,* 1980; Chang *et al.,* 1981). This division was dependent on the presence of monocytes in the cultures and the interaction of the Fc regions of OKT3 with them (Tax *et al.,* 1983, 1984). It was also learned that the mitogenic effects of OKT3 and other anti-CD3 antibodies were dependent on the isotype of the antibody; mouse IgG2a was mitogenic for all people, mouse IgG1 for some, and mouse IgG2b for only a few individuals (Smith *et al.*, 1986). In addition to its mitogenic properties, OKT3 was found to be capable of inhibiting T cell function in two other *in vitro* assays: the mixed lymphocyte reaction (MLR) and the cytotoxic lymphocyte reaction (CTL) (Chang *et al.*, 1981; Reinherz *et al.*, 1980). In the first of these reactions, T cells from one individual are mixed with non-T cells from another. Within a week of initiation of the assay, the T cells will begin to divide, stimulated by recognition of the foreign histocompatibility antigens of the non-T cells. In the second type of assay, cells are stimulated in an MLR culture and then exposed to fresh cells from the donor of the non-T cells. Cytotoxic T cells, generated during the MLR culture, and directed against the histocompatibility antigens of the non-T cells, will then kill the fresh cells during the second culture. It was this ability to inhibit the response of T cells against foreign histocompatibility antigens that suggested the use of OKT3 as therapy for transplant rejection.

2. BIOLOGICAL EFFECTS OF DEGRADATION OF OKT3

2.1. Effects on Binding Affinity

One of the quality assurance (QA) assays that is used to follow Orthoclone OKT3 during stability testing periods has been termed the potency assay. In this assay, the test lot of OKT3 is diluted and tested, at each dilution, for its ability to competitively inhibit the binding of FITC-conjugated, Orthomune OKT3 FITC (Orthomune OKT3 contains antibody from the same clone as Orthoclone OKT3). A reference standard lot of OKT3 is similarly diluted and tested and the competition curves for both lots of OKT3 are compared statistically. The displacement of the linear portion of the competition curve of test samples from the linear portion of the reference standard curve is expressed as a percentage. It was observed that as the test lot of OKT3 aged, either under elevated temperature storage to accelerate reactions (37°C), or under labeled storage conditions (5°C), its competition curve would be displaced relative to the curve generated by the reference standard. It was assumed that this displacement of the competition curve resulted from a decrease in the binding affinity of the test lot of OKT3.

Studies were then carried out to measure directly the binding affinity of new and aged lots of OKT3. Because it was known that the CDR regions of OKT3 contained several residues sensitive to oxidative degradation (see below), we felt it would be counterproductive to measure the binding affinity using OKT3 labeled with ^{125}I under oxidizing conditions. We therefore chose to label the lots of OKT3 to be tested by direct conjugation of FITC. Since FITC binds through free amino groups and would not expose the antibody to oxidizing conditions, it was hoped that OKT3 labeled with FITC would retain its original binding affinity. All lots of antibody labeled with FITC were compared with the same lot, unlabeled, in a rough dilutional analysis for saturation staining of cells. In all cases, flow cytometry indicated that the same amount of antibody was necessary to stain cells to saturation. From these results we concluded that the FITC labeling had not significantly impaired the binding of OKT3 to cells. The F/P (moles of fluorescein per mole of antibody) ratio was determined for each lot of antibody using Simply Cellular beads and quantitative microbead standards from Flow Cytometry Standards.

Analyzing cells stained with increasing amounts of each FITC-OKT3 by flow cytometry, it was possible to derive data suitable for Scatchard analysis. Binding affinities derived in this manner indicated that aged OKT3 did, in fact, have a decreased binding affinity for the CD3 antigen of human peripheral blood lymphocytes. Fresh lots of OKT3 were found to have a binding affinity of 2.4×10^9 M^{-1}. Lots of OKT3 that had been stored at 5°C for 12–15 months were found to have binding affinities ranging from 5×10^8 M^{-1} to 1.2×10^9 M^{-1}. These decreases in affinity correlated with shifts in the competition curves observed in the potency assay (Fig. 2).

Although the two- to fourfold losses in binding affinity over a year at 5°C were quite modest and the final affinities were still indicative of strong, specific binding, we wanted to determine whether the reduction of binding affinity would affect the function of OKT3 in *in vitro* assays. Two assays of OKT3 function were performed: induction of mitogenesis and inhibition of the MLR.

2.2. Ability to Induce Mitogenesis

OKT3 binds to the human T cell receptor and is able to induce a mitogenic response in those cells. At low antibody concentrations, the magnitude of this mitogenic response can be proportional to the amount of antibody used. The lower binding affinity might have resulted in delivery of a lower effective dose of OKT3 to the cells which would have been reflected in a less intense mitogenic signal at the lower antibody concentrations. Human peripheral blood lymphocytes (PBL) were incubated with a range of concentrations of reference standard OKT3 and OKT3 from the aged lots tested in the binding affinity tests. The results are shown in Fig. 3. All lots of OKT3 tested,

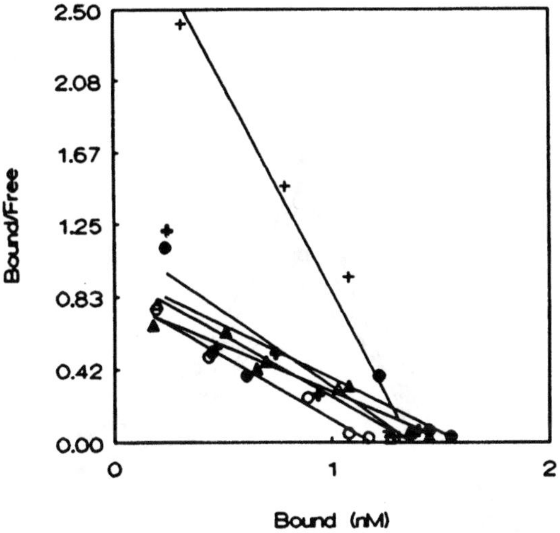

	Sample	Age	Potency
+	Ref. Std.		100
△	OKT3 Lot A	14	70
○	OKT3 Lot B	15	57
✦	OKT3 Lot C	12	72
▲	OKT3 Lot D	12	83
●	OKT3 Lot E	11	79

Figure 2. Scatchard plot, age and potency assay results from five aged lots of OKT3. Production lots of OKT3, formulated in PBS with 0.02% Tween were stored at 5°C for the indicated number of months. Samples were then tested in the potency assay and binding affinities were measured as described in the text.

including the reference standard, appeared to deliver the same mitogenic signal at all concentrations, indicating that the two- to fourfold decrease in binding affinity had no detectable effect on *in vitro* T cell mitogenesis.

2.3. Ability to Inhibit MLR

OKT3 is also able to inhibit the response of human T cells to foreign histocompatibility antigens in the MLR. This assay can be progressively

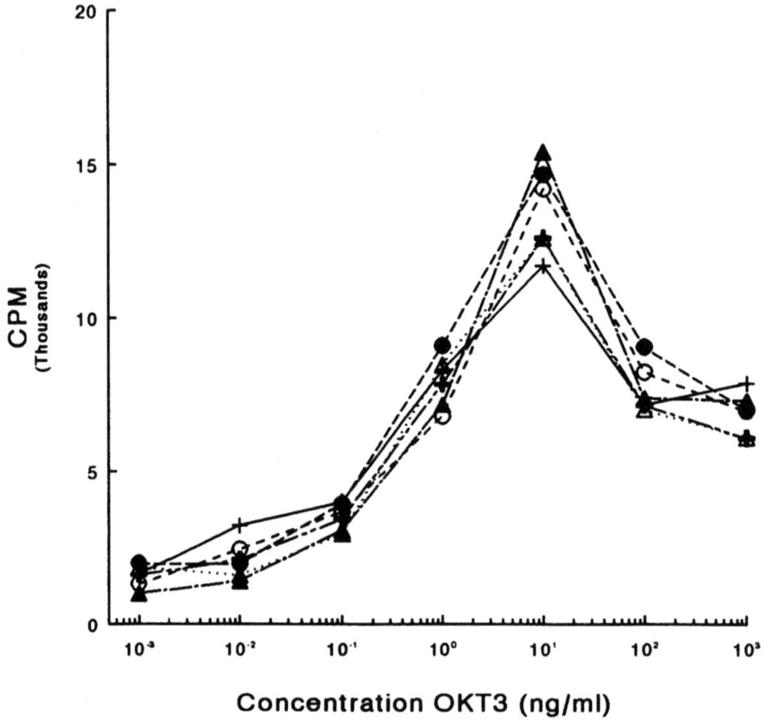

Figure 3. Ability to induce human T-cell mitosis by lots of OKT3. Five aged lots of OKT3 (see Fig. 1) and reference standard were added at the concentrations indicated to cultures of 2×10^5 human periheral blood lymphocytes, and the resulting cell division was measured by uptake of [^{3}H]thymidine. (+, OKT3 reference standard; ○, OKT3 lot B; ●, OKT3 lot E; △, OKT3 lot A; ▲, OKT3 lot D; **+**, OKT3 lot C.)

inhibited by increasing concentrations of OKT3. To determine whether the loss of binding affinity would significantly reduce the ability of a given amount of OKT3 to inhibit the MLR, reference standard OKT3 and the aged lots tested above were tested for their abilities to inhibit the MLR at various concentrations. As shown in Fig. 4, the decrease in binding affinity was not sufficient to affect the inhibitory abilities of OKT3.

3. BIOCHEMICAL EVIDENCE OF DEGRADATION

Several routine analytical techniques for proteins were found useful for detecting changes in OKT3 with time. In addition, as more was learned of

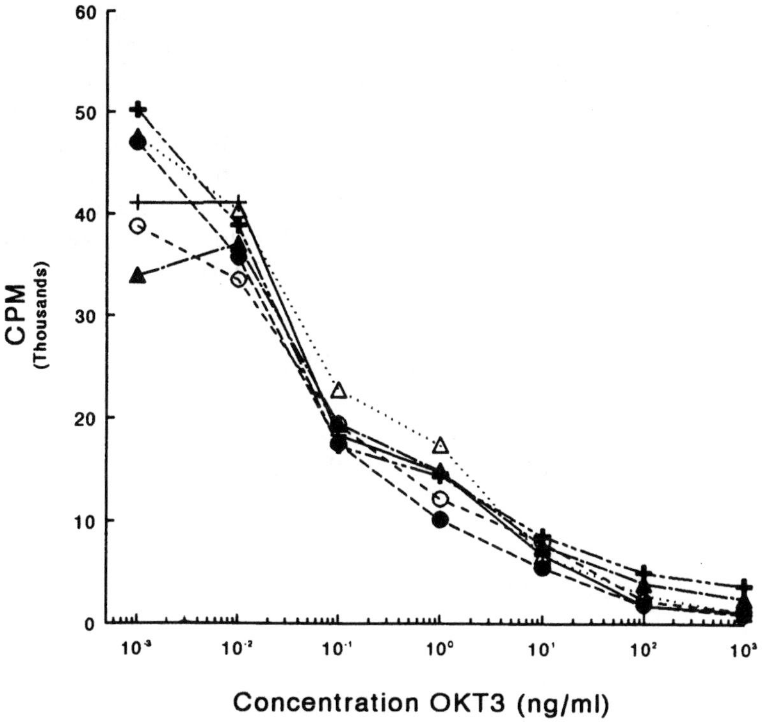

Figure 4. Ability of aged lots of OKT3 to inhibit the human mixed lymphocyte response. Human T cells, purified by rosetting with sheep erythrocytes, were mixed with irradiated, non-T cells from two other donors, in the presence of the indicated concentrations of reference standard OKT3 or samples from aged lots of OKT3 (see Fig. 1). After 7 days, the cultures were fed with media containing [^{3}H]thymidine and the incorporated radioactivity was counted after 6 h. (See Fig. 3 for explanation of symbols.)

the chemical mechanisms of degradation it was apparent that different techniques detected different degradation products.

3.1. Shifts in IEF Pattern

Alterations in the charge of the OKT3 molecule were followed using both isoelectric focusing (IEF) and cation exchange chromatography (HPLC-IEC). IEF, although difficult to quantitate, provided the clearest and most sensitive evidence of slight alterations in the charge of OKT3 as it aged. Reference standard OKT3 consistently produces a pattern of four strong, distinct bands ranging in pI from 6.8 to 7.2. As OKT3 ages, this pattern progressively shifts toward more acidic pH's (Fig. 5). As the most basic band begins to lose intensity, a new band appears just below the most acidic band of the reference standard. As the degradation process continues, the most basic band is lost and additional more acidic bands appear. The pattern retains four or five sharp bands until the entire pattern is more acidic than any of the reference standard bands. Beyond this point, the bands become indistinct and the pattern is lost (Fig. 5A). This sequence of events occurs regardless of the temperature at which OKT3 is stored, but takes longer to occur at 5°C (more than 1 year) than at 37°C (2 months).

3.2. Alteration in HPLC-IEC Retention Times

Changes are also seen in the retention times and number of peaks eluted from HPLC-IEC as OKT3 ages. As with the changes in the IEF profiles, the

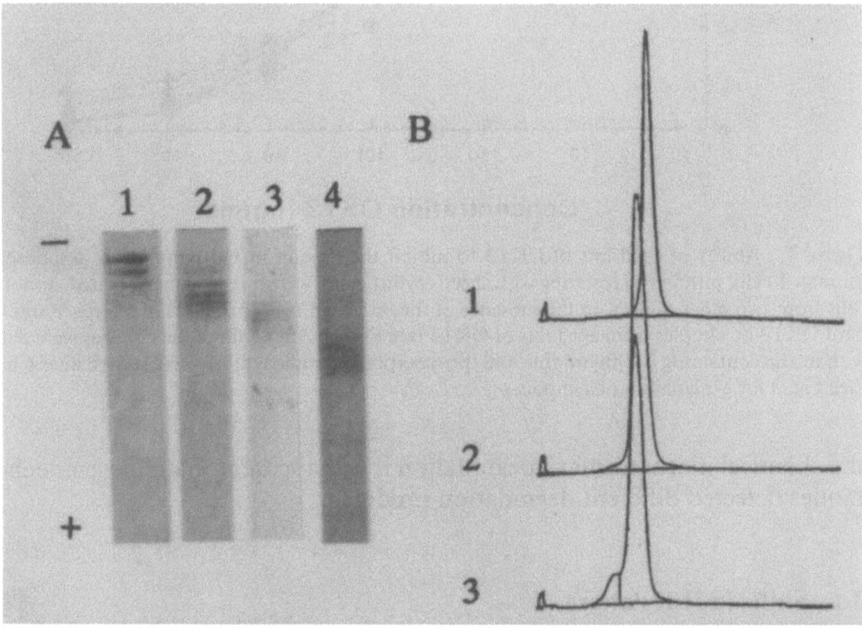

Figure 5. Changes in IEF and HPLC-IEC results as OKT3 ages. (A) Reference standard OKT3 (lane 1) or OKT3 held at 37°C for 2, 4, or 6 weeks (lanes 2, 3, and 4, respectively) was separated by IEF. Progressive shifting of the pattern toward the anode can be seen. (B) Reference standard OKT3 (1), OKT3 stored at 5°C for 11 months under nitrogen (2) or air (3), were analyzed by HPLC-IEC. Samples were applied to a Synchropak CM-300 column at 32°C in 0.01 M K_2HPO_4 (buffer A). Samples were eluted with an increasing gradient of buffer B, 0.5 M KCl in buffer A, to 26% and OKT3 was eluted isocratically at 26% B. Progressive shifting of the major peak and formation of a second peak can be seen.

same changes occur at different temperatures, only the kinetics differ. When analyzed with an isocratic elution, even small differences in charge will alter the retention time of the peaks. Using this type of method it was possible to clearly distinguish changes in OKT3 by comparison with the retention of the reference standard material. Initially, degradation results in earlier elution of the entire OKT3 peak when compared with the reference standard. The peak remains quite sharp at this point and in other tests, such as the potency test described above, appears indistinguishable from reference standard (Fig. 5B). A second peak, eluting even earlier, develops as the material continues to degrade. When material is stressed at high temperatures or OKT3 several years old is used, enough of the second peak can be obtained for additional studies. OKT3 in this second peak is found to compete poorly for binding with FITC-OKT3 in comparison with reference standard. In addition, it is enriched in OKT3 containing nonreducible cross-links between the heavy and light chains (see below).

3.3. Protein Chain Alterations Detected by SDS-PAGE

Both of the above methods detect primarily degradation events leading to alterations in the charge of the protein. When SDS-PAGE is used to follow material during stability studies, additional information is obtained on the physical condition of the protein chains. OKT3 stored at 5°C for over a year and a half was compared with reference standard OKT3 on SDS-PAGE under reducing conditions. Protein bands were identified with Coomassie staining and origin of the bands was determined in Western blots using heavy- and light-chain-specific antisera (Fig. 6). Using heavily loaded gels, reference standard OKT3 was found to contain primarily a heavy chain of 52 kDa and a light chain of 26 kDa. There were also faintly visible protein bands at 46, 43, and 31 kDa, all derived from heavy chains. Aged and degraded OKT3 was still composed primarily of the normal heavy and light chains, but in addition, contained a definite band at 88 kDa as well as lighter bands at 46, 43, 31, 24, and 17 kDa. All fragments except the 24- and 17-kDa bands were derived from heavy chains; 24- and 17-kDa fragments were from light chains (Fig. 6). Additional, even smaller fragments were also produced, but were not resolved on the gels used in these experiments. Interestingly, the band that appeared at 88 kDa was stained by both anti-heavy- and anti-light-chain antisera in the Western blot. This observation and the molecular mass suggested that this new band represented a nonreducible, covalently linked heavy and light chain dimer.

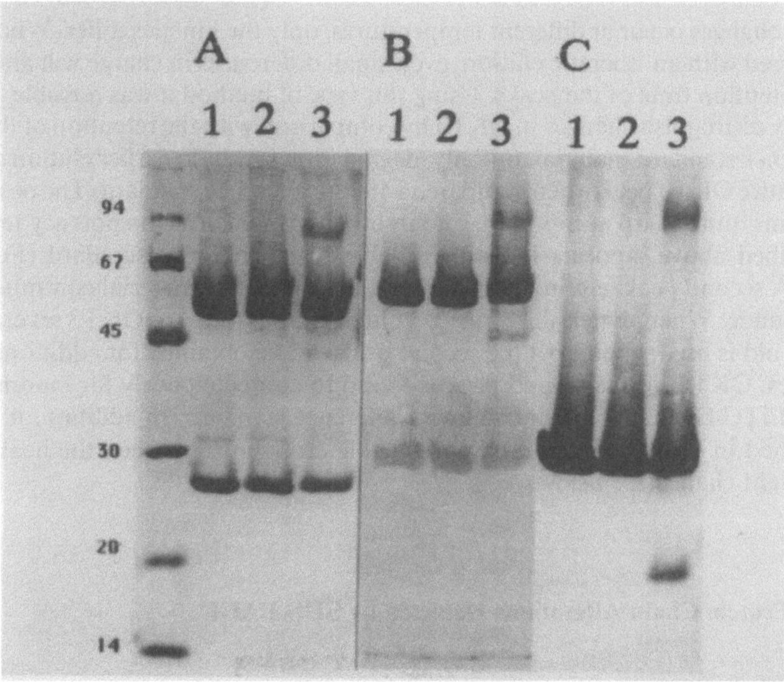

Figure 6. Western blot of samples of OKT3. Reference standard OKT3 (lane 1), a fresh lot of OKT3 (lane 2), or 22-month-old OKT3 (lane 3) were treated with 10% 2-mercaptoethanol and run on 12.5% acrylamide SDS-PAGE gels using the Laemmli buffers. The gel was then stained with Coomassie blue to detect all protein bands (A) or electroblotted onto nitrocellulose and probed with either anti-mouse IgG heavy-chain sera (B) or anti-mouse light-chain sera (C).

The 31-kDa fragment was partially sequenced by cutting the band from a Western blot membrane (Aebersold *et al.,* 1987; Matsudaira, 1987) and its NH_2-terminus determined to be P-P-C-K-C-P-A-P. This sequence occurs at residues H (heavy chain residues) 227–234 of OKT3, the hinge region of the antibody. In antibodies, the constant portion of the heavy chain contains a highly flexible region, between two constant domains, which provides the antibody with the ability to flex and bind otherwise spatially inaccessible parts of antigen molecules. Because of its structure, this region is exposed to solvent, and perhaps is more accessible to chemical interactions leading to degradation. The size of the fragment implies that it extends to, or nearly to, the COOH-terminus. Chain breakage at this point would have produced another fragment composed of the NH_2-terminal portion of the heavy chain. This fragment of the heavy chain would begin with the NH_2-terminus of OKT3, which previous work had shown to be blocked by pyroglutamate and difficult to sequence by ordinary means. The only other cleavage fragment we have positively identified by sequencing thus far is a small fragment not observed on SDS gels because of the molecular mass limitation of the gels used. This fragment was isolated by gel filtration chromatography and determined by protein sequencing to be the COOH-terminal 7-kDa fragment of heavy chain. The counterpart of this fragment could be the 43- or 46-kDa fragment observed on the Western blots.

4. PRELIMINARY STUDIES OF DEGRADATION MECHANISMS

Effects of Removing Oxygen on Stability of OKT3

The results of the various biochemical studies of degraded OKT3 suggested that multiple mechanisms of degradation would be required to account for the effects observed. Acidic shifts in the charge of the molecules are most easily explained by deamidation of amino acids such as glutamine and asparagine. This type of degradation has been reported for other proteins (Johnson *et al.*, 1989; Chazin *et al.*, 1989; DiAugustine *et al.*, 1987). Hydrolysis of peptide bonds would account for the production of smaller molecular-weight fragments. Oxidation of labile amino acids is another common event in the degradation of proteins, and we believe this mechanism to be most likely involved in the formation of covalent cross-links between protein chains.

In an effort to determine whether oxidation was, in fact, responsible for the cross-linking which apparently occurred between the heavy and light chains of OKT3, some samples of OKT3 were formulated in buffers we believed would inhibit oxidation. OKT3 was formulated in either the current phosphate buffer supplemented with 0.1% sodium thiosulfate, the current phosphate buffer overlaid with argon, or the current phosphate buffer overlaid with air. These samples were stored at 37°C for up to 3 months, and at appropriate points, samples were removed and tested by IEF, by the potency assay and by SDS-PAGE (Fig. 7). SDS-PAGE clearly showed that the formation of the 88-kDa species was delayed by either the argon overlay or inclusion of thiosulfate in the formulation buffer. In the potency assay as well, the binding affinity of OKT3 was preserved by the argon overlay or the thiosulfate additive. However, neither argon nor the addition of thiosulfate affected the degradation detected by IEF. All samples of OKT3 shifted at the same rate toward more acidic pI's. These results are consistent with the existence of at least two mechanisms of degradation for OKT3, oxidation and deamidation. In addition, it was shown using the potency assay that material that had shifted noticeably to more acidic pI's was unaf-

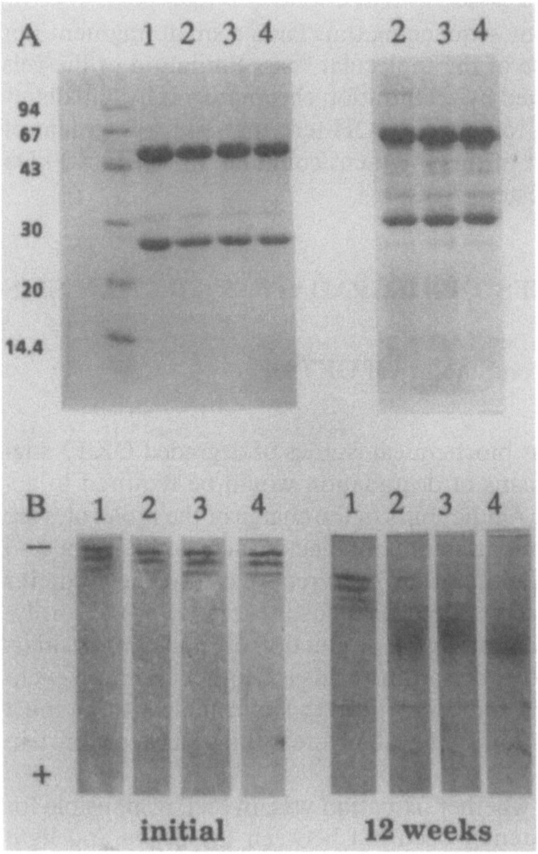

Figure 7. Effect of elimination of oxidation on OKT3 degradation. Samples of OKT3 were formulated in either current phosphate buffer overlaid with air (2), current buffer overlaid with argon (3), or current buffer supplemented with 0.1% Na thiosulfate and overlaid with air (4) and stored for 12 weeks at 37°C. Samples were examined by SDS-PAGE under reducing conditions (A) or IEF (B). All test samples were compared to an OKT3 reference standard (1).

fected in its ability to bind human T cells. The ability to bind to cells appeared most affected by oxidative events.

5. IDENTIFICATION OF DEGRADATION SITES BY PEPTIDE MAPPING

5.1. Peptide Mapping Method

To further delineate the specific sites and chemical mechanisms of OKT3 degradation, a peptide mapping procedure was developed. The procedure involved digesting the protein with enzyme, separating the resulting peptide fragments by HPLC, and identifying the peptides that changed as OKT3 degraded. Native antibodies are not very susceptible to enzyme digestion. Denaturation, especially reduction and alkylation, exposes enzyme-sensitive sites, but has a tendency to make the antibody insoluble. In our technique, OKT3 was denatured, the disulfide bonds were cleaved, and the cysteine *S*-sulfonated according to the procedure of Thannhauser *et al.* (1984). This method enabled soluble heavy and light chains of OKT3 to be separated by gel permeation chromatography. The heavy and light chains were then separately digested with enzymes which simplified the chromatography needed to resolve the more than 60 fragments generated by trypsin digestion of the two chains. Other enzymes, endoproteinases such as Glu-C and Lys-C, gave fewer peptide fragments, but the larger size of these peptides made complete sequence determination problematic, and changes in the structure of one amino acid in large peptides did not alter the HPLC retention times significantly. Although most digests were produced with trypsin, Glu-C and Lys-C endoproteinase digestions were still useful to confirm the identity of certain peptides and amino acid modifications. The enzyme digests were analyzed by reversed-phase HPLC with gradient elution to generate the peptide maps. Peptide fragments were identified on the maps by sequencing the isolated component, coelution with a chemically synthesized known peptide, and/or molecular weight determination by ion-spray mass spectrometry. A number of lots of OKT3 from various stability studies were compared with reference standard by this peptide mapping procedure, and our results are summarized in Table I.

Table I
Sites of Degradation Identified in OKT3

Amino acid	Antibody chain, region	Chemical change
Asn_{156}	L, constant	Deamidation
Asn_{386}	H, constant	Deamidation
Asn_{423}	H, constant	Deamidation
Met_{34}	H, variable	Oxidation
Met_{316}	H, constant	Oxidation
Met_{360}	H, constant	Oxidation
Met_{408}	H, constant	Oxidation
Met_{174}	L, constant	Oxidation
Trp_{47}	H, variable	Unknown
Amino acids 46–52	L, variable	Cross-link
Amino acids 99–121	H, variable	Cross-link
Cys_{226}–Pro_{227}	H, constant	Hydrolysis
Lys_{388}-Thr_{389}	H, constant	Hydrolysis

5.2. Sites of Deamidation

Initially, OKT3 from accelerated (37°C) stability studies was examined by peptide mapping. The earliest obvious changes in the peptide maps of this material were decreases in the relative size of the HPLC peaks corresponding to the amino acid 155–168 tryptic fragment of the light chain and the amino acid 364–388 fragment of the heavy chain (Fig. 8). Two small peaks in the light chain map correspondingly grew in size, as did a peak in the heavy chain map. L (light chain peptide fragment) 155–168 contains an Asn–Gly sequence at residues 156–157, and H (heavy chain peptide fragment) 364–388 has Asn–Gly at residues 386–387. The Asn–Gly sequence in proteins is known to be particularly labile to deamidation of the Asn side chain, resulting in production of Asp and isoAsp residues via hydrolysis at alternate sites of the cyclic imide intermediate (Stephenson and Clarke, 1989; Yuksel and Gracy, 1986; Capasso *et al.,* 1989). The isolated light chain degradation peptides were sequenced and one was found to be the Asp^{156} analogue of L155–168, the other sequence terminated after Gln^{155}. This termination indicates that this peptide is the isoAsp-containing fragment which cannot undergo Edman degradation at isoAsp, and therefore resists sequencing. The isoAsp peptide was the predominant deamidation product. In the heavy chain, the peak that grew in size at the expense of H364–388 was found to be related to that peptide in sequence, but the yields in the sequencing and possible content of isoAsp peptide made unequivocal establishment of Asn^{386} deamidation difficult. We confirmed that this residue does deamidate readily in OKT3 at 37°C by construction of synthetic peptides of the H383–390 fragment containing Asn or Asp. These synthetic peptides coeluted with the HPLC peaks that exhibited the greatest difference in size in the Glu-C endoproteinase map of stressed and reference standard OKT3. These are the only two Asn residues in OKT3 which are contained within Asn–Gly sequences; however, there are several Asn–Ser sequences, also known to be prone to deamidation (Kossiakoff, 1988; Violand *et al.,* 1990) and one of these Asn, H423, was also shown to be deamidated to a small extent in some samples of OKT3.

The conversion of four Asn (one in each H and L chain) in intact OKT3 to Asp, and small amounts of deamidation at other sites, could account for much of the observed acidic shift in IEF bands and shift in ion exchange HPLC retention time of OKT3 with time. This decomposition pathway is greatly accelerated with increasing temperature; samples of OKT3 held at 37°C for 8 weeks showed more Asn deamidation than batches kept at 4–8°C over 2 years, although a number of other degradation products were identified in the highly aged batches. The predominant sites of Asn deamidation in

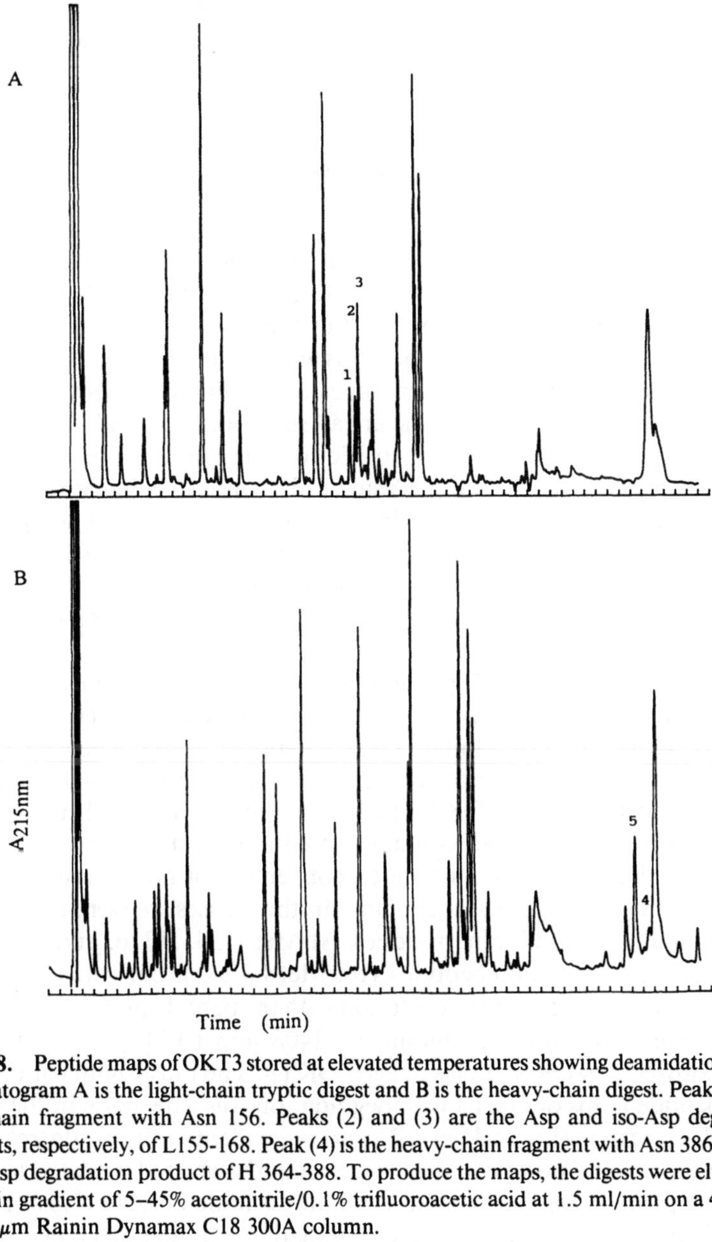

Figure 8. Peptide maps of OKT3 stored at elevated temperatures showing deamidation. HPLC chromatogram A is the light-chain tryptic digest and B is the heavy-chain digest. Peak (1) is the light-chain fragment with Asn 156. Peaks (2) and (3) are the Asp and iso-Asp degradation products, respectively, of L155-168. Peak (4) is the heavy-chain fragment with Asn 386. Peak (5) is the Asp degradation product of H 364-388. To produce the maps, the digests were eluted with a 60-min gradient of 5–45% acetonitrile/0.1% trifluoroacetic acid at 1.5 ml/min on a 4.6 × 250 mm, 5-μm Rainin Dynamax C18 300A column.

OKT3 are in the constant region of the antibody and therefore would not be expected to have significant effect on antigen binding, a result confirmed by the finding that samples which showed greater than 50% deamidation still had acceptable potency in the competition assay. Potency was also found not to be correlated with the shift in IEF pattern.

5.3. Potential Sites of Oxidative Degradation

In OKT3 aged under normal storage conditions at 5°C, it appears that oxidation is a major factor resulting in loss of binding affinity (see above). Determination of oxidation products in proteins is not always straightforward. Although a number of amino acids are susceptible to oxidation, detection of the products of these reactions is not trivial. Methionine, cysteine, tryptophan, histidine, and tyrosine are all susceptible to oxidation. Methionine sulfoxide is converted back to Met during conventional amino acid analysis and protein sequencing. Cysteine gives multiple oxidation products which are also problematic in amino acid analysis and sequencing. In our process of preparing OKT3 for peptide mapping, the Cys were S-sulfonated, so the status of this amino acid could not be determined. Trp, His, and Tyr residues are also susceptible to oxidation, producing a range of products (Manning *et al.,* 1989).

The peptide maps of degraded OKT3, stored at 5°C, (Figs. 9 and 10) showed reduced relative peak sizes for some peptides and several new peaks not found in the maps of standard OKT3. Peptides from the new peaks were isolated and analyzed by protein sequencing. In many cases the sequence data matched those of an OKT3 fragment already identified in the maps. Since most of these peptides contained methionine, this observation could result from the previously mentioned conversion of methionine sulfoxide back to methionine during sequencing. In three cases we were able to confirm oxidation at methionine residues. At Met H34, H316, and H360, the identity of the degradation peptide as the Met sulfoxide derivative was established by mass spectroscopy or coelution with the peptide prepared by chemical synthesis. Two other methionines, H408 and L174, gave evidence of being oxidized in some samples of aged OKT3. The most significant change in the peptide maps of OKT3 with degradation was a decrease in the size of the peak corresponding to H99–121 and coincident with this decrease was the appearance of a new peak that sequencing showed to be related to H99–121. This peptide fragment contains a non-disulfide-bonded Cys at residue 105, a likely candidate for oxidative degradation. Another degraded peptide that was sequenced, H41–55, contained Trp as the only labile residue. In a

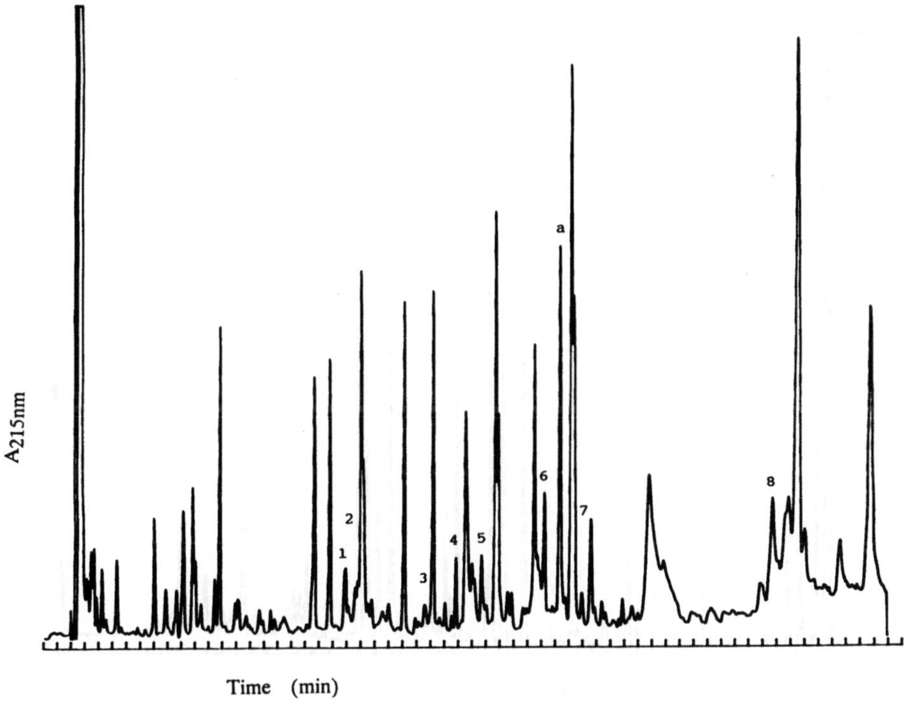

Figure 9. Peptide map of the heavy chain of OKT3 aged at 5°. Numbered peaks are degradation peptides not present, or in very small amounts, in the reference standard OKT3 map. They are related in sequence to the undegraded fragments as follows: (1) H32-38, (2) H423-441, (3) H347-363, (4) H395-411, (5) H304-322, (6) H99-121, (7) H41-55, and (8) H364-388. Peak (a) is the H99-121 fragment that disappears in the 88-kD degradation product. HPLC conditions are the same as in Fig. 8.

comparative study with OKT3 formulated with and without a nitrogen overlay, the nitrogen-overlaid material showed reduced amounts of the peaks in the peptide maps attributed to oxidations as described above.

5.4. Formation of Interchain Cross-links

A major degradation product that was observed in reduced SDS-PAGE gels of aged OKT3 is a component with an apparent molecular mass of 88 kDa. This component did not separate cleanly from the heavy chain in systems used for size exclusion chromatography in the preparation of materials for peptide mapping, but a sufficiently enriched sample was obtained to

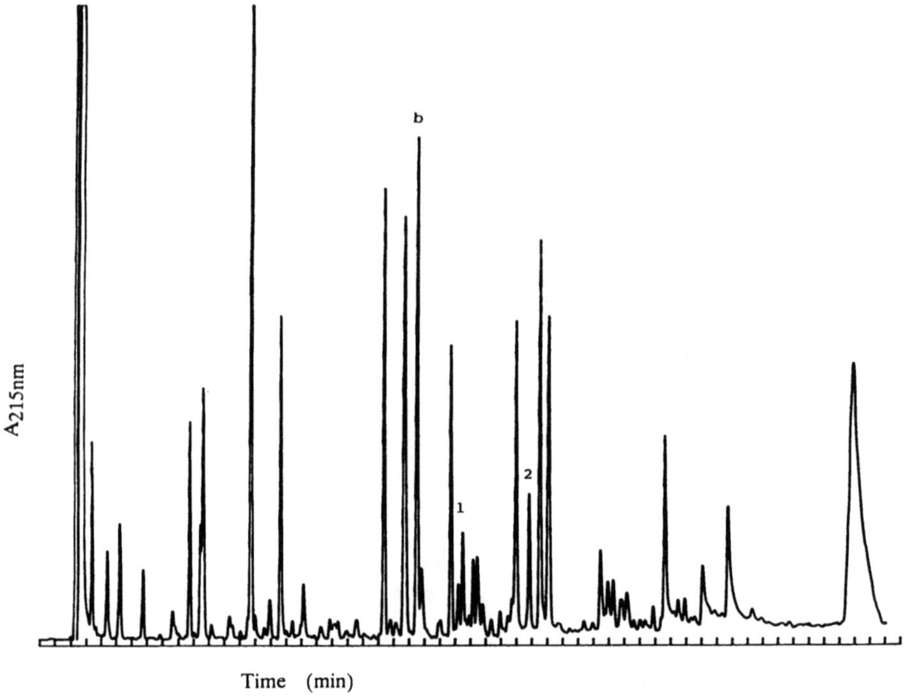

Figure 10. Peptide map of the light chain of OKT3 aged at 5°. Numbered peaks are: (1) deamidated L155-168 and (2) degraded L169-187. Peak (b), L46-52, is not present in the map of the 88-kD degradation product. HPLC conditions are the same as in Fig. 8.

confirm by peptide mapping that the 88-kDa protein consists of both heavy and light chain. In the map of the sample, all of the light chain peptide fragments were present except for L46–52, implying that this region is part of a cross-link joining the chains. Since it appeared that the 88-kDa species results from a cross-link between the heavy and light chains in the variable regions, $F(ab')_2$ fragments were prepared for further study. The $F(ab')_2$ fragment is an enzymatically cleaved fragment of immunoglobulin consisting of intact light chains disulfide bonded to heavy chain fragments which are also disulfide bonded together. The heavy chains are cleaved in the hinge region. $F(ab')_2$ fragments of degraded OKT3 were prepared by pepsin digestion for isolation of a cross-linked species. Reduction and S-sulfonation of the $F(ab')_2$ fragment gave one product that was nearly twice the molecular mass of the light and heavy chain fragments and separable by gel filtration. The peptide map of this product showed equal amounts of light and heavy chain peptide fragments and an absence of L46–52 and H99–121, establishing that these two regions form the cross-link. Examination of the known crystal structures of several antibodies and F(ab) fragments reveals that portions of the analogous regions in the antibody structures are in close proximity. Although the exact nature of the cross-link has not been determined, a mechanism involving oxidation of Trp L46 and nucleophilic addition of the sulfhydryl of Cys H105 to the Trp indole ring is proposed. Literature precedent for formation of similar cross-links can be found in the oxidation of lysozyme by iodine (Imoto and Rupley, 1973) and in the reaction of cysteine with tryptophan during acid hydrolysis (Ohta and Nakai, 1978) and under oxidative conditions (Savige, 1976). Increased amounts of the 88-kDa product were observed in OKT3 exposed to mild oxidizing conditions. Since both peptides involved in the cross-link are part of the antigen binding regions of the antibody, it is expected that this degradation pathway would result in loss of binding affinity as measured by the potency assay. In fact, when OKT3 enriched in the 88-kDa species was prepared by HPLC-IEC and tested in the potency assay, it was found to have very poor binding affinity.

6. CONCLUSION

Studying the degradation of a molecule as large and complex as an immunoglobulin is difficult. We have therefore aimed most of our studies to indicate general mechanisms of degradation which affect the molecule, rather than concentrating on a single type or site of degradation. These studies have clearly indicated that two types of chemical degradation appear to account for most of the physical changes which occur as OKT3 degrades: oxidation and deamidation. Our biochemical and functional studies suggest that oxidative mechanisms may more seriously affect the binding affinity of OKT3, while deamidation may account for most of the charge shifts which are visible by IEF. At the approved storage temperature of 5°C, both mechanisms apparently occur at about the same rate and failure of the potency assay and shift of IEF patterns occur together. At elevated temperatures, however, IEF changes appear to precede the loss of binding activity. The same mechanisms of degradation are operative at low and high temperatures; however, the deamidation pathway is accelerated preferentially with increasing temperature. In light of this, one should interpret high-temperature stability studies with caution.

A molecule as complex as OKT3 contains multiple sites where either of these reactions could occur. However, we have identified a number of Asn–Gly and Asn–Ser sequences which spontaneously deamidated during storage. Detection of oxidation products is more complicated, but we successfully determined that five methionine residues were becoming oxidized.

Most interestingly, we were also able to deduce the location of an interchain covalent linkage between the heavy and light chain of OKT3. Although we are still attempting to identify the precise residues involved in the cross-link, the location of this bond, between the CDR regions of the heavy and light chains, would be expected to alter the binding characteristics of this antibody.

Combining the results of our functional testing with the chemical analysis of some of the material tested has also yielded important information about the functional significance of this degradation. Most of the OKT3 which was found to be altered in charge, and presumably deamidated, still retained most of its binding affinity and performed well in *in vitro* assays of function. This would be expected if, as our chemical analysis suggests, most of the sites of deamidation were located in the constant regions of the immunoglobulin. Tests of binding of the OKT3 enriched in the cross-linked, 88-kDa species indicated that binding of this degradation product was impaired. Our peptide mapping suggests that the cross-link accounting for the 88-kDa species is located in the antigen binding regions of OKT3, and would explain the results of the binding study.

In addition to providing interesting information on the chemistry of spontaneous protein degradation, the results of these studies have been utilized in the development of new, more stable formulations of OKT3. Based on these findings, Orthoclone OKT3 is now filled in ampules under nitrogen. Flushing with nitrogen has been shown, like argon, to greatly slow the oxidative degradation of OKT3, and preserve the binding ability of the antibody. Additional studies on possible functional effects of deamidation and methods of prevention of these reactions, as well, are under way.

REFERENCES

Aebersold, R. H., Leavitt, J., Saavedra, R. A., Hood, R. E., and Kent, S. B. H., 1987, N-terminal amino acid sequence analysis of proteins separated by one or two dimensional gel electrophoresis after in situ digestion on nitrocellulose, *Proc. Natl. Acad. Sci. USA* **84**:6970–6974.

Amzel, L. M., and Poljak, R. J., 1979, Three-dimensional structure of immunoglobulins, *Annu. Rev. Biochem.* **48**:961–997.

Bristow, M. R., Glibert, E. M., O'Connell, J. B., Renlund, D. G., Watson, F. S., Hammond, E., Lee, R. G., and Menlove, R., 1988, OKT3 monoclonal antibody in heart transplantation, *Am. J. Kidney Dis.* **11**:135–147.

Capasso, S., Mazzarella, L., Sica, F., and Zagari, A., 1989, Deamidation via cyclic imide in asparaginyl peptides, *Pept. Res.* **2**:195–200.

Chang, T. W., Kung, P. C., Gingras, S. P., and Goldstein, G., 1981, Does OKT3 monoclonal antibody react with an antigen-recognition structure on human T cells? *Proc. Natl. Acad. Sci. USA* **78**:1805–1808.

Chatenoud, L., Baudrihaye, M., Kreis, H., Goldstein, G., Schlindler, J., and Bach, J. F., 1982, Human in vivo antigenic modulation induced by the anti-T cell OKT3 monoclonal antibody, *Eur. J. Immunol.* **12**:979–982.

Chazin, W. J., Kordel, J., Thulin, E., Hofmann, T., Drakenberg, T., and Forsen, S., 1989, Identification of an isoaspartyl linkage formed upon deamidation of bovine calbindin D_{9k} and structural characterization by 2D ^1H NMR, *Biochemistry* **28**:8646–8653.

Cosimi, A. B., Jenkins, R. L., Rohrer, R. J., Delmonico, F. L., Hoffman, M., and Monaco, A. P., 1990, A randomized clinical trial of prophylactic OKT3 monoclonal antibody in liver allograft recipients, *Arch. Surg.* **125**:781–784.

DiAugustine, R. P., Gibson, B. W., Aberth, W., Kelly, M., Ferrua, C. M., Tomooka, Y., Brown, C. F., and Walker, M., 1987, Evidence for isoaspartyl (deamidated) forms of mouse epidermal growth factor, *Anal. Biochem.* **165**:420–429.

Edelman, G. M., 1971, Antibody structure and molecular immunology, *Ann. N.Y. Acad. Sci.* **190**:5–25.

Imoto, T., and Rupley, J. A., 1973, Oxidation of lysozyme by iodine: Identification and properties of an oxindolyl ester intermediate; evidence for participation of glutamic acid 35 in catalysis, *J. Mol. Biol.* **80**:657–667.

Johnson, B. A., Shirokawa, J. M., Hancock, W. S., Spellman, M. W., Basa, L. J., and Aswad, D. W., 1989, Formation of isoaspartate at two distinct sites during *in vitro* aging of human growth hormone, *J. Biol. Chem.* **264**:14262–14271.

Kossiakoff, A. A., 1988, Tertiary structure is a principal determinant to protein deamidation, *Science* **240**:191–194.

Kung, P., Goldstein, G., Reinherz, E., and Schlossman, S., 1979, Monoclonal antibodies defining distinctive human T cell surface antigens, *Science* **206**:347–349.

Manning, M. C., Patel, K., and Borchardt, R. T., 1989, Stability of protein pharmaceuticals, *Pharm. Res.* **6**:903–918.

Matsudaira, P., 1987, Sequences from picomole quantities of proteins electroblotted onto polyvinylidene difluoride membranes, *J. Biol. Chem.* **262**:10035–10038.

Nisonoff, A., Hopper, J. E., and Spring, S. B., 1975, *The Antibody Molecule,* Academic Press, New York.

Norman, D. J., Shield, C. F., Barry, J., Hennell, K., Funnell, M. B., and Lemon, J., 1987, A U.S. clinical study of Orthoclone OKT3 in renal transplantation, *Transplant. Proc.* **19**:21–27.

Ohta, T., and Nakai, T., 1978, The reaction of tryptophan with cystine during acid hydrolysis of proteins, *Biochim. Biophys. Acta* **533**:440–445.

Ortho Multicenter Transplant Study Group, 1985, A randomized clinical trial of OKT3 monoclonal antibody for acute rejection of cadaveric renal transplants, *N. Engl. J. Med.* **313**:337–342.

Reinherz, E. L., Hussey, R. E., and Schlossman, S. F., 1980, A monoclonal antibody blocking human T cell function, *Eur. J. Immunol.* **10**:758.

Savige, W. E., 1976, New oxidation and photo-oxidation products of tryptophan, *Proc. Int. Wolltextil-Forschungskonf., 5th 1975* **2**:570–579.

Smith, K. G. C., Austyn, J. M., Hariri, G., Beverly, P. C. L., and Morris, P. J., 1986, T cell activation by anti-T3 antibodies: Comparison of IgG1 and IgG2b switch variants and direct evidence for accessory function of macrophage Fc receptors, *Eur. J. Immunol.* **16**:478–483.

Stephenson, R. C., and Clarke, S., 1989, Succinimide formation from aspartyl and asparaginyl peptides as a model for the spontaneous degradation of proteins, *J. Biol. Chem.* **264**:6164–6170.

Tax, W. J. M., Willems, H. W., Reekers, P. P. M., Capel, P. J. A., and Koene, R. A. P., 1983, Polymorphism in mitogenic effect of IgG1 monoclonal antibodies against T3 antigen on human T cells, *Nature* **304**:445–447.

Tax, W. J. M., Hermes, F. F. M., Willems, H. W., Capel, P. J. A., and Koene, R. A. P., 1984, Fc receptors for mouse IgG1 on human monocytes: polymorphism and role in antibody-induced T cell proliferation, *J. Immunol.* **133**:1185–1189.

Thannhauser, T. W., Konishi, Y., and Scheraga, H. A., 1984, Sensitive quantitative analysis of disulfide bonds in polypeptides and proteins, *Anal. Biochem.* **138**:181–188.

Thistlethwaite, J. R., Gaber, A. O., Haag, B. W., Aronson, A. J., Broelsch, C. E., Stuart, J. K., and Stuart, F. P., 1987, OKT3 treatment of steroid-resistant renal allograft rejection, *Transplantation* **43**:176–184.

VanWauwe, J. P., DeMay, J. R., and Goossens, J. G., 1980, OKT3: A monoclonal anti-human T lymphocyte antibody with potent mitogenic properties, *J. Immunol.* **124**:2708–2713.

Violand, B. N., Schlittler, M. R., Toren, P. C., and Siegel, N. R., 1990, Formation of isoaspartate 99 in bovine and porcine somatotropins, *J. Prot. Chem.* **9**:109–117.

Yuksel, K. U., and Gracy, R. W., 1986, *In vitro* deamidation of human triosephosphate isomerase, *Arch. Biochem. Biophys.* **248**:452–459.

Chapter 5

Leuprolide and Other LH-RH Analogues

Akwete Lex Adjei and L. Hsu

1. INTRODUCTION

Luteinizing hormone-releasing hormone (LH-RH) is the blood-borne messenger between the hypothalamus and the anterior pituitary which controls reproductive function. It is synthesized and stored in the hypothalamus in neurons which project to the median eminence and is released in periodic bursts into the hypophyseal portal circulation. The isolation (Schally *et al.*, 1971a,b), elucidation of structure (Matsuo *et al.*, 1971a), and synthesis of porcine LH-RH (Matsuo *et al.*, 1971b) prompted intense activity by various laboratories around the world. This chapter briefly describes some of the structural chemistry work that has been done to stabilize and to improve the activity of LH-RH analogues in order to make clinical application possible. Discussions of marketed and developmental drugs for diseases mediated by gonadotropins are included.

1.1. Pharmacology of LH-RH

Porcine LH-RH, whose structure is identical to ovine, bovine, human, and rat LH-RH, is a decapeptide (Matsuo *et al.*, 1971a) and chemically designated as follows:

Akwete Lex Adjei and L. Hsu • Product Development, Pharmaceutical Products Division, Abbott Laboratories, North Chicago, Illinois 60064.

Stability and Characterization of Protein and Peptide Drugs: Case Histories, edited by Y. John Wang and Rodney Pearlman. Plenum Press, New York, 1993.

(pyro)Glu-His-Trp-Ser-Tyr-Gly-Leu-Arg-Pro-GlyNH_2
1 2 3 4 5 6 7 8 9 10

Native LH-RH

When introduced to the systemic circulation, LH-RH stimulates secretion of the gonadotropins, luteinizing hormone (LH) and follicle-stimulating hormone (FSH), by the anterior pituitary. These two hormones exert trophic and steroidogenic effects upon the gonadal tissues. However, the responsiveness of rats to LH-RH varies during the estrous cycle, being greatest in proestrus and estrus (Martin *et al.*, 1974). This variation in sensitivity may be related to circulating estrogen levels since 2 μg/kg of LH-RH increased plasma LH 250-fold in rabbits while ovulation induction was achieved at 250 μg (Kanematsu *et al.*, 1974). The essential role played by endogenous LH-RH in regulating reproductive function has been demonstrated by both active and passive immunization studies. Rabbits which produced an antibody to LH-RH developed gonadal atrophy (Arimura *et al.*, 1973); in addition, administration of rabbit anti-LH-RH serum to normally cycling rats blocked ovulation and prevented the preovulatory surge of LH and FSH (Arimura *et al.*, 1974).

1.2. Leuprolide Acetate and Other LH-RH Analogues

Studies with synthetic ^3H-labeled LH-RH revealed that the decapeptide is rapidly degraded in blood by enzymatic cleavage of the (pyro)Glu-His group and excreted along with intact peptide by the kidney (Redding *et al.*, 1973). The plasma half-life is about 57 min in man and 7 min in the rat (Redding and Schally, 1973). Studies also demonstrated that homogenates of rat and pig hypothalamus degrade LH-RH by cleaving the decapeptide between amino acid residues 6 and 7 (Koch *et al.*, 1974). This fact suggested that therapeutic usefulness of LH-RH would be limited because of its short biological half-life and also because of rapid inactivation *in vivo.* Several laboratories, therefore, aimed to modify the decapeptide by chemical substitution. These investigations have led to two directions in LH-RH research, namely (1) highly active agonists which stimulate the release of LH and FSH and (2) competitive antagonists which block the effect of LH-RH.

1.2.1. Agonists

The *in vitro* activity of analogues which mimic the activity of LH-RH is determined by the release of LH and FSH from monolayer cultures of rat pituitary cells. LH-RH and its agonists stimulate the release of LH and FSH from cells in a dose-related manner (Vale *et al.*, 1972). Binding affinity to pituitary receptor sites by most LH-RH analogues may be related closely to their ability to trigger gonadotropin release. The binding of 20 LH-RH analogues, including Leuprolide, to the pituitary receptor has been investigated (Heber and Odell, 1978). Results revealed high-affinity binding (K_a = 10^9 M^{-1}) for agonists with strong *in vivo* activity. The *in vivo* activity was determined either by: (1) incremental levels in serum LH by radioimmunoassay (RIA) after injection of test materials in ovariectomized estrogen- and progesterone-treated rats (Ramirez and McCann, 1963), or in immature male rats; or by (2) measuring either the induction of ovulation in proestrous rats blocked by pentobarbital after administration of LH-RH agonists, or the inhibition of ovulation in preestrous rats after administration of LH-RH antagonists (Pedroza *et al.*, 1978). Studies (Rippel and Johnson, 1976; Johnson *et al.*, 1976) on the inhibitory effect of Leuprolide on human chorionic gonadotropin (HCG)-induced ovarian and uterine weight increase in immature rats showed that high subcutaneous (s.c.) doses of the drug inhibit the effect in intact and hypophysectomized rats.

1.2.1. Antagonists

The study of *in vitro* and *in vivo* antagonism to LH-RH requires assays that rely on blocking the effect of exogenous LH-RH or on suppressing the endogenous surge of LH-RH preceding LH release and ovulation. Subcutaneous injection of an antagonist, [D-Phe^2, D-Phe^6]LH-RH, with exogenous LH-RH at a molar ratio of 2500 : 1 antagonist to LH-RH blocked the effect of the exogenous LH-RH for up to 8 hr (Pedroza *et al.*, 1978; Rippel and Johnson, 1976; Johnson *et al.*, 1976). Other studies showed that single or multiple doses of antagonists effectively inhibit ovulation in normally cycling rats at proestrus (Nishi *et al.*, 1976).

1.3. Chemistry and Structural Properties of LH-RH Analogues

The first three amino acids at the NH_2-terminus of LH-RH—p-Glu (also known as 5-oxoproline, a lactam), His, and Trp—play a functional role in the biological activity of LH-RH and simple substitutions or deletions in these positions decrease or abolish LH-RH activity. Yet, considerable activity can be retained by substitution with structures resembling them. His and Trp can be replaced by structures possessing similar acid–base and hydrogen bonding capacity, or suitably oriented aromatic nuclei capable of generating similar electronic interactions. Substitutions in positions 4–10 can lead to quite active analogues or even, in the case of positions 6 and 10, to superactive peptides which are more potent than LH-RH and also possess prolonged activity (Monohan *et al.*, 1973). Examples include: desGly^{10}-LH-RH ethylamide (Fujino *et al.*, 1972), 3–5 times more active and much longer-acting than LH-RH; D-Phe^{6}-LH-RH and D-Trp^{6}-LH-RH, 10 and 13 times more potent and longer-acting, respectively, than LH-RH.

Fujino *et al.* and chemists at Abbott (Fujino *et al.*, 1974) incorporated changes at both positions 6 and 10. [D-Ala^{6}-desGly^{10}]LH-RH ethylamide was noted to be 30–40 times more potent than LH-RH over a 6-hr period. FSH releasing activity of this compound was 15 times greater than that of LH-RH resulting in about 50–80 times higher ovulation induction in rats compared with the parent hormone. Subcutaneous injection of [D-Leu^{6}-desGly^{10}]LH-RH ethylamide, known as Leuprolide, into immature male rats released 53 times more LH and 15 times more FSH over a 6-hr period than a similar dose of LH-RH. Rat ovulation studies revealed Leuprolide to be 50–80 times more active than LH-RH. Leuprolide (molecular weight 1269.46) is chemically defined as:

p-Glu-L-His-L-Trp-L-Ser-L-Tyr-D-Leu-L-Leu-L-Arg-L-Pro-ethylamide

Leuprolide (Abbott 43818; TAP 144)

Ovulation studies with s.c. administration of Buserelin, Hoechst's LH-RH analogue, [D-Ser(Bu^{t})^{6}-desGly^{10}]LH-RH ethylamide, released 19 times as much LH and 17 times as much FSH as did a similar dose of LH-RH (Sandow *et al.*, 1978). Studies with anterior pituitary cells in monolayer cultures showed that *in vitro* LH and FSH-releasing activities of Leuprolide, [D-Ala^{6}-desGly^{10}]LH-RH ethylamide, [D-Phe^{6}-desGly^{10}]LH-RH ethylamide, and [D-Trp^{6}-desGly^{10}]LH-RH ethylamide were respectively 30, 30, 15, and 15 times greater than LH-RH. Zoladex, [D-Ser(Bu^{t})^{6}-AzGly^{10}]LH-RH (developed by ICI), was found to be as active as Leuprolide in the modified Steelman–Pohley assay (Rippel and Johnson, 1976). This test is a bioassay for measuring the potential FSH activity of an LH-RH analogue in immature rats resulting from dose-dependent increases in ovarian and uterine weights compared with NIH-FSH standards coadministered with HCG.

The increase in biological activity of the superactive LH-RH analogues with substitutions in positions 6 and 10 has been attributed to either enhanced binding to pituitary receptors, slower inactivation, or a combination of both factors (Fujino *et al.*, 1972). Thus, [D-Ala^{6}-desGly^{10}]LH-RH ethylamide is less readily degraded by brain enzymes than is LH-RH because of substitutions at positions 6 and 10. Monohan *et al.* (1973) suggested that the greater activity of [D-Ala^{6}]LH-RH might be due to changes in conformation;

the stabilized type II' B-bend results in greater affinity to pituitary receptors than is found with LH-RH. This hypothesis was validated by NMR (Deslauriers and Somorjai, 1976) and conformational energy analysis (Momany, 1976). The hydrophobic character of the amino acid in position 6 and, possibly, its size and aromaticity may contribute to the activity. For example, [D-Nal(2)^{6}]LH-RH, Syntex's compound, Nafarelin, contains a hydrophobic naphthyl D-amino acid in position 6. It was shown to be 200 times more potent than LH-RH in the estrus suppression assay when administered for 14 consecutive days (Nestor *et al.*, 1982).

The first competitive inhibitors of LH-RH were discovered soon after the disclosure of the structure of the native hormone (Matsuo *et al.*, 1971a). The early inhibitors, [Gly^{2}]LH-RH and [des-His^{2}]LH-RH, required high doses in order to be effective *in vitro*. Substitutions in [des-His^{2}]LH-RH with D-Ala at position 6 that increased potency in the agonist series gave more effective *in vitro* antagonists (Monohan *et al.*, 1973). Workers in several groups achieved increases in antagonist potency by further substitutions in positions 1, 2, 3, and 6, one of the most potent being [*N*-Ac-D-Phe^{1}, D-p-Cl-Phe^{2}, D-Trp^{3,6}]LH-RH, which inhibited ovulation by 60% at a 62-μg dose in the rat antiovulatory assay (Coy *et al.*, 1979). The mechanism of LH-RH antagonism has not been fully defined, but antagonistic peptides have been reported to bind to pituitary receptors (Spona, 1974) thus blocking endogenous LH-RH by competitive antagonism or allosteric inhibition. The antagonist, [*N*-Ac-D-p-Cl-Phe^{1,2}-D-Trp^{3}-D-Arg^{6}-D-Ala^{10}]LH-RH, was shown to be effective in blocking preovulatory surge of LH in proestrous rats after either s.c. injection of 10-μg dose or oral gavage of 2-mg dose (Nekola *et al.*, 1982). The findings are encouraging because they indicate that with increased enzymatic resistance of the peptide (due to the five D-amino acids in the molecule) oral activity can be achieved. However, the required oral dose of this antagonist is 200 times greater than the s.c. one, and thus is clinically attractive, particularly if cost of the peptide is not prohibitive.

1.3.1. Amino Acid Sequence of Leuprolide Acetate

There are a variety of techniques used to investigate the amino acid sequence of protein and peptide drugs. The functional components of Leuprolide acetate and structural elucidation of this peptide were confirmed using the following methods:

1. Hydrolysis of the peptide followed by quantitative analyses of the amino acid residues by ion exchange chromatography
2. Ionization by fast atom bombardment (FAB) and mass spectrometry

3. High-performance liquid chromatography (HPLC) following acid/alkaline exposure of the peptide and enzymatic hydrolysis

Leuprolide acetate is synthesized sequentially in solution using the classical method of blocking, coupling, and deblocking of the functional amino acids all of which are levorotatory (L) except for the Leu in position 6 which is dextrorotatory (D). Being a peptide produced by a well-defined process, the amino acid sequence is known.

The termini of Leuprolide are "blocked" by lactam formation in the case of the NH_2-terminal amino acid (p-Glu) and amide formation in the case of the COOH-terminal amino acid (Pro-ethylamide). Thus, methods for determining amino acid sequencing in peptides and proteins (Edman degradations, Sanger reaction, enzymatic hydrolysis) are not directly applicable. Although various ways of circumventing the terminal blockages have been described in the literature (methanolysis, acid or base exposure), these are often associated with racemization and peptide bond cleavage resulting in the formation of free amino acids. Once the terminal blockages are cleared, theoretically, sequential Edman degradation can be applied to the molecule. The unblocked peptides can also be sequentially cleaved at the peptide bonds using specific enzymes. Such enzymes, e.g., aminopeptidase M (Jones *et al.*, 1981), are generally ineffective for linkages involving D-amino acids (-D-D- or -L-D-).

Since the determination of peptide sequence by mass spectrometry offers considerable advantages over chemical approaches, this method was also used to investigate the functional components of Leuprolide acetate. Volatility of large molecules such as peptides and proteins, originally a problem using mass spectrometric techniques, has been successfully overcome with newer soft ionization techniques such as field desorption and FAB. The terminal blockages are inconsequential to ionization techniques as are the numerous purification steps associated with manual degradation of the Edman and Sanger methods (Simmons and Johnson, 1976). However, differentiation of the optical isomers of the various amino acid fragments cannot be achieved by this technique. Thus, problems associated with amino acid sequencing in Leuprolide acetate are unique since the molecule is blocked at both ends, and has a D-amino acid in the sixth position. Therefore, studies were conducted to:

1. Establish the presence of the various amino acids in the molecule at their purported ratios (within experimental errors)
2. Establish the NH_2-terminal blockage by the absence of release of any amino acid by direct enzymatic hydrolysis
3. Open the NH_2-terminus by acid/alkali treatment and follow the course of enzymatic hydrolysis to locate the L–D linkage

4. Establish the sequence by mass spectrometry
5. Explore the possibility of racemizing the D-amino acid by prolonged alkaline treatment followed by enzymatic hydrolysis.

1.3.2. Amino Acid Analysis

Approximately 0.5 μmole of Leuprolide acetate was hydrolyzed in 6 N HCl for 24 hr at 110°C. The resulting mixture of amino acids was freed of HCl by evaporation under vacuum and reconstituted in 1 ml of pH 2.2 citrate buffer. The solution was analyzed directly on a Durrum (Dionex) D-500 Amino Acid Analyzer. Mole ratios of the amino acids and ethylamine were obtained by comparison with a standard mixture of pure amino acids run concurrently with the sample hydrolysate. Mole ratios of the amino acids determined by ion exchange chromatography following acid hydrolysis of Leuprolide acetate are shown in Table I. Although no information on amino acid sequence or enantiomeric identity could be derived from this analysis, the data are useful in establishing the presence and molecular ratios of the amino acids in a protein/peptide structure. Generally, Trp and Ser are degraded during acid hydrolysis by HCl. The pyrrolidone ring opens up during the acid hydrolysis and the p-Glu is determined as Glu. Other hydrolytic media such as alkali and alkyl or aryl sulfonic acids may be used but these could be reactive to other amino acids such as Arg and Ser. The presence of 1 mole each of Trp and Tyr in Leuprolide acetate was established by ultraviolet spectrophotometry (Section 1.4.5).

Table I
Amino Acid Analysis of the Acid Hydrolysate of Leuprolide Acetate

Amino acid	Mole ratio: Found	Mole ratio: Theory	Specifications for the bulk drug
Ser	0.91	1.0	Present
Glu	1.02	1.0	0.85–1.1
Pro	1.00	1.0	0.85–1.1
Leu	2.01	2.0	1.8–2.2
Tyr	1.02	1.0	0.85–1.1
His	1.02	1.0	0.85–1.1
Arg	1.04	1.0	0.85–1.1
Trp	0.77	1.0	Present
Ethylamine[a]	0.99	1.0	

[a] Not an amino acid, but derived from proline ethylamide of Leuprolide acetate.

1.3.3. Mass Spectrometry

Mass spectra were obtained on a Kratos MS50 spectrometer with a FAB source. Low-resolution scans were run at 2000 resolution. Magnet scan rate was set at 100 sec/decade. Spectrometer calibration was achieved with tris(perfluoroheptyl)-*S*-triazine. Approximately 1 mg of solid material was dissolved in 5 μl of ethanol and added to about 100 μl of glycerin. A trace of dilute HCl was used to acidify the solution and about 0.5 μl placed in the source on a copper tip. In the FAB source, argon ions of kinetic energy 4–6 keV were converted to atoms of similar energy by charge exchange during passage through argon gas. The 4- to 6-keV beam of atoms was made to impact onto the sample. High-resolution mass measurements were made at 10,000 resolution using manual peak matching techniques. A reference compound with a peak of known empirical formula whose exact mass was within 10% of the mass of the unknown ion was used. The magnetic field was adjusted so that the peak due to one of the ions (reference or unknown) was in the center of the oscilloscope. The ratio of the accelerating voltage necessary to bring the signal from the other ion to coincide with the first ion (on alternate sweeps of the oscilloscope) was used to calculate the mass of the unknown ion.

The mass spectra obtained for a typical batch of Leuprolide acetate (Tables II and III, Fig. 1) exhibited a distinct protonated ion at m/z 1209. Fragment ion peaks, although of lower intensity, agreed well with the assigned amino acid sequence. In many cases, a peak of greater intensity was observed two mass units higher than that of the peak due to simple cleavage probably a result of back transfer of a proton to the protonated end of the molecule during bond cleavage. Verification of empirical formulas for the protonated molecular ion and several of the assigned fragment ion peaks was obtained by high-resolution mass measurement. The exact masses which were measured are listed in Table III along with the calculated masses for those fragments. The peaks occurring at m/z 466 and 353 (Fig. 1) are typical of the presence of Leu in the peptide. These might occur by ion decomposition (Barber *et al.*, 1980) in the gas phase with the loss of olefinic fragments, analogous to other ionization methods (electron impact and chemical ionization).

1.3.4. Hydrolysis with Aminopeptidase M and HPLC

An HPLC method was developed to analyze component amino acids in Leuprolide acetate (except Pro) using *o*-phthaldialdehyde derivatization and fluorometric detection. The dialdehyde reacts with primary amino acids

Table II
Structurally Significant Fragment Peaks Predicted and Observed in the Spectrum of Leuprolide Acetate

Fragment	Mass/formula	+2H	−CO	+NH	
A	1121/$C_5H_6NO_2$	114	(84)[a]	127	
AB	249/$C_{11}H_{13}N_4O_3$[b]	251	221	264	
ABC	435/$C_{22}H_{23}N_6O_4$[b]	(437)	(407)	(450)	
ABCD	522/$C_{25}H_{28}N_7O_6$	(524)	(494)	(537)	
ABCDE	(685)/$C_{34}H_{37}N_8O_8$	(687)	(657)	(700)	
ABCDEF	(798)/$C_{40}H_{48}N_9O_9$	(800)	(770)	(813)	
ABCDEFG	(911)/$C_{46}H_{59}N_{10}O_{10}$	(913)	(883)	(926)	
ABCDEFGH	(1067)/$C_{52}H_{71}N_{14}O_{11}$	(1069)	(1039)	(1082)	
Fragment	Mass/formula	+2H	+CO	−NH	−NH + 2H
I	141/$C_7H_{13}N_2O$[b]	143	169	—	—
HI	297/$C_{13}H_{25}N_6O_2$[b]	299	325	282	284
GHI	410/$C_{19}H_{36}N_7O_3$[b]	412	438[b]	(395)	397
FGHI	523/$C_{19}H_{47}N_8O_4$[b]	525	551	508	510
EFGHI	686/$C_{34}H_{56}N_9O_6$[b]	688[b]	714	671	673
DEFGHI	773/$C_{37}H_{61}N_{10}O_8$	775	801	758	760
CDEFGHI	959/$C_{48}H_{71}N_{12}O_9$	961	987	(944)	946
BCDEFGHI	(1096)/$C_{54}H_{78}N_{15}O_{10}$	(1098)	(1124)	1081	(1083)

[a] Peaks not observed are in parentheses.
[b] Empirical formulas verified by peak matching.

(amines) in the presence of 2-mercaptoethanol to produce fluorescent reaction products known as 1-alkylthio-2-alkyl-substituted isoindoles (Simmons and Johnson, 1976). The fluorescent derivatives were separated under normal chromatographic conditions and compared with several amino acid standards. To determine the enantiomeric identity of Leu in position 6, enzymatic hydrolysis of the molecule was investigated. Aminopeptidase M, also known as leucine aminopeptidase, isolated from microsomal fraction of swine kidney, is generally used for such studies since this enzyme does not need any further activation. The enzyme hydrolyzes all amino acids at the NH_2-terminus except blocked ones such as bonds involving X-Pro (probably because of the tertiary amide linkage) or L-X–D-X linkages (Beynum *et al.*, 1977). Leuprolide is not susceptible to this enzyme because of the p-Glu at the NH_2-terminus. Thus, if the p-Glu is ring opened, only the first four amino acids will be released. Therefore, treatment of drug with dilute HCl (Nishi *et al.*, 1980) was used to accomplish the ring opening.

The concentrations of the peptide and enzyme, as well as the derivatization procedure employed, were similar to those reported elsewhere (Jones *et*

Table III
The Atomic Composition of the Prominent Fragment Peaks in the Mass Spectrum of Leuprolide Acetate

Measured mass	Calculated mass	Empirical formula	Assignment
1209.6508	1209.6532	$C_{59}H_{85}N_{16}O_{12}$	$(M + H)^+$
959.5461	959.5466	$C_{48}H_{71}N_{12}O_9$	CDEFGHI
688.4509	688.4510	$C_{34}H_{58}N_9O_6$	EFGHI + 2H
686.4332	686.4353	$C_{34}H_{56}N_9O_6$	EFGHI
523.3742	523.3720	$C_{25}H_{47}N_8O_4$	FGHI
438.2814	438.2829	$C_{20}H_{56}N_4$	GHI + CO
435.1788	435.1781	$C_{22}H_{23}N_6O_4$	ABC
410.2862	410.2879	$C_{19}H_{36}N_7O_3$	GHI
297.2056	297.2039	$C_{13}H_{25}N_6O_2$	HI
249.0971	249.0987	$C_{11}H_{13}N_4O_3$	AB
141.1035	141.1028	$C_7H_{13}N_2O$	I

al., 1981). The responses of the various amino acids to the fluorometer were found to be linear to the concentrations between 100 and 800 pmole. Under acid conditions, only one major transformation seemed to have occurred, implying that if the p-Glu were to ring-open, the resulting L-Glu analogue of Leuprolide acetate would elute earlier. The alkaline exposure results in several components generally eluted as triplets. Based on other work (Nishi *et al.*, 1980), these peaks were assigned to [D-Ser^4] and [D-His^2] analogues of Leuprolide acetate. The respective alkali and further enzyme-hydrolyzed ring-opened products are shown in Fig. 2. When Leuprolide acetate was incubated with aminopeptidase M and analyzed for the presence of amino acids after derivatization, no free amino acids were detected. This confirmed the NH_2-terminal residue to be p-Glu. When the peptide was exposed to 0.1 N HCl at 37°C for 48 hr, and the resulting solution was derivatized and analyzed, only trace amounts of free amino acids were seen in the chromatograms, showing the lack of any significant peptide bond cleavage. However, these conditions are capable of ring opening the p-Glu. The acid-treated peptide on further enzymatic hydrolysis released the first four amino acids in the peptide chain almost quantitatively. If the peptide chain were comprised only of L-amino acids, one would expect all of the amino acids (except Arg and Pro) to be released according to their mole ratios. To prove this hypothesis, Abbott-45509 (Glu-Trp-Ser-Tyr-Gly-Leu-Arg-Pro-GlyNH_2), a synthetic peptide with an X-Pro linkage and all L-amino acids, was treated with the enzyme and analyzed by HPLC along with a standard mixture of the amino acids. The resulting chromatograms (Fig. 3) and mole ratios (Table IV) of the

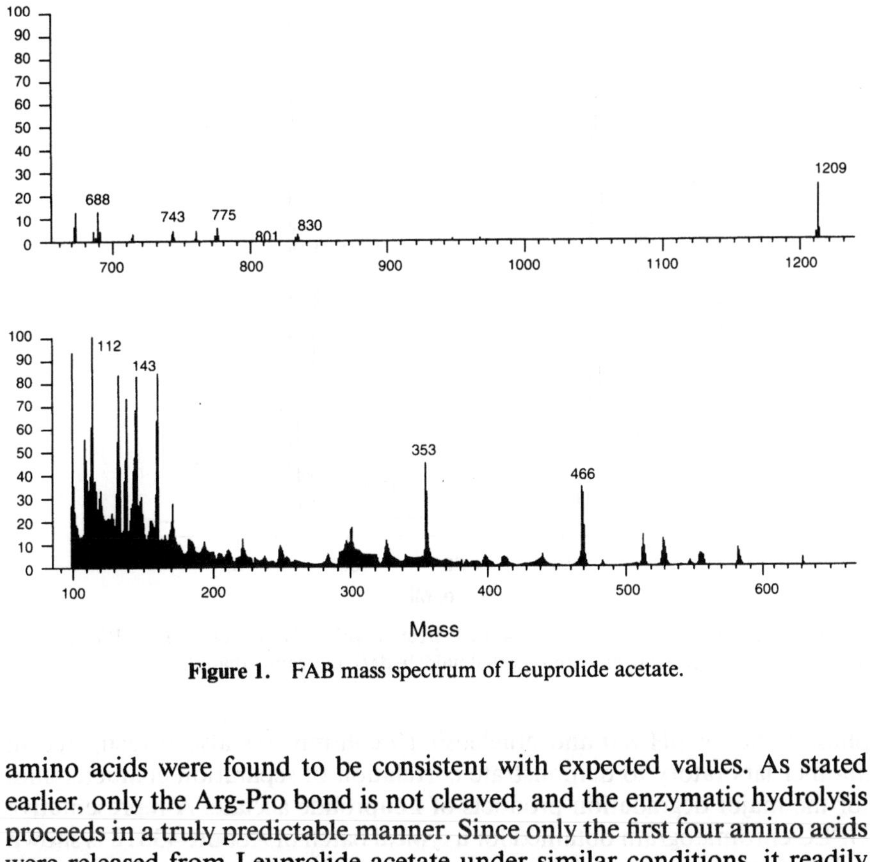

Figure 1. FAB mass spectrum of Leuprolide acetate.

amino acids were found to be consistent with expected values. As stated earlier, only the Arg-Pro bond is not cleaved, and the enzymatic hydrolysis proceeds in a truly predictable manner. Since only the first four amino acids were released from Leuprolide acetate under similar conditions, it readily followed that the next bond would involve an L–D linkage. With the mass spectrum already establishing Tyr as the fifth amino acid, it was concluded that the sixth amino acid was D-Leu.

1.4. Analytical Characterization

1.4.1. Chromatography

Chromatographic analysis of Leuprolide for purity is based on an impurity limit of <4% by TLC and HPLC. Both procedures use ammonium phosphate/acetonitrile mobile phase, pH 7.0. The reversed-phase HPLC method uses a BioSil ODS-5S column. A second test system using a mobile

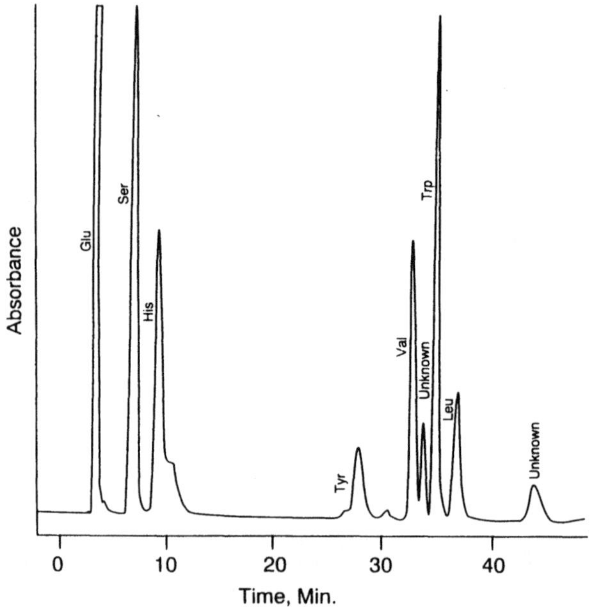

Figure 2. Chromatogram of Abbott-43818, treated with 1 N NaOH, 4 days, RT, enzyme hydrolyzed 24 hr, derivatized with *o*-phthaldialdehyde/2-mercaptoethanol.

phase buffer at pH 8.0 and Nucleosil 10 column was also investigated in another laboratory to demonstrate correlation of separation characteristics for the major degradation products of Leuprolide acetate. A representative HPLC chromatogram obtained for a typical batch of Abbott-43818 is shown in Fig. 4. A comparison of the separation and relative retention times for the two HPLC systems is presented in Table V. The data demonstrated that the two HPLC test methods and the respective separation effects obtained by the two laboratories were similar. Results from a typical batch of leuprolide are as follows:

Component	Relative retention time (based on condition B)	% estimate (based on peak area)
[des-Leu^7] TAP-144	0.26	<0.1
Unknown	0.62	<0.1
TAP-144 (modified)	0.78	0.45
[des-ProNHEt^9] TAP-144	0.93	0.25
Abbott-43818	1.00	
[N^G-Ac-Arg^8] TAP-144	1.64	<0.1
Total impurities		<1.0%

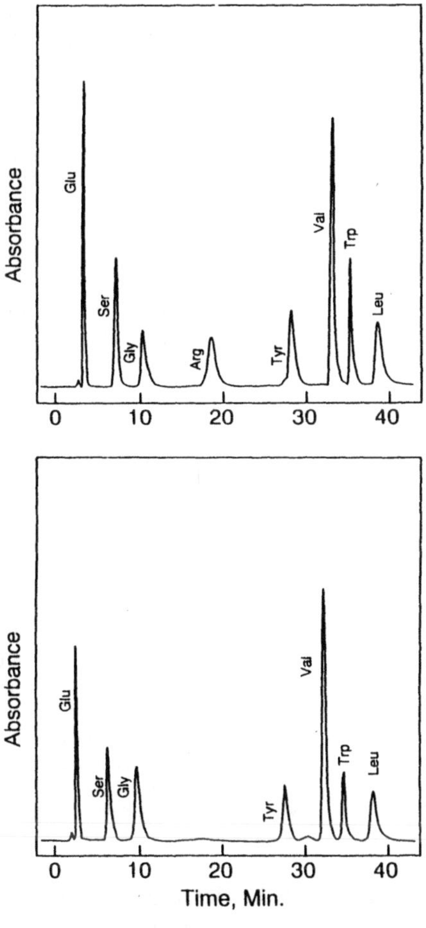

Figure 3. (Top) Chromatogram of standard mixture of amino acids derivatized with *o*-phthaldialdehyde/2-mercaptoethanol, for analysis of Abbott-45509 ([des-pGlu[1],Glu[2]]LH-RH). (Bottom) Chromatogram of Abbott-45509 treated with aminopeptidase M (4 hr) and derivatized with *o*-phthaldialdehyde/2-mercaptoethanol.

Correlation of the detectable impurities with the knowns listed in Tables V and VI gives a tentative identity of CH-1 (Leuprolide acetate or TAP-144 modified, a form of the peptide different than Abbott-43818) for the impurity with a relative retention time of 0.78. The terms CH, X_A, and HPB are nomenclatures assigned to the respective impurities. The possibility exists that CH-5 is also present in the sample but is not resolved from CH-1. The TLC data indicate CH-5 is present in trace amounts. Separation characteristics of the HPLC system are summarized in Tables V and VI for isolated impurities and available stereoisomers of the drug. In order to maximize resolution of the leading and tailing impurity peaks adjacent to Leuprolide acetate, the amount of bulk drug injected was set at 10 μg. The method is

Table IV
Mole Ratios of Amino Acids from the Enzymatic Hydrolysis of Acid-Treated Abbott-43818

	Mole ratio	Theory
Glu	0.62[a]	1.0
His	0.96	1.0
Trp	0.96	1.0
Ser	1.08	1.0
Tyr	<0.1	2.0
Leu	<0.1	1.0
Arg	ND[b]	1.0
Pro	ND	1.0
Abbott-45509 (des-p-Glu[1], Glu[2]) LH-RH		
Glu	1.04	1.0
Ser	0.96	1.0
Gly	1.83	2.0
Tyr	1.05	1.0
Trp	1.01	1.0
Leu	1.11	1.0
Arg	ND	1.0
Pro	ND	1.0

[a] The slightly lower than expected value for Glu may be due to degradation occurring in the solution.
[b] ND, data not generated.

sensitive for the detection of impurities at 0.1 to 0.2% especially for those eluting prior to and immediately following Leuprolide acetate.

1.4.2. TLC

The TLC procedure used a solvent system consisting of chloroform : methanol : 32% acetic acid (60 : 45 : 20). Silica gel GF_{254} plates were obtained from Analtech. Spotted solvent volume was 20 μl of a 10 mg/ml sample in methanol. Standards consisting of 1-, 2-, 4-, 8-, and 12-μl portions of a 0.5 mg/ml solution of Leuprolide acetate reference standard were spotted. These amounts corresponded to 0.25, 0.5, 1.0, 2.0, and 3% relative to the sample size spotted. Plates were heated at 100°C for 10 min, cooled, treated with chlorine gas, and sprayed with a freshly prepared solution of 1% potassium iodide/1% soluble starch solution. Summary data from

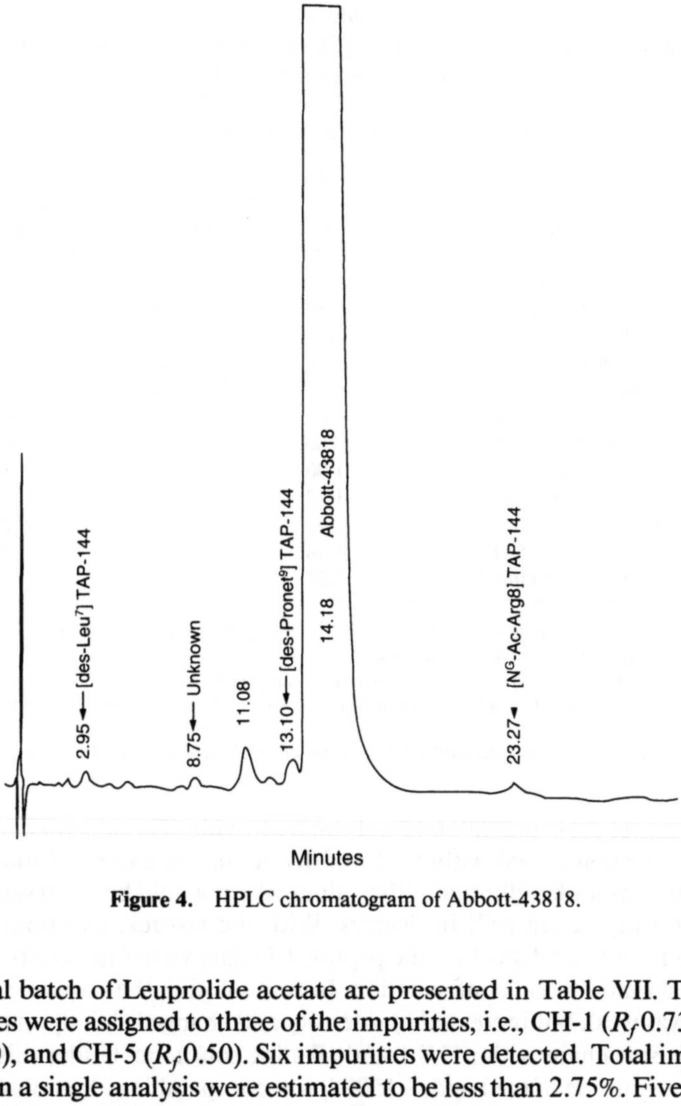

Figure 4. HPLC chromatogram of Abbott-43818.

a typical batch of Leuprolide acetate are presented in Table VII. Tentative identities were assigned to three of the impurities, i.e., CH-1 (R_f0.73), CH-7 (R_f0.60), and CH-5 (R_f0.50). Six impurities were detected. Total impurities based on a single analysis were estimated to be less than 2.75%. Five impurities totaling <1.0% were described earlier by HPLC.

1.4.3. Circular Dichroism

The CD spectra of a typical batch of Leuprolide acetate under various solvent conditions from 214 to 310 nm are shown in Figs. 5 and 6. The solutions were about 3×10^{-4} *M* (0.4 mg/ml) in 0.01 *N* sulfuric acid, 0.01 N

Table V
Correlation of Potential Manufacturing Impurities in Abbott-43818 by HPLC: Comparison of Two Laboratory HPLC Systems

		Relative retention time between Laboratories		
		Laboratory No. 1		
Code[a]	Compound identity[a]	Condition A[b]	Condition B[c]	Laboratory No.2 Condition B
HPB-4	p-Glu-His-Trp	0.02	0.01	0.06
CH-6	[des-Leu⁷] TAP-144	0.37	0.26	0.24
CH-13	Unknown	0.35	0.25	
CH-12	Unknown	0.74	0.65	0.26; 0.62; 0.66
CH-11	Unknown	0.77	0.70	
CH-1	TAP-144 (modified)	0.69	0.58	0.77
CH-5	[des-ProNHEt⁹] TAP-144	0.80	0.79	0.74
TAP-144	Abbott-43818	1.00	1.00	1.00
X_A	[D-His²] TAP-144	1.12	1.17	1.12
	[L-Leu⁶] TAP-144	1.43	1.58	1.42
CH-3	[N^G-Ac-Arg⁸] TAP-144	1.48	1.70	2.05
CH-7	TAP-144 (modified)	1.57	1.93	1.99

[a] The term "modified" represents a form of the peptide different than Abbott-43818. HPB, CH, and X_A are nomenclatures assigned to the respective impurities.
[b] Condition A—mobile phase: acetonitrile : 0.03 M ammonium phosphate, dibasic (295 : 705); column: Nucleosil 10 C_{18}, 4 mm × 30 cm; flow rate: 1.5 ml/min; injection volume: 12 μl (1% solution); detector: UV (220 nm).
[c] Condition B—same as A except acetonitrile : 0.03 M ammonium phosphate, dibasic (27 : 73) mobile phase.

Tris buffer (pH 8.1), and 0.016 N sodium hydroxide. The spectra in alkaline solution were completed within 2 hr of dissolving the sample. Final values were plotted as molar ellipticity (the values which would be observed for a 1 M solution in a 1-cm cell) in degrees. With the possible exception of the spectrum in *tert*-butyl alcohol, the peptide CD data were found to be consistent with a random coil conformation. In acid solution the aromatic band at 280 nm was weak but increased in intensity and changed sign upon titration of the His side chain with pH 8.1 and alkaline solutions. Under this condition the magnitude of the band, −1500°, was comparable to that observed in small molecules with fairly rigid conformations of the aromatic ring. There was a relatively small redshift of the 280-nm band in going from pH 8.1 to 0.016 N NaOH. For Tyr alone, a doubling in intensity upon ionization would be expected if there was no conformational change. This indicated that the 280-nm band was due to Trp.

There also was a moderately strong band at 228 nm (20,000°) in acid solution. The intensity of this band was reduced upon titration of His. In

Table VI
Relative Retention Times for Various Stereoisomers of Leuprolide Acetate

Isomer	Relative retention time
[L-Leu^6] Abbott-43818	1.42
[L-Leu^6, D-Leu^7] Abbott-43818	1.20
[D-Pro^9] Abbott-43818	1.16
[D-Tyr^5] Abbott-43818	1.15
[D-His^2] Abbott-43818	1.12
[D-Leu^7] Abbott-43818	1.11
[D-Trp^3, L-Leu^6] Abbott-43818	1.10
Abbott-43818	1.00
[D-Ser^4, L-Leu^6] Abbott-43818	0.88

alkaline solution the positive shoulder at 250 nm was due to Tyr. The 228-nm band was probably due to Trp absorbance in alkaline solution and a combination of Tyr and Trp absorbance at pH 8.1 and in acid solution. This assignment seemed plausible since changes in the 228-nm band paralleled those in the 280-nm band while there were smaller changes at shorter wavelengths. It was not possible to completely rule out contribution of the peptide *n*–pi* transition to the 228-nm band. His does not absorb light at either 228 or 280 nm as discussed earlier; however, its titration does influence the CD spectrum probably due to the conformational changes in the molecule. The data also suggested that in *tert*-butyl alcohol, some ordered peptide structure may have been induced in the peptide. The per residue ellipticity of ca. $-4000°$ was about one-third that expected for a β structure and about one-fifth that expected for an α-helix. The contribution of the aromatic transitions and the ionization state of the peptide were uncertain. Thus, the conformational inferences regarding this alcohol must be considered tentative.

1.4.4. Infrared Spectroscopy

The IR spectrum (Fig. 7) showed a strong broad band between 3600 and 2500 cm^{-1} due to superimposing O—H stretching vibrations of the phenolic hydroxyl and water (presumably for crystallization) in the molecule. The spectrum also showed superimposing N—H stretching vibrations attributed to the γ-lactam, the guanidine, the various secondary amides and amines, and the protonated amine. There was a complex of weak to medium-intensity bands from 3000 to 2800 cm^{-1} due to superimposing C—H stretching vibrations of the various methyl and methylene groups. A strong

Table VII
TLC Analysis of Impurities in Leuprolide Acetate

Component	R_f	% estimate
Minor spot	0.51	<0.5
Leuprolide acetate	0.56	—
Minor spot	0.60	<1
Minor spot	0.73	<0.5
Trace spots	0.50, 0.67, 0.69	<0.25 each
Total impurities		<2.75

Preparative TLC of Leuprolide Acetate (Abbott-43818): Coding of Isolated Manufacturing Impurities

Code name	R_f
CH-11	0.81
CH-12, CH-13	0.75
CH-1	0.72
CH-3	0.64
CH-7	0.62
Abbott-43818	0.57
CH-5	0.49
CH-6	0.36

band at 1660 cm^{-1} was attributed to superimposing of the C=N stretching vibration and N—H bending vibrations of the guanidine, to the C=O stretching vibrations of the acetate salt, to the γ-lactam and the various secondary amides, and finally to the skeletal stretching modes of the benzene rings. Thus, the strong band at 1540 cm^{-1} was felt to be due to superimposing of the amide II band of the various secondary amides, and to skeletal stretching modes of the benzene rings.

1.4.5. Ultraviolet Spectroscopy

UV spectra of Leuprolide acetate gave absorption maxima at 240, 280, and 289 nm with molar absorptivities 14,360, 6662, and 6536, in 0.1 N NaOH, respectively (see Fig. 8). The absorption maxima in 0.1 N HCl were at 278 and 288 nm with molar absorptivities of 6347 and 4713, respectively. The shift in the UV spectrum between acid and basic solution was felt to be due to the ionization of the phenolic -OH group in tyrosine. Figure 9 shows that the absorption spectrum for Leuprolide acetate above 240 nm could have been due to the summation of the absorption spectra from the Tyr and Trp segments of the molecule.

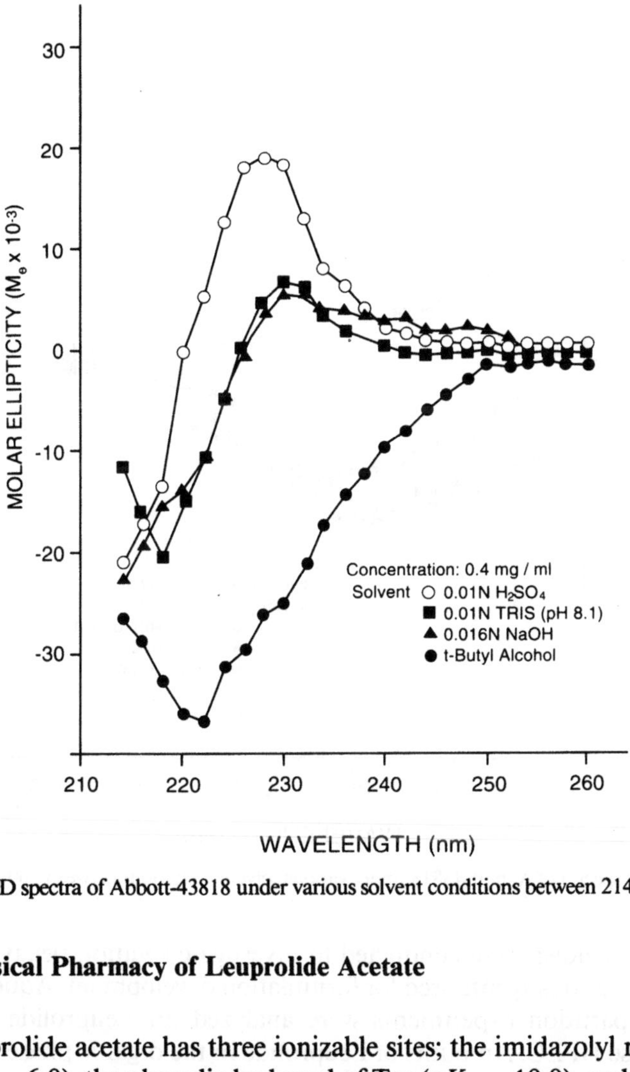

Figure 5. CD spectra of Abbott-43818 under various solvent conditions between 214 and 260 nm.

1.5. Physical Pharmacy of Leuprolide Acetate

Leuprolide acetate has three ionizable sites; the imidazolyl nitrogen of His (p$K_a \approx$ 6.0), the phenolic hydroxyl of Tyr (p$K_a \approx$ 10.0), and the guanidine nitrogen of Arg (p$K_a \approx$ 13.0). Since the guanidine nitrogen is extremely basic, this peptide as synthesized exists in the protonated form and is generally associated with at least 1 mole of acetic acid. The drug is a hydrophilic salt ($-NH_3^+ \cdot {}^-OOCCH_3$) because of the short hydrocarbon backbone of acetic acid. Leuprolide also is mostly ionized across a wide pH range of physiological interest, and it is not absorbed orally. For this reason it was felt that salts with improved lipophilicity might be more bioavailable nasally or orally than the acetate form of the drug.

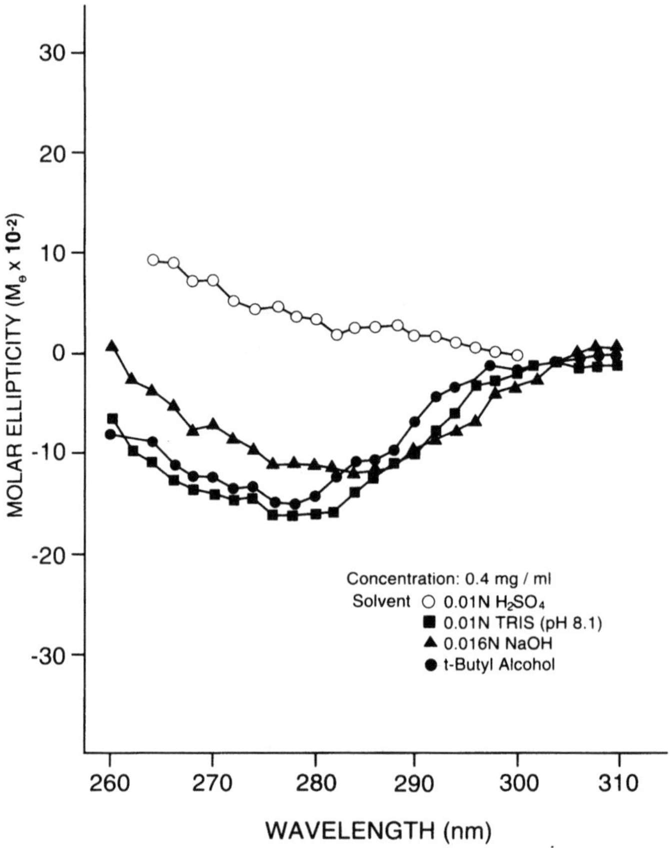

Figure 6. CD spectra of Abbott-43818 under various solvent conditions between 260 and 310 nm.

Several studies were conducted to investigate solution behavior in the presence of various agents used for formulation development. Aqueous samples from partition experiments were analyzed for Leuprolide using reversed-phase HPLC. Amounts of Leuprolide in the organic phase were generated by mass difference based on initial concentrations of drug used in the partitioning medium. Ion chromatography was used to determine *n*-alkyl sulfonate concentrations in the partitioning media. A solution of 0.005 M sodium carbonate with 2 to 5% methanol was used as mobile phase for the separation of decanesulfonate, octanesulfonate, and hexanesulfonate. A solution of 0.015 M sodium bicarbonate was used for the separation of methanesulfonate and butanesulfonate. Mobile phase flow rate was 69 ml/hr at an injection volume of 100 μl. In studies where salicylate was used as a counterion, salicylate levels in the aqueous phase were determined using

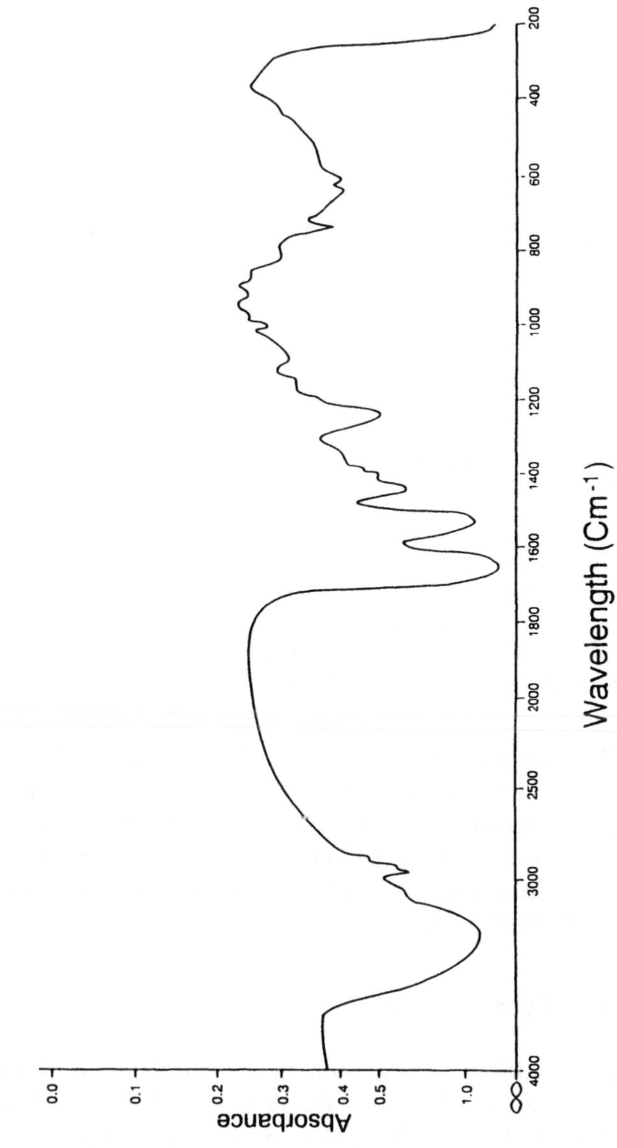

Figure 7. Infrared spectrum of Abbott-43818.

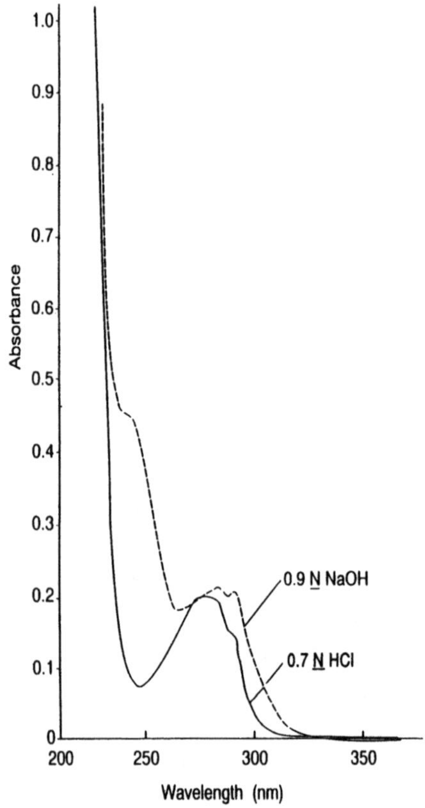

Figure 8. UV spectra of Abbott-43818 in acid and alkali at a concentration of 0.0404 mg/ml.

HPLC assay. Separation of the anion was achieved on a Bondapak C_{18} column at 313 nm with a 0.1% acetic acid (30 : 70% v/v) mobile phase. Dehydrocholate was determined by HPLC at 205 nm on a Nucleosil C_{18} column. The mobile phase consisted of 0.014 M sodium phosphate monobasic monohydrate, acetonitrile, methanol, triethylamine, and 85% phosphoric acid (50 : 35 : 15 : 0.02 : 0.2% v/v).

1.5.1. Theoretical Considerations

The Appendix contains a schematic of the partition model used to investigate solution kinetics of Leuprolide. In this model [Leup1] represents the concentration of Leuprolide species possessing two positive charges (guanidine and histidine). [Leup2] represents the concentration of Leuprolide species with one positive charge (guanidine). Species designated as Leup1 may contain two counterions of "interest" to form Leup1–2X where

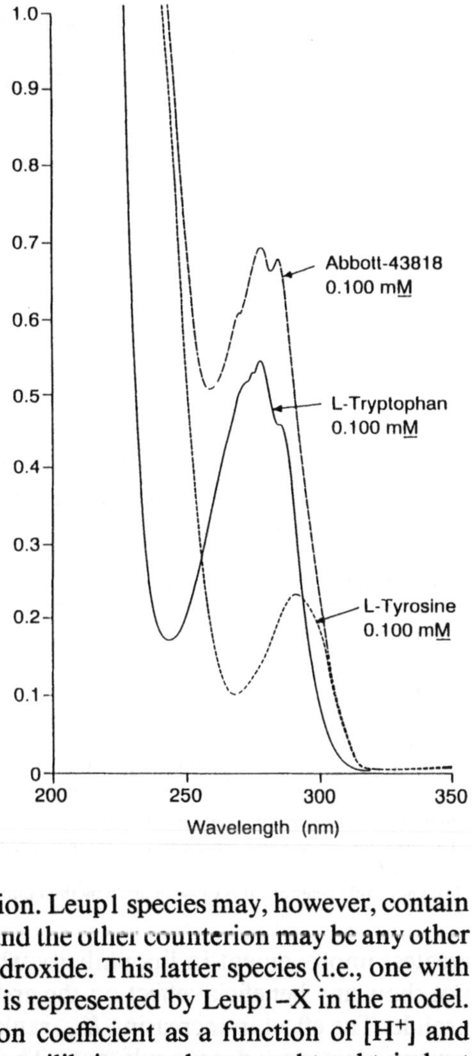

Figure 9. UV spectra of Abbott-43818, tryptophan, and tyrosine, each at 0.100 mM in 0.1 N sodium hydroxide.

X represents the lipophilic counterion. Leup1 species may, however, contain only one counterion of "interest" and the other counterion may be any other anion, e.g., acetate, chloride, or hydroxide. This latter species (i.e., one with only one counterion of "interest") is represented by Leup1–X in the model. Equation (7) describes the partition coefficient as a function of $[H^+]$ and counterion concentration $[X]_{aq}$ at equilibrium and was used to obtain best fits for K_{obs}. Values of $[H^+]$ and $[X]_{aq}$ were known experimentally. Estimates of K_1, K_2, and K_3 were obtained by nonlinear regression fit to Eq. (7).

1.5.2. Alkyl Sulfonates

Results of K_{obs} for Leuprolide using several alkyl sulfonates are reported in Table VIII. The data demonstrate significant increase in lipophilicity of the ion pairs with carbon number of the anion. The decyl salt shows the

Table VIII
Effect of pH and Concentration of Alkyl Sulfonic Acids on Apparent Partition Coefficient of Leuprolide at 30°C

Ratio (pH)	Decane	Octane	Hexane	Butane	Methane
1 : 2 (3)	4.01	0.86	0.03	0.02	0.01
1 : 2 (5)	18.67	1.87	0.19	0.02	0.03
1 : 2 (7)	38.36	4.14	0.55	0.11	0.15
1 : 1 (3)	0.72	0.30	0.02	0.02	0.02
1 : 1 (5)	2.25	0.55	0.09	0.02	0.03
1 : 1 (7)	6.49	1.57	0.32	0.16	0.15
2 : 1 (3)	0.27	0.10	0	0.01	0.02
2 : 1 (5)	0.66	0.33	0.05	0.03	0.02
2 : 1 (7)	1.11	0.60	0.17	0.15	0.15

Equilibrium Partition Coefficients of Ion Pairs of Leuprolide: Effect of Alkyl Chain Length on Lipophilicity of Leuprolide[a]

Carbon number	$\log K_1$	$\log K_2$	$\log K_3$
C-10	3.84	7.75	4.98
C-8	3.33	6.35	3.89
C-6	1.76	4.83	2.88
C-4	ND[c]	ND	ND
C-1	ND	ND	ND
Slope	0.52 (±0.15)	0.98 (±0.13)[b]	0.52 (±0.01)

[a] Note: K_{obs} for Leuprolide acetate at pH 3, 4, and 5 are respectively 2.0×10^{-6}, 3.5×10^{-5}, and 2.0×10^{-2}.
[b] The slope for the K_2 curve represents a di-anionic salt. Therefore the effect of carbon number is half this value, i.e., 0.49.
[c] Data not easily determined because of assay method limitation.

greatest improvement in lipophilicity while the C-1 and C-4 analogues do not show any beneficial effect on the apparent partition coefficient of Leuprolide. The effect in increasing K_{obs} ranks according to the following order: hexane < octane < decane sulfonate. The data also demonstrate a consistent increase in K_{obs} as a function of pH and counterion concentration, suggesting that deprotonation of the imidazolyl nitrogen of histidine and the guanidine nitrogen may both be participating in the equilibrium reaction.

Mathematical treatment of the data to provide estimates of partition coefficients for the respective ion pairs was done using Eq. (7). Results for the equilibrium constants K_1, K_2, and K_3 are tabulated in Table VIII. For the given counterion, lipophilicities of different Leuprolide species increase in the following order: Leup1–X < Leup2–X < Leup1–2X (i.e., $K_1 < K_3 < K_2$). Although Leup2–X species (which exists under acid conditions) has the highest partition coefficient, its contribution to observed partition coefficients is relatively low in the range of counterion concentration studied. This may be due to the fact that mass law is operative under these experimental conditions. Although not reported, the theoretically derived log K results agreed well with the observed data.

Log K_{obs} values of each species were plotted as a function of number of carbon atoms in the alkyl side chain of the counterion. This treatment yielded straight lines with a slope of 0.52 for K_1 and K_3 and 0.98 for K_2. Results for K_1, K_2, and K_3 are summarized in Table VIII. The slope, K_1 and K_2 represent the Hansch π constant, and the values agree well with the literature value of 0.5 for methylene. The slope of 0.98 for K_2 represents contributions from two methylene groups.

1.5.3. Salicylate and Dehydrocholic Acid

Summarized in Table IX are K_{obs} data obtained with salicylate counterion for Leuprolide. Mathematical treatment of the data did not provide a meaningful correlation, indicating that different and perhaps additional factors may be involved in ion pair formation of Leuprolide with salicylate. As with the alkyl sulfonates, there is an increase in K_{obs} with a corresponding increase in pH and counterion concentration. The magnitude of increase in partition coefficient is also about the same as hexane sulfonate but is dependent more on pH than on counterion concentration, probably due to the low pK_a of His, i.e., 2.97.

Table IX
Effect of pH and Concentration of Other Ion-Pairing Agents on Apparent Partition Coefficient of Leuprolide at 30°C

Ratio (pH)	Salicylate	Dehydrocholate
1 : 2 (3)	0	0.01
1 : 2 (5)	0.16	0.01
1 : 2 (7)	0.45	0.10
1 : 1 (3)	0.01	0
1 : 1 (5)	0.11	0
1 : 1 (7)	0.39	0.07
2 : 1 (3)	0	0
2 : 1 (5)	0.06	0
2 : 1 (7)	0.25	0.06

Apparent partition coefficients obtained for Leuprolide in the presence of dehydrocholic acid are also summarized in Table IX. This counterion did not provide any meaningful improvement of K_{obs} even at pH 7. Steric hindrance from the bulky nature of this molecule may be largely responsible for the low ion pair formation with Leuprolide at the pH range (i.e., 3–7) under which the studies were carried out. Low K_{obs} with cholic acid derivatives at lower pH values should be expected from theoretical considerations. Notably, the pK_a of the counterion is greater than 6 and should not be available for ion pairing at pH values lower than 7. Salicylate also did not satisfactorily predict results consistent with K_{obs} in the model, suggesting inapplicable assumptions or oversimplification of the model.

Summarizing, the partitioning results indicate that the extent of increase in lipophilicity of Leuprolide may be proportional to the extent of deprotonation of the imidazolyl nitrogen. Thus, lipophilicity of these ion pairs may be proportional to the pK_a of the acid from which the anion is derived, i.e., sulfonic acid (p$K_a < 2.0$) > salicylic acid (p$K_a \approx 3.0$) > dehydrocholic acid (p$K_a > 6.0$).

1.6. Bioanalysis

Bioassay methods have been used to monitor the pharmacological end points of LH-RH analogues. The methods are based on receptor binding techniques (Marshall and Odell, 1975). To assay Leuprolide in biological media, rat pituitary plasma membrane fraction is prepared as a source of LH-RH receptors. ^{125}I-labeled [Tyr^5]-Leuprolide, made via chloramine-T oxidation and purified by ion exchange chromatography, is coincubated at 4°C with varying drug levels and LH-RH receptors. At equilibrium, bound tracer is separated from free by centrifugation. Data are reported as pK_I, the negative log of the equilibrium dissociation constant. Binding is saturable and reversible, and the affinities agree with pharmacological potencies.

1.6.1. LH Release

The ability of compounds to release LH from cultured pituitary cells (Vale and Grant, 1974) is the primary means of establishing their activities. Pituitaries are dispersed with collagenase and hyaluronidase, and the cells are planted in culture plates for 72 hr. Cells are incubated with compounds for 3 hr at 37°C. Released LH is determined by RIA using ^{125}I-labeled LH (ovine) as tracer with immunoaffinity-purified anti-rat-LH IgG, and rat LH (NIHRP2) as standard. Bound and free tracer are separated with immobilized protein A.

1.6.2. Agonists

The LH releasing potency is expressed as the pD_2 which is the negative log of the molar concentration to produce a half-maximal release of LH. Intrinsic activity is the maximum effect produced by a compound relative to 10^{-8} M Leuprolide acetate. Compounds with greater than 70% intrinsic activity are considered full agonists; those with less than 70% are partial agonists. The potencies of agonists are usually in good agreement with their binding affinities. Compounds with high potencies ($pD_2 > 9$) in general produce a half-maximal effect at a lower concentration than their equilibrium dissociation constant (i.e., $pD_2 > pK_I$). This may be due to the fact that compounds possessing high efficacy might not need to occupy the whole receptor population to produce a maximal response.

1.6.3. Antagonists

The ability of a compound to produce a rightward shift in the dose-dependent LH release by a typical agonist such as Leuprolide is a measure of its antagonistic potency, or pA_2. The pA_2 is the negative log of the concentration which shifts the Leuprolide curve twofold (i.e., requires a doubling of the Leuprolide concentration to produce an equivalent response). Antagonist potencies also show good agreement with binding affinities; but, in general, receptor affinities of antagonists exceed *in vitro* potencies by about tenfold. It is not certain whether this phenomenon is related to a property of the LH-RH receptor, or if it is an artifact of the differences in methodology.

2. FORMULATIONS AND DOSIMETRY

Parenteral administration of LH-RH analogues is a major concern in therapeutics. Besides being invasive, these routes are not well accepted by patients requiring long-term drug therapy. Various companies have explored other routes. For example, Hoechst AG used the nasal route (i.n.) for Buserelin in man (Nillius *et al.*, 1984). This route has the drawback of low and erratic bioavailability. TAP Pharmaceuticals has developed and marketed a sustained-release i.m. formulation of Leuprolide, Lupron Depot. The formulation contains biodegradable microspheres of poly(*d,l*-lactide-glycolide) which slowly release Leuprolide as the matrix degrades to soluble lactate and glycolate (Sanders *et al.*, 1984). Syntex, ICI, Ortho, and Wyeth also developed sustained-release formulations of their LH-RH agonists. It has been reported that ICI 118630 (Zoladex) successfully reduced circulating LH and testosterone in prostatic cancer patients with repeated injections only once every 28 days (Ahmed *et al.*, 1985). Since the material disappears from the injection site, no long-term irritation was detected. TAP Pharmaceuticals also markets sterile aqueous solutions of Leuprolide (Lupron 1 mg/ml and 5 mg/ml) for s.c. use in prostatic cancer.

2.1. Parenteral Product Forms

Generic composition of TAP's injectable Leuprolide formulations is as follows:

Lupron Injection, 5 mg/ml

Leuprolide acetate powder		20.0 mg
Benzyl alcohol, NF		36.0 mg
Sodium chloride, USP		25.2 mg
Water for injection, USP	q.s.	4.0 ml
Acetic acid, glacial, USP	q.s.	pH 5.2

Lupron injection is available in type I glass vials and packaged in multiple fill volumes. It is also available in a unit dose prefilled syringe for self-administration.

Lupron Depot, for Intramuscular Injection, 15 mg/ml

Leuprolide acetate powder	30.0 mg
Polylactic/polyglycolide	264.7 mg
Gelatin, USP	5.3 mg
D-Mannitol, USP	52.9 mg
Vehicle for Lupron Depot	2.0 ml

Lupron Depot is available in two strengths, a 7.5-mg dosage for monthly administration to prostatic cancer patients, and a 3.75-mg dosage for use in endometriosis. Both dosages are sustained-release preparations and release Leuprolide over 30 days.

2.2. Nasal Dosage Forms

Nasal formulations of Leuprolide acetate have been investigated for absorption in humans by TAP Pharmaceuticals, but unfortunately, like Buserelin (Nillius *et al.*, 1984) and Nafarelin (Chan *et al.*, 1988), absorption was low ($\simeq$2% relative to i.v. administration) and erratic (Adjei *et al.*, 1992) with an intersubject coefficient of variation approximately 90%. The nasal formulations of Leuprolide acetate were packaged in 1-ounce amber glass bottles each closed with a metered mechanical spray pump. All solutions were adjusted to pH 5.2 with acetic acid and made isotonic with sodium chloride. In animal model studies, nasal absorption of Leuprolide revealed a significant beneficial effect of an absorption promoter. The enhancer system used in prototype formulation was 5% α-cyclodextrin and 25 mM EDTA. Formulations were administered with a mechanical spray pump to each nostril so that the total dosages administered were 1, 1, and 3 mg of Leuprolide acetate for formulations A, B, and C, respectively. Generic composition of three clinical formulations is:

Formula	Material identification	Material composition	Leuprolide concentration	Dose level	Bioavailability (% of i.v.)
A	NaCl	0.9%	5 mg/ml	1 mg	0–1.8
B	α-Cyclodextrin	5.0%	5 mg/ml	1 mg	0–2.7
C	α-Cyclodextrin	5.0%	15 mg/ml	3 mg	0–6.0

2.3. Inhalation Dosage Forms

TAP Pharmaceuticals is conducting clinical trials with inhalation aerosols of Leuprolide acetate for the palliative treatment of endometriosis. Several formulations were screened in preclinical studies three of which showed significant absorption of the drug via lung. Generic composition of the three metered dose inhaler formulations are as follows:

Composition	Solution (A)	Suspension (B)	Suspension (C)
Water, purified, USP	1.7%	—	—
Decanesulfonic acid	0.7%	—	—
Alcohol, dehydrated, USP	20.0%	—	—
Leuprolide acetate	10 mg/ml	10 mg/ml	20 mg/ml
Sorbitan monooleate, NF	5.0%	—	—
Sorbitan trioleate	—	0.5%	0.5%
Trichlorofluoromethane	—	25.0%	25.0%
Dichlorodifluoromethane	q.s.	q.s.	q.s.
Bioavailability (% of i.v.)	6.2	26.3	20.6

Formulation A was a solution aerosol and contained 10 mg/ml of Leuprolide acetate. Formulations B and C were suspension aerosols and contained 10 and 20 mg/ml of Leuprolide acetate, respectively. Two sprays ($2 \times 50\ \mu l$) of each of the aerosols were administered so that a total of 1, 1, and 2 mg of Leuprolide acetate was administered via formulation A, B, and C, respectively. All three aerosol formulations utilized mixtures of chlorofluorocarbons (Union Carbide) as propellant systems for the drug.

3. PRECLINICAL AND CLINICAL STUDIES WITH LEUPROLIDE ACETATE

Studies were conducted to determine the bioavailability of Leuprolide acetate in healthy human males when administered i.n. by spray, as an inhalation aerosol (i.h.), and also when administered parenterally by s.c. and i.v. injection. As indicated earlier, i.n. bioavailability relative to i.v. injection was approximately 2% with significant subject-to-subject variability. Inhalation delivery showed extensive absorption of the drug with slightly lower degree of intersubject variability. The i.v. and s.c. were bioequivalent to each other but the volume of distribution of the central compartment following s.c. administration was nearly threefold greater than for i.v. administration, suggesting extensive distribution of the drug. Bioavailability of a suspension aerosol, 27.9% (see formulation B, Section 2.3), using s.c. injection as control, was approximately fourfold greater than a solution aerosol of the drug (i.e., 6.6%, formulation A). Mean bioavailability of the suspension aerosols indicated superior performance of i.h. over i.n. delivery of the drug, suggesting that other LH-RH analogues may be systemically delivered via lung for effective treatment of diseases controlled by gonadotropins.

3.1. Pharmacokinetics following Acute Administration

The clinical studies showed that upon nasal delivery of Leuprolide to humans, plasma peak concentrations (C_{max}) with 1- and 3-mg doses ranged between 0.24–1.6 ng/ml and 0.10–11.0 ng/ml, respectively. Inhalation delivery gave a mean C_{max} with a 1-mg dose solution aerosol of 0.97 ng/ml, and 4.4 and 11.4 ng/ml for suspension aerosols given at 1- and 2-mg doses, respectively.

The recommended daily dose to prostatic cancer patients is 1 mg s.c. of Leuprolide acetate. Analysis of variance of plasma concentration–time profiles for 1-mg doses of this drug following i.n., i.h., and s.c. administration revealed that mean plasma levels for the suspension aerosols were significantly higher ($p < 0.01$) than either the solution aerosol or the two i.n. formulations. Plasma concentration–time profiles for the nasal and inhalation formulations compared with the marketed product (s.c.) are shown in Fig. 10. Plasma $t_{1/2}$ following i.n. and i.h. administration of Leuprolide (average 3.0 and 4.0 hr, respectively) were also significantly different ($p \leq 0.05$) which may partially account for the higher AUCs obtained for the 1-mg i.h. aerosol relative to the 1-mg i.n. formulations. The elimination rates for the three routes were not different ($p < 0.05$). Thus, AUC and plasma level differences between formulations might be in part attributed to undefined differences at the absorption site.

The relevance of the pharmacokinetics of LH-RH analogues in assessing clinical efficacy of these drugs is imperative. However, peptide metabolism in body fluids as well as chemical degradation during normal use life and storage are equally relevant. For this reason, a discussion of clinical pharmacokinetics of these compounds should be complemented with *in vivo* efficacy models.

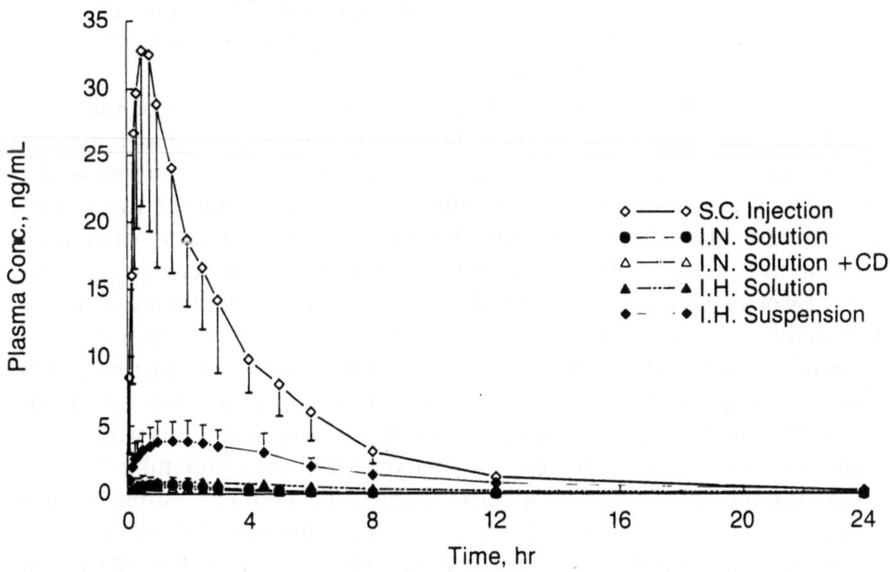

Figure 10. Bioavailability of Leuprolide in humans via various routes of administration of 1-mg bolus doses.

3.2. Pharmacodynamics following Chronic Administration

Johnson *et al.* (1976) studied the suppressant effect of Leuprolide in immature female rats following s.c. injection of 10 μg of the peptide daily for 77 days. A suppression of ovarian and uterine growth weights was observed. Organ weights in Leuprolide-treated rats were similar to those seen in prepubertal untreated animals. Four weeks after cessation of the treatment, the weights of the sex organs increased slowly reaching values near those of control animals. During the course of the treatment the animals did not exhibit normal cycling. Normal cycling was resumed within 5–10 days after treatment was stopped.

Labrie *et al.* (1978, 1980) demonstrated that treatment of adult male rats with Leuprolide or [D-Ala^6, desGly-NH_2^{10}]LH-RH ethylamide leads to a loss of the testicular LH and prolactin receptors accompanied by decreased weight of the testes, seminal vesicle, and ventral prostate. Other studies in animals indicated that chronic administration of Leuprolide resulted in inhibition of the pituitary release of LH and FSH following an initial stimulation. Consequently, plasma testosterone level and spermatogenesis were suppressed, and degenerative changes in the seminiferous tubules were observed. All of these physiological changes were reversible within 1 month after cessation of the treatment. This ability of Leuprolide to induce "reproductive atrophy" pharmacologically, initially suggested its potential usefulness in treating androgen-dependent neoplasia.

Huggins and Hodges (1941) demonstrated the androgen dependence of most prostatic cancers. Since then, hormonal manipulation has remained the primary treatment for most patients with metastatic carcinoma of the prostate. Bilateral orchiectomy or oral estrogens have been the most common treatment used and have been found to induce temporary disease remission in a majority of patients. However, the possible adverse psychological impact of castration and the frequently observed cardiovascular and feminizing side effects of estrogen therapy have prompted clinical investigations of alternative methods of androgen suppression. The endocrine effects observed in animals after chronic administration of Leuprolide, as previously described, have encouraged the use of the drug in patients with metastatic prostatic cancer. The drug is effective and offers a preferable form of endocrine therapy in view of adverse effects associated with estrogens. The following possible drug-related side effects were observed: (1) an initial increase in serum testosterone and dihydrotestosterone during the first week of therapy was observed; and (2) many patients experienced vasomotor hot flashes.

4. STABILITY STUDIES AND DEGRADATION KINETICS

Because Leuprolide acetate contains nine different amino acids, an amide grouping at the COOH-terminus, and a lactam at the NH_2-terminus, the molecule undergoes degradation reactions typical of a peptide. Using reversed-phase HPLC, Nishi *et al.* (1980) observed that in acid solution (i.e., 0.1 N HCl, 40°C for 2 days) the primary degradation product resulting from Leuprolide acetate is [Glu^1]Abbott-43818. Under similar stress conditions (0.1 N HCl, 37°C for 2 days), a single major degradation peak at the relative retention time of 0.45 was found (Fig. 11). The sample, on neutralization, can be enzyme hydrolyzed with amino peptidase as described earlier. Since Leuprolide acetate cannot be enzyme hydrolyzed directly, the reaction product must be the lactam ring-opened product. A shorter relative retention time of the acid by-product compared with Abbott-43818 is thus plausible following replacement of p-Glu with glutamic acid.

Nishi *et al.* (1980) also reported that the principal reaction observed in aqueous alkaline solution (0.1 N NaOH, 5°C, 48 hr) was racemization. Racemization was greatest at positions 4 and 2. Under mild alkali stress conditions, two major degradation products (labeled D and F in Fig. 12) were observed. Peak F was identified to be [D-His^2] Abbott-43818 by comparison with a known. A tentative identification of [D-Ser^4] Abbott-43818 was assigned to peak D. Peak B has the same relative retention time as [Glu^1] Abbott-43818 from acid stress studies. This is consistent with the fact that the alkali-stressed sample, after neutralization, could be enzyme hydrolyzed. Thus, if the [Glu^1] Abbott-43818 assignment is correct, then peak A would be [Glu^1, D-Ser^4] Abbott-43818, and peak C would be [Glu^1, D-His^2] Abbott-43818.

Syntex scientists (Strickley *et al.*, 1990) studied the aqueous stability of RS-26306, an LH-RH antagonist, using HPLC-MS and thermospray ionization. This LH-RH analogue is a decapeptide containing five D-amino acids. It is acetylated at the NH_2-terminus and amidated at the COOH-terminus. Results indicated that at pH 3 or lower, the primary degradation product is formed by acid-catalyzed deamidation of the COOH-terminal amide. Above pH 7, epimerization of the individual amino acids to form diastereomers is the principal reaction pathway. Results also indicated that between pH 4 and 6, the degradation occurs by intramolecular serine-catalyzed hydrolysis to yield two primary products, a tripeptide and a heptapeptide. The pH–rate study demonstrated a U-shaped profile with a projected shelf life of at least 10 years at 25°C. The pH-dependent stability profile for this peptide is similar to results obtained with the superagonist LH-RH analogue Nafarelin

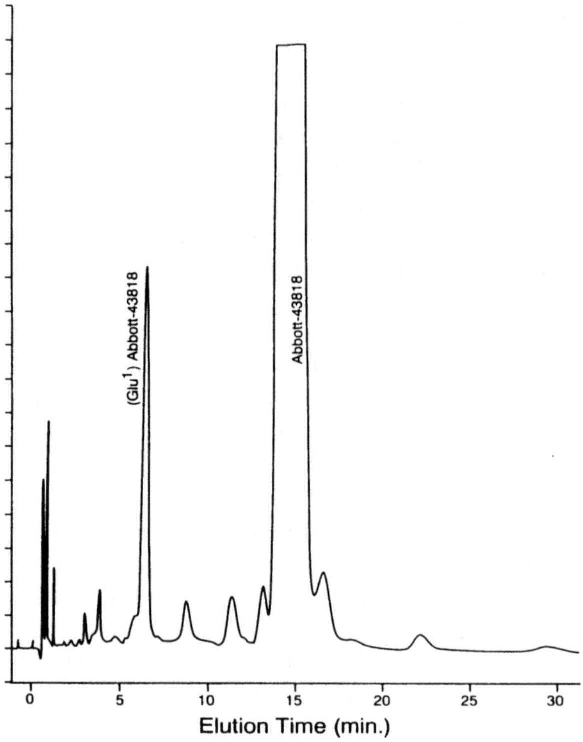

Figure 11. HPLC chromatogram of Abbott-43818 after stress in acid (0.1 N HCl, 40°C, 2 days).

acetate, [D-Nal(2)[6]]LH-RH, in aqueous solution (Johnson *et al.*, 1986) using reversed-phase HPLC. Maximum stability is obtained at pH 4–6 and in acid a specific acid-catalyzed hydrolysis of the glycinamide group to yield the free acid as primary degradation product is observed. Also, several smaller degradation products are noted in the pH range 4–10 but the overall impact is small. Results indicated a lower degree of degradation at lower temperatures with optimal stability at pH 5.3 and a predicted T_{90} of 3.8 years at 25°C.

4.1. Drug Substance Stability

Stability of bulk Leuprolide acetate was examined by HPLC and TLC. Room temperature stress studies using UV irradiation at approximately 100° for up to a year did not reveal any significant conversion of Leuprolide to any of its recognized degradation products.

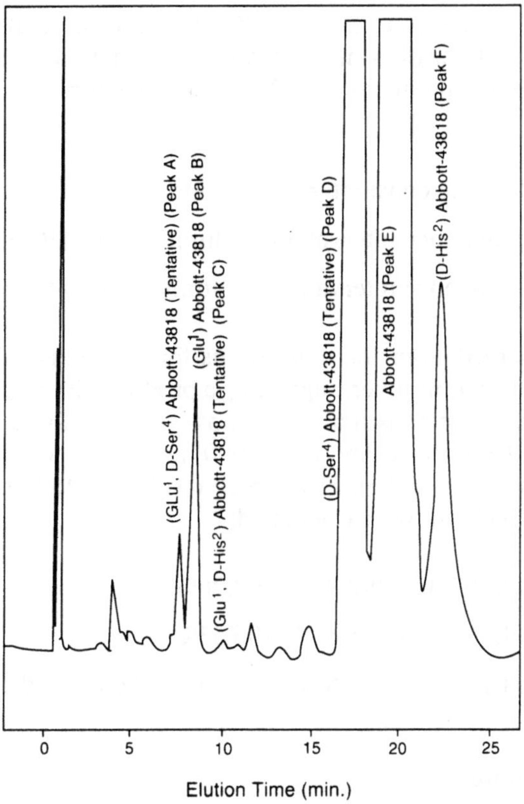

Figure 12. HPLC chromatogram of Abbott-43818 after stress in alkali (0.1 N NaOH, 25°C, 2 hr).

4.2. Formulation Stability

The major degradation products in aqueous solution were discussed earlier. Analysis of data for several product lots of Lupron injection under various stress (temperature, light, and headspace storage) conditions revealed that Leuprolide is both heat and light sensitive. Stability under refrigerated storage, however, is satisfactory and does not show measurable drug losses to below the specification limit of 90% over at least 2 years.

4.3. Shelf-Life Projections

Stability data through 24 months for multiple lots of Lupron were generated by HPLC (condition B). No measurable drug losses occurred at the recommended storage condition of 5°C. The data were also found to be supportive of a shelf life of 24 months. Mean degradation rates (5°C to 50°C) based on Arrhenius predictions at 25°C for typical pilot scale lots are as follows:

Stability rate constant = 0.0017/month

95% confidence interval, lower limit = −0.0036/month

95% confidence interval, upper limit = 0.0069/month

As may be observed from the data, limits for the predicted rate constant include zero as a valid number, suggesting a product with negligible degradation. These results are consistent with those reported for Syntex's LH-RH antagonist RS-26306 (Strickley *et al.*, 1990) and the superagonist Nafarelin acetate (Johnson *et al.*, 1986). From these results, the following profiles for the drug (% label claim) were calculated:

12 months 97.47% (92.92%–102.26%)

24 months 95.53% (85.66%–106.53%)

36 months 93.62% (78.87%–111.13%) when stored at 25°C

5. CONCLUSIONS

Several studies have been discussed which demonstrate that LH-RH activity and stability to enzymatic processes can be achieved with structural modification. Studies also show that *in vivo* activity of these peptides can be improved by orders of magnitude relative to the parent. Based on work by chemists around the world, several LH-RH analogues are now in various phases of clinical development. Leuprolide acetate (Abbott-43818), a potent LH-RH agonist, is currently on the market for use in palliative treatment of prostatic cancer and endometriosis. Methods to detect stability problems as well as impurities during bulk drug manufacture have been discussed. Solution kinetics data were also discussed in terms of the effect of lipophilic counterions on partition characteristics of this drug. A theoretical model was presented for predicting the partition coefficient of various salts of this drug which may be applicable to other LH-RH analogues. A variety of clinical formulations were discussed which might provide solutions to the compliance problems associated with daily s.c. injection of this peptide.

ACKNOWLEDGMENTS

The authors are thankful to Dr. J. Greer and the LH-RH team, Dr. G. Menon and his analytical research group, and Dr. T. Reiland for technical support. The authors also are very grateful to Ms. M. Herrin and Dr. J. Seely of TAP Pharmaceuticals, Division of Abbott Labs, for continued support and funding of LH-RH development work.

6. APPENDIX: SCHEMATIC REPRESENTATION OF ION-PAIR PARTITION MODEL

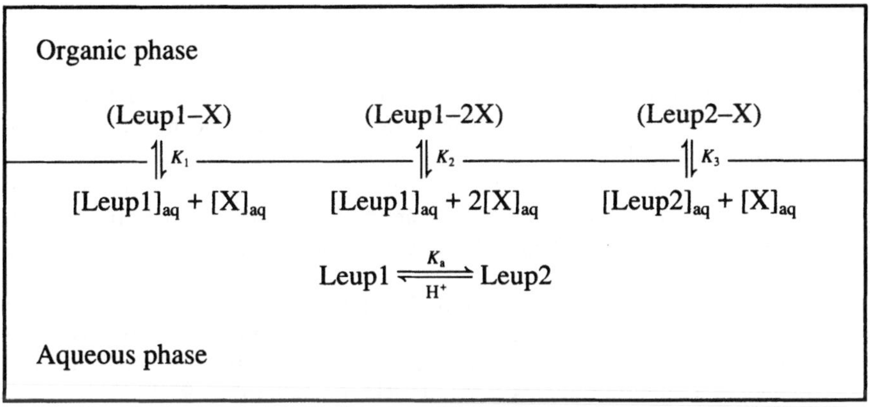

$$K_1 = \frac{[\mathrm{Leup1\ X}]_{org}}{[\mathrm{Leup1}]_{aq}[\mathrm{X}]_{aq}} = \frac{[\mathrm{Leup1{-}X}]_{org}}{\dfrac{\mathrm{H}^+}{\mathrm{H}^+ + K_a}[\mathrm{Leup}]_{Taq}[\mathrm{X}]_{aq}} \quad (1)$$

$$K_2 = \frac{[\mathrm{Leup1{-}2X}]_{org}}{[\mathrm{Leup1}]_{aq}[\mathrm{X}]^2_{aq}} = \frac{[\mathrm{Leup1{-}2X}]_{org}}{\dfrac{\mathrm{H}^+}{\mathrm{H}^+ + K_a}[\mathrm{Leup}]_{Taq}[\mathrm{X}]^2_{aq}} \quad (2)$$

$$K_3 = \frac{[\mathrm{Leup2{-}X}]_{org}}{[\mathrm{Leup2}]_{aq}[\mathrm{X}]_{aq}} = \frac{[\mathrm{Leup2{-}X}]_{org}}{\dfrac{K_a}{\mathrm{H}^+ + K_a}[\mathrm{Leup}]_{Taq}[\mathrm{X}]_{aq}} \quad (3)$$

$$\begin{aligned}[\mathrm{Leup}]_{Taq} &= [\mathrm{Leup1}]_{aq} + [\mathrm{Leup2}]_{aq} \\ &= \text{total Leuprolide concentration in aqueous phase}\end{aligned} \quad (4)$$

$$K_{obs} = \text{partition coeff.} = \frac{\text{total Leuprolide concn in organic phase}}{\text{total Leuprolide concn in aqueous phase}}$$

$$K_{obs} = \frac{[\text{Leup1–X}]_{org} + [\text{Leup1–2X}]_{org} + [\text{Leup2–X}]_{org}}{[\text{Leup}]_{Taq}} \tag{5}$$

$$K_{obs} = K_1\left(\frac{H^+}{H^+ + K_a}\right)[X]_{aq} + K_2\left(\frac{H^+}{H^+ + K_a}\right)[X]^2_{aq} + K_3\left(\frac{K_a}{H^+ + K_a}\right)[X]_{aq} \tag{6}$$

$$K_{obs} = \frac{[X]_{aq}}{H^+ + K_a}\{K_1[H^+] + K_2[H^+][X]_{aq} + K_3K_a\} \tag{7}$$

REFERENCES

Adjei, A., Sundberg, D., Miller, D., and Chun, A., 1992, Bioavailability of Leuprolide acetate following nasal and inhalation delivery to rats and healthy humans, *Pharm. Res.* **9**:244–249.

Ahmed, S. R., Grant, J., Shalet, S. M., Howell, A., Chowdhury, S. D., Weatherson, T., and Blacklock, N. J., 1985, Preliminary report on use of depot formulation of LH-RH analogue ICI 118630 (Zoladex) in patients with prostatic cancer, *Br. Med. J.* **290**:185.

Arimura, A., Sato, H., Kumasaka, T., Worobe, R. B., Debeljuk, L., Dunn, J. D., and Schally, A. V., 1973, Production of antiserum to LH-RH associated with marked atrophy of gonads in rabbits; characterization of the antibody and development of a radioimmunoassay for LH-RH, *Endocrinology* **93**:1092.

Arimura, A., Debeljuk, L., and Schally, A. V., 1974, Blockade of preovulatory surge of gonadotropins LH and FSH and of ovulation by anti-LH-RH serum in rats, *Endocrinology* **93**:323.

Barber, M., Bordoli, R. S., and Sedwick, R. D., 1980, Fast atom bombardment mass spectrometry, in: *Soft Ionization Biological Mass Spectrometry* (H. R. Morris, ed.), Heyden, London.

Beynum, G. M. A. van, DeGraff, J. M., Castel, A., Kraal, B., and Bosch, L., 1977, Structural studies on the coat protein of alfalfa mosaic virus, *Eur. J. Biochem.* **72**:63–78.

Chan, R. L., Henzl, M. R., LePage, M. E., LaFargue, J., Nerenberg, S. A., Anik, S., and Chaplin, M. D., 1988, Gonadotropin-releasing hormone, *Clin. Pharmacol. Ther.* **44**(3):275–282.

Coy, D. H., Mezo, I., Pedroza, E., Nekola, M. V., Vilchez-Martinez, J. A., Piyachaturawat, P., Schally, A. V., Seprodi, J., and Teplan, I., 1979, LH-RH antagonists with potent antiovulatory activity, in: *Peptides, Proceedings of the 6th American Peptide Symposium* (E. Gross and J. Meienhofer, eds.), Pierce Chemical Co., Rockville, Ill., p. 775.

Deslauriers, R., and Somorjai, R. L., 1976, Internal rotations of side chains and backbone in luteinizing hormone-releasing hormone (LHRH). Analysis of carbon-13 spin–lattice relaxation times, *J. Am. Chem. Soc.* **98**:1931.

Fujino, M., Kobayashi, S., Obayashi, M., Shinagawa, S., Fukuda, T., Kitada, C., Nakayama, R., Yamazaki, I., White, W. F., and Rippel, R. H., 1972, Structure–activity relationships in the C-terminal part of luteinizing hormone releasing hormone (LH-RH), *Biochem. Biophys. Res. Commun.* **49**:863.

Fujino, M., Fukuda, T., Shinagawa, S., Kobayashi, S., Yamazaki, I., Nakayama, R., Seely, J. H., White, W. F., and Rippel, R. H., 1974, Synthetic analogs of luteinizing hormone releasing hormone (LH-RH) substituted in position 6 and 10, *Biochem. Biophys. Res. Commun.* **60**:406.

Heber, D., and Odell, W. D., 1978, Pituitary receptor binding activity of active, inactive, superactive, and inhibitory analogs of gonadotropin-releasing hormone, *Biochem. Biophys. Res. Commun.* **82**:67.

Huggins, C., and Hodges, C. V., 1941, Studies on prostatic cancer. I. The effect of castration, of estrogen and androgen injection on serum phosphatases in metastatic carcinoma of the prostate, *Cancer Res.* **1**:293.

Johnson, D. M., Pritchard, R. A., Taylor, W. F., Conley, D., Zuniga, G., and McGreevy, K. G., 1986, Degradation of the LH-RH analog Nafarelin acetate in aqueous solution, *Int. J. Pharm.* **31**:125–129.

Johnson, E., Gendrich, R., and White, W., 1976, Delay of puberty and inhibition of reproductive processes in the rat by a gonadotropin-releasing hormone agonist analog, *Fertil. Steril.* **27**:853.

Jones, B. N., Pääbo, S., and Stein, S., 1981, Amino acid analysis and enzymatic sequence determination of peptides by an improved o-phthalaldehyde precolumn labeling procedure, *J. Liq. Chromatogr.* **4**:565.

Kanematsu, S., Scaramuzzi, R. J., Hilliard, J., and Sawyer, C. H., 1974, Patterns of ovulation-inducing LH release following coitus, electrical stimulation and exogenous LH-RH in the rabbit, *Endocrinology* **95**:247.

Koch, Y., Baram, T., Chobsieng, P., and Fridkin, M., 1974, Enzymic degradation of luteinizing hormone-releasing hormone (LH-RH) by hypothalamic tissue, *Biochem. Biophys. Res. Commun.* **61**:95.

Labrie, F., Auclair, C., Cusan, L., Kelly, P., Pelletier, G., and Ferland, L., 1978, Inhibitory effect of LH-RH and its agonists on testicular gonadotrophin receptors and spermatogenesis in the rat, *Int. J. Androl. Suppl.* **2**:303.

Labrie, F., Belanger, A., Cusan, L., Seguin, C., Pelletier, G., Kelly, P. A., Reeves, J. J., Lefebvre, F. A., LeMay, A., Gourdeau, Y., and Raynaud, J. P., 1980, Antifertility effects of LH-RH agonists in the male, *J. Androl.* **1**:209.

Marshall, J. C., and Odell, W. D., 1975, Preparation of biologically active [^{125}I] LH-RH suitable for membrane binding studies, *Proc. Soc. Exp. Biol. Med.* **149**:351.

Martin, J. E., Turey, L., Everett, J. W., and Fellows, R. E., 1974, Variations in responsiveness to synthetic LH releasing factor (LRF) in proestrous and diestrous-3 rats, *Endocrinology* **94**:556.

Matsuo, H., Baba, Y., Nair, R. M. G., Arimura, A., and Schally, A. V., 1971a, Structure of the porcine LH and FSH-releasing hormone. I. The proposed amino acid sequence, *Biochem. Biophys. Res. Commun.* **43:**1334.

Matsuo, H., Arimura, A., Nair, R. M. G., and Schally, A. V., 1971b, Synthesis of the porcine LH and FSH-releasing hormone by the solid phase method, *Biochem. Biophys. Res. Commun.* **45:**822.

Momany, F. A., 1976, Conformational energy analysis of the molecule, luteinizing hormone-releasing hormone. I. Native decapeptide, *J. Am. Chem. Soc.* **98:**2990.

Monohan, M. W., Amoss, M. S., Anderson, H. A., and Vale, W., 1973, Synthetic analogs of the hypothalamic luteinizing hormone releasing factor with increased agonist or antagonist properties, *Biochemistry* **12:**4616.

Nekola, M. V., Horvath, A., Ge, L. J., Coy, D. H., and Schally, A. V., 1982, Suppression of ovulation in the rat by an orally active antagonist of luteinizing hormone-releasing hormone, *Science* **218:**160.

Nestor, J. J., Ho, T. L., Simpson, R. A., Horner, B. L., Jones, G. H., McRae, G. I., and Vickery, B. H., 1982, Synthesis and biological activity of some very hydrophobic superagonist analogues of luteinizing hormone-releasing hormone, *J. Med. Chem.* **25:**795.

Nillius, S. J., Bergquist, C., Gudmunsson, J., and Wide, L., 1984, Superagonists of LH-RH for contraception in women, *J. Steroid Biochem.* **20:**1973.

Nishi, K., Ito, H., Shinagawa, S., Hatanaka, C., Fujino, M., and Hattori, M., 1980, Racemization of LH-RH analogs in alkaline solution, in: *Peptide Chemistry* (H. Yonehara, ed.), Protein Research Foundation, Osaka.

Nishi, N., Coy, E. J., Arimura, A., and Schally, A. V., 1976, Suppression of LH-RH-induced ovulation in hamsters and rats by synthetic analogues of LH-RH, *J. Reprod. Fertil.* **48:**119.

Pedroza, E., Vilchez-Martinez, J. A., Coy, D. H., Arimura, A., and Schally, A. V., 1978, Correlation between in vivo inhibition of gonadotropin release induced by luliberin and the blockade of ovulation by synthetic analogs of luliberin, *Int. J. Fertil.* **23:**294.

Ramirez, V. D., and McCann, S. M., 1963, A highly sensitive test for LH releasing activity: The ovariectomized, estrogen progesterone-blocked rat, *Endocrinology* **73:**193.

Redding, T. W., and Schally, A. V., 1973, Synthesis of luteinizing hormone releasing hormone containing tritium-labeled pyroglutamic acid, *Life Sci.* **12:**23.

Redding, T. W., Kastin, A. J., Gonzalez-Barcena, D., Coy, D. H., Coy, E. J., Schalch, D. S., and Schally, A. V., 1973, The half-life, metabolism and excretion of tritiated luteinizing hormone-releasing hormone (LH-RH) in man, *J. Clin. Endocrinol. Metab.* **37:**626.

Rippel, R. H., and Johnson, E. S., 1976, Inhibition of HCG-induced ovarian and uterine weight augmentation in the immature rat by analogs of GnRH, *Proc. Soc. Exp. Biol. Med.* **152:**432.

Sanders, L. M., Kent, J. S., McRae, G. I., Vickery, B. H., Tice, T. R., and Lewis, D. H., 1984, Controlled release of a luteinizing hormone-releasing hormone analogue from poly(d,l-lactide, glycolide) microspheres, *J. Pharm. Sci.* **73:**1294.

Sandow, J., Rechenberg, W. V., Konig, W., Hahn, M., Jerzabek, G., and Fraser, H., 1978, Physiological studies with highly active analogues of LH-RH, in: *Hypothalamic Hormones: Chemistry, Physiology, and Clinical Applications* (D. Gupta and W. Voelter, eds.), Verlag Chemie, p. 307.

Schally, A. V., Arimura, A., Baba, Y., Nair, R. M. G., Matsuo, H., Redding, T. W., Debeljuk, L., and White, W. F., 1971a, Isolation and properties of the FSH and LH-releasing hormone, *Biochem. Biophys. Res. Commun.* **43**:393.

Schally, A. V., Nair, R. M. G., Redding, T. W., and Arimura, A., 1971b, Isolation of the LH and FSH-releasing hormone from porcine hypothalami, *J. Biol. Chem.* **246**:7230.

Simmons, S. S., and Johnson, D. F., 1976, The structure of the fluorescent adduct formed in the reaction of o-phthalaldehyde and thiols with amines, *J. Am. Chem. Soc.* **98**:7098.

Spona, J., 1974, LH-RH–receptor interaction is inhibited by des-His-2 des-Gly-10-LH-RH-ethylamide, *FEBS Lett.* **48**:88.

Strickley, R. G., Brandl, M., Chan, K. W., Straub, K., and Gu, L., 1990, High-performance liquid chromatographic (HPLC) and HPLC-mass spectrometric (MS) analysis of the degradation of the luteinizing hormone-releasing hormone (LH-RH) antagonist RS-26306 in aqueous solution, *Pharm. Res.* **7**:530–536.

Vale, W., and Grant, G., 1974, In vitro pituitary hormone secretion assay for hypophysiotropic substances, *Methods Enzymol.* **37**:82.

Vale, W., Grant, G., Amoss, M., Blackwell, R., and Guilemin, R., 1972, Culture of enzymatically dispersed pituitary cells: Functional validation of a method, *Endocrinology* **91**:562.

Chapter 6

Stability of Adrenocorticotropic Hormone (ACTH) and Pathways of Deamidation of Asparaginyl Residue in Hexapeptide Segments

Kamlesh Patel

1. INTRODUCTION

1.1. Pharmacology

Adrenocorticotropic hormone (corticotropin or ACTH), secreted by the anterior pituitary, stimulates the adrenal cortex to produce and secrete cortisol, corticosterone, aldosterone, and a number of weakly androgenic substances by activating adenyl cyclase in the cell membranes (Haynes and Larner, 1975). The activity of ACTH is destroyed by the proteolytic enzymes of the gastrointestinal tract; therefore, the hormone is ineffective when given orally. It is readily absorbed from parenteral sites and is usually administered by intramuscular injection and occasionally by intravenous infusion (Haynes and Larner, 1975). ACTH rapidly disappears from the circulation following its intravenous administration in man; the plasma half-life is about 15 min. No significant biological activity is present in the urine (Haynes and Larner, 1975).

Kamlesh Patel • SmithKline Beecham Pharmaceuticals, King of Prussia, Pennsylvania 19406.

Stability and Characterization of Protein and Peptide Drugs: Case Histories, edited by Y. John Wang and Rodney Pearlman. Plenum Press, New York, 1993.

1.2. Clinical Use

At present, the most important use of ACTH is for diagnosis in adrenal insufficiency. It has limited therapeutic value in conditions responsive to corticosteroid therapy. ACTH may also be used in the treatment of pituitary insufficiencies, special forms of arthritis, allergies, multiple sclerosis, acute encephalitis and myelitis, tuberculous meningitis, childhood epilepsies, brain edema, acute polyradiculoneuritis, temporal arteritis, herpes zoster, idiopathic paresis of the nervous facialis, myasthenia gravis, and others (Haynes and Larner, 1975; Schwyzer, 1977).

1.3. Structure and Physicochemical Properties

ACTH is a single-chain polypeptide containing 39 amino acids. The first 24 amino acids are identical in all species. Ovine, porcine, and bovine ACTHs differ from human ACTH only at amino acid positions 25, 31, and 33 (Li, 1972). The primary structure of human ACTH is shown in Fig. 1.

ACTH is freely soluble in water, partly precipitated at the isoelectric point (pH 4.65–4.8), and appreciably soluble in 60 to 70% alcohol or acetone (Merck Index, 1989).

ACTH has not been crystallized and a crystal structure is lacking. ACTH is a flexible molecule in aqueous solutions, and is able to adopt preferred conformations in specialized surroundings (Schwyzer, 1977). According to a classic work by Squire and Bewley (1965), a small region close to the NH_2-terminus appears to have the tendency to form 1 to 1½ turns of a helical structure. Schiller (1972) was able to show that the adjacent region, residues 10 to 21, assumes a flexible conformation. Low *et al.* (1975) have confirmed these findings by showing that a CD spectrum indicative of α-helix conformation is produced in trifluoroethanol, a solvent favoring ordered structures.

1

Ser Tyr Ser Met Glu His Phe Arg Trp Gly Lys Pro Val Gly Lys Lys Arg Arg Pro Val

39

Lys Val Tyr Pro Asn Gly Ala Glu Asp Glu Ser Ala Glu Ala Phe Pro Leu Glu Phe

Figure 1. Amino acid sequence of human ACTH.

The NH_2-terminal part of the ACTH molecule contains the essential molecular information for corticotropic effects and the COOH-terminus is responsible for its stabilization in the body. The high *in vivo* corticotropic activity of segments containing the NH_2-terminal amino acid residues 1–19, 1–20, 1–23, and 1–24 have been demonstrated (Li *et al.*, 1960; Hofmann *et al.*, 1961, 1962; Schwyzer *et al.*, 1960). Reduced activity was observed for the sequences 1–18 and 1–17. However, at least one-half of the molar potency is lost by omitting the COOH-terminal stretches 25–39 and 24–39 (Schwyzer, 1964). Also, the duration of corticotropic action is diminished by such deletions. Potency and duration of action can be enhanced by making substitutions in the amino acid chain, e.g., D-Ser for L-Ser in position 1, norleucine for Met in position 4, and Val for Asn in position 25 (Jubiz *et al.*, 1968; Friedman, 1969).

1.4. Preparations and Routes of Administration

Corticotropin Injection, U.S.P. (ACTH), is available as a sterile solution or as a lyophilized powder for intramuscular or intravenous use. The preparation is derived from the pituitaries of mammals used for food.

Repository Corticotropin Injection, U.S.P. (H.P. ACTHAR GEL, CORTROPHIN GEL), is administered either intramuscularly or subcutaneously. It is a highly purified ACTH in gelatin solution.

Sterile Corticotropin Zinc Hydroxide Suspension, U.S.P. (CORTROPHIN-ZINC), is a preparation of purified corticotropin adsorbed on zinc hydroxide, intended for intramuscular injection.

Cosyntropin for injection (CORTROSYN) is a synthetic peptide corresponding to amino acid residues 1 to 24 of human ACTH. This preparation, approved for diagnostic purposes, is given intramuscularly or intravenously.

2. CHEMICAL STABILITY

2.1. Deamidation of ACTH

In recent years, deamidation of Asn residues has received considerable attention as one of the major chemical pathways of peptide/protein degradation (Manning *et al.*, 1989). In the deamidation reaction, the side chain amide linkage in a Gln or Asn residue is hydrolyzed to form a free carboxylic acid. Depending on the conditions, the deamidation of an Asn residue can

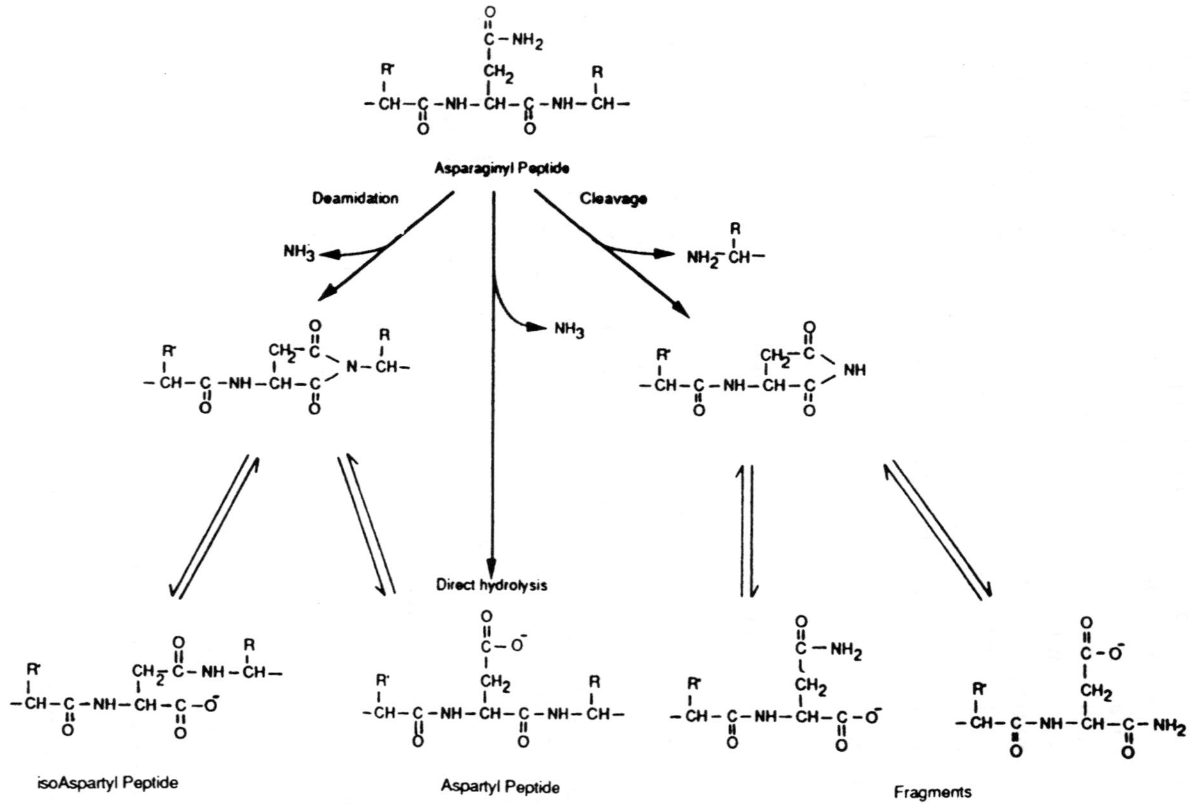

Figure 2. Proposed pathways for deamidation of Asn residues in peptides/proteins.

lead to the formation of the normal α-linked Asp residue or a mixture of the α-linked Asp residue and a β-linked Asp residue (isoAsp) (Fig. 2). ACTH contains a single Asn residue at position 25 which is known to deamidate readily under mildly alkaline conditions (Dixon and Stack-Dunne, 1955; Li *et al.*, 1955; Shepherd *et al.*, 1956; Pickering *et al.*, 1963; Graf *et al.*, 1971; Riniker *et al.*, 1972; Bhatt *et al.*, 1990). Graf *et al.* (1973) have shown that the deamidation of porcine ACTH results in a considerable decrease of the corticosteroidogenic potency. Recently, Bhatt *et al.* (1990) have investigated the deamidation of the Asn residue in a porcine ACTH. Separation of native ACTH from the deamidated ACTH (mixture of Asp- and isoAsp-peptides) was achieved by cation exchange HPLC (Fig. 3) and isoelectric focusing (IEF) (Fig. 4). However, these methods cannot separate Asp and isoAsp products. Nonetheless, IEF was used to determine the effects of pH and buffer concentration on the rate of deamidation of ACTH. The production of ammonia was also detected using a coupled enzymatic assay involving the reductive amination of 2-oxoglutarate to form glutamate and the conversion of NADH to NAD. At all pHs and buffer concentrations studied, the disappearance of ACTH, the appearance of deamidated ACTH, and the appearance of ammonia followed pseudo-first-order kinetics. The rate of disappearance of ACTH was observed to increase with increasing pH (Table I). The deamidation of ACTH was also found to depend on buffer (glycine) concen-

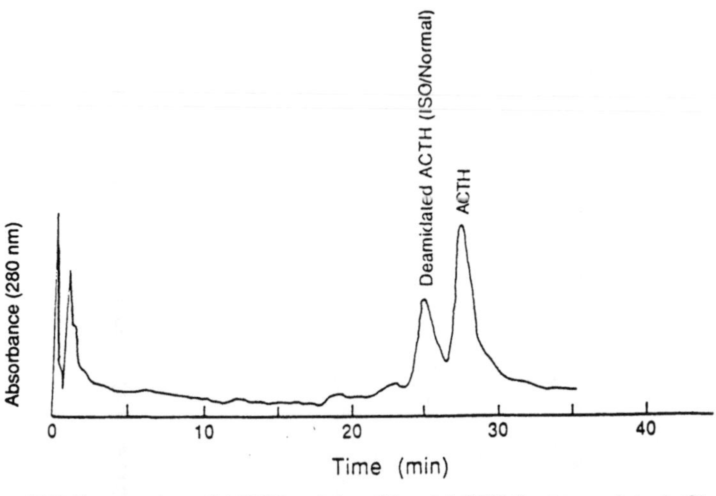

Figure 3. HPLC separation of ACTH and deamidated ACTH (isoAsp and Asp). Chromatographic conditions were as follows. Cation exchange HPLC column (Aquapore CX-300, 4.6 × 100 mm). Mobile phase: solvent A (10 mM ammonium acetate, pH 6.7), solvent B (500 mM ammonium acetate, pH 6.7). Linear gradient: 0–5 min, 100% solvent A; 5–35 min, 0 to 100% solvent B. Flow rate: 1.5 ml/min. UV detection: 280 nm.

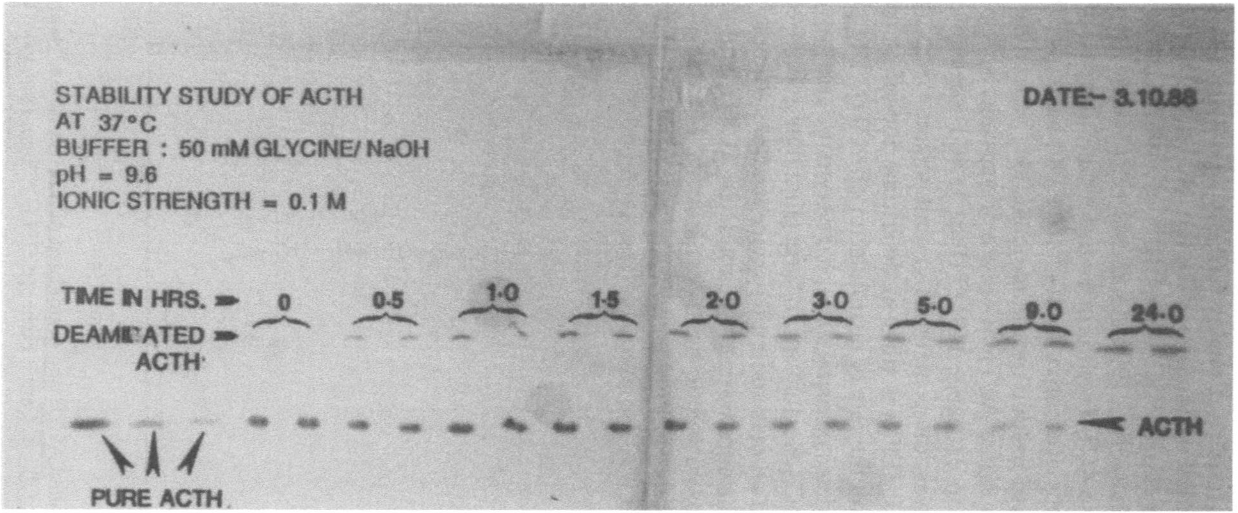

Figure 4. Representative photograph of an isoelectric focusing gel showing separation of ACTH from deamidated ACTH.

Table I
Pseudo-First-Order Rate Constants for the Disappearance of ACTH and ACTH[22–27] at 37°C and Various pH and Buffer Concentrations ($I = 0.1$)[a]

			k_{obs} (hr^{-1})	
pH	Buffer	buffer concn (M)	ACTH	ACTH[22–27]
1.9	HCl	0.01	3.8×10^{-4}	1.6×10^{-4}
5.0	Acetate	0.1	7.5×10^{-4}	N.D.[b]
7.0	Phosphate	0.005	1.9×10^{-3}	6.8×10^{-4}
9.6	Gycine	0.01	6.6×10^{-2}	N.D.
9.6	Glycine	0.05	1.4×10^{-1}	3.2×10^{-1}
9.6	Glycine	0.1	2.3×10^{-1}	N.D.

[a] Each value is the mean of two experimental values, the values being within 5% of each other.
[b] ND, not determined.

trations as shown in Table I. Since separation of the deamidated ACTH (isoAsp- and Asp-peptides) was not achieved using HPLC and IEF, formation of the isoAsp product at pH 7.0 and 9.6 was verified by the protein carboxymethyltransferase (PCM)-catalyzed methylation of the deamidated ACTH (Aswad, 1984; Johnson and Aswad, 1985; Bhatt *et al.*, 1990). No such methylation was observed when ACTH was incubated at pH 1.9 and 37°C. These data indicate that the deamidation of ACTH at alkaline and neutral pHs proceeds via formation of a cyclic imide, which hydrolyzes to give the Asp- and isoAsp-peptides. Under acid conditions the amide moiety of the Asn residue undergoes direct hydrolysis, without assistance by intramolecular catalysis (Bhatt *et al.*, 1990).

2.2. Deamidation of ACTH[22–27]

In order to understand the factors that influence deamidation of Asn in ACTH, several investigators have used a hexapeptide [Val-Tyr-Pro-Asn-Gly-Ala (Asn-Gly hexapeptide)], based on residues 22–27 of ACTH, as a model peptide (Geiger and Clarke, 1987; Patel and Borchardt, 1990a). In the case of hexapeptides, it was possible to separate the Asn-Gly hexapeptide, Asp-peptide, and isoAsp-peptide using HPLC (Fig. 5).

2.2.1. Effects of pH and Buffer Species

At a fixed pH, the overall loss of the Asn-Gly hexapeptide exhibited pseudo-first-order kinetics. In the neutral and alkaline pH range (pH 6–12),

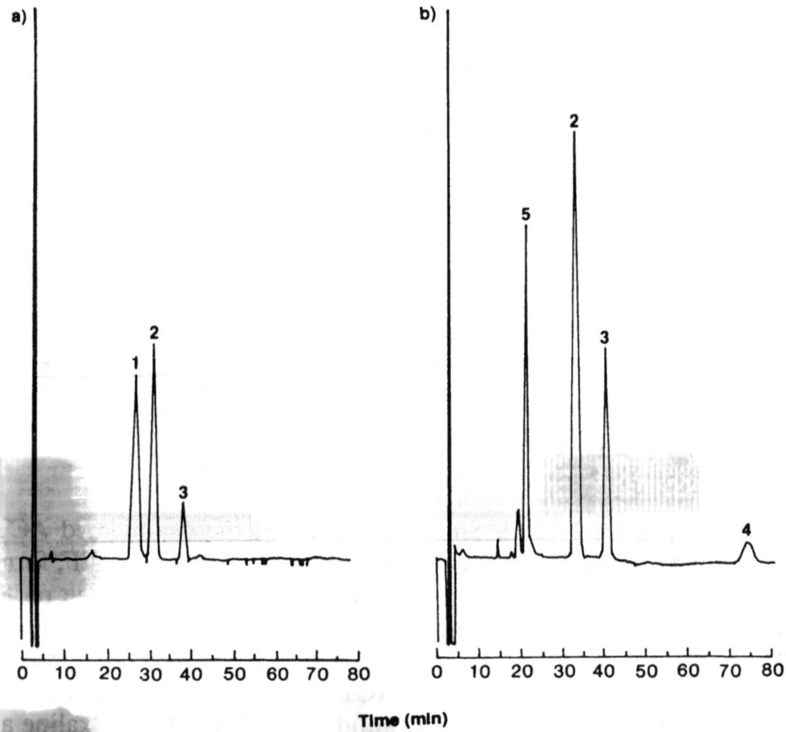

Figure 5. Typical HPLC chromatograms for the deamidation of $ACTH^{22-27}$ at (a) neutral to alkaline pH and (b) acid pH. Chromatographic conditions were as follows: Alltech Econosphere C_{18} column (4.6 × 250 mm); Mobile phase: 7% acetonitrile, 0.1% TFA in water; Flow rate: 1 ml/min; UV detection: 214 nm. Peaks: 1. Val-Tyr-Pro-IsoAsp-Gly-Ala; 2. Val-Tyr-Pro-Asn-Gly-Ala; 3. Val-Tyr-Pro-Asp-Gly-Ala; 4. Val-Tyr-Pro-Asu-Gly-Ala; 5. Val-Tyr-Pro-Asp.

both Asp- and isoAsp-peptides were formed upon deamidation (Patel and Borchardt, 1990a). The catalytic effect of buffer was observed in the pH range 7–11. However, no buffer catalysis was observed at pH 5–6.5. At pH 5, using acetate buffer, the reaction was found to be slow at 37°C; thus, the rate constants at 37°C were calculated from 90°C data and an activation energy of 22.2 kcal/mole. The pH dependency of the deamidation of the Asn-Gly hexapeptide in the neutral and basic range at 37°C is summarized in Table II.

As illustrated with this peptide, deamidation exhibits a strong dependency in pH, with the rate increasing with increasing pH in the pH range 5–12. In contrast, the distribution of the deamidated products (ratio of isoAsp- to Asp-peptides) remains essentially constant at approximately 3.3–4.2 over the pH range 6–12 as shown in Table III.

Table II
Observed Pseudo-First-Order Rate Constants for Deamidation of ACTH[22–27] in Aqueous Buffer Solutions of Various pH Values (37°C, I = 0.5)

pH (buffer)	Buffer concn (M)	k_{obs}[a] (hr^{-1}) ± S.D.
1.0 (HCl)	~0.1	$(2.33 \pm 0.06) \times 10^{-3}$
1.5 (HCl)	~0.032	$(0.71 \pm 0.04) \times 10^{-3}$
2.0 (HCl)	~0.01	0.23×10^{-3}[b,c]
5.0 (acetate)[d]	0.01	$(4.51 \pm 0.11) \times 10^{-2}$
	0.05	$(4.54 \pm 0.03) \times 10^{-2}$
	0.1	$(5.24 \pm 0.06) \times 10^{-2}$
6.0 (phosphate)	0.005	$(1.53 \pm 0.12) \times 10^{-3}$
	0.05	$(1.67 \pm 0.06) \times 10^{-3}$
	0.1	$(1.63 \pm 0.06) \times 10^{-3}$
6.5 (phosphate)	0.005	$(4.44 \pm 0.04) \times 10^{-3}$
	0.05	$(4.49 \pm 0.31) \times 10^{-3}$
	0.1	5.0×10^{-3}[c]
7.0 (phosphate)	0.005	$(0.63 \pm 0.04) \times 10^{-2}$
	0.05	$(1.13 \pm 0.06) \times 10^{-2}$
	0.1	$(1.47 \pm 0.05) \times 10^{-2}$
7.5 (phosphate)	0.005	$(1.51 \pm 0.08) \times 10^{-2}$
	0.05	$(2.81 \pm 0.02) \times 10^{-2}$
	0.1	$(3.48 \pm 0.13) \times 10^{-2}$
8.0 (Tris)	0.05	$(3.99 \pm 0.33) \times 10^{-2}$
	0.2	$(5.91 \pm 0.05) \times 10^{-2}$
	0.4	$(7.65 \pm 0.14) \times 10^{-2}$
9.0 (Tris)	0.05	0.116 ± 0.016
	0.2	0.201 ± 0.006
	0.4	0.268 ± 0.009
10.0 (bicarbonate)	0.005	0.294 ± 0.016
	0.05	0.617 ± 0.019
	0.1	0.871 ± 0.024
11.1 (phosphate)	0.01	1.09 ± 0.2
	0.05	1.54 ± 0.04
	0.08	1.85 ± 0.09
12.0 (NaOH)	~0.01	1.85 ± 0.09

[a] N = 3.
[b] Only 25% deamidation observed at pH 2.0, 37°C in 80 days.
[c] N = 1.
[d] Reactions carried out at 90°C.

Table III
Effect of pH and Buffer Concentration on the Ratio of isoAsp/Asp-peptides at 37°C, $I = 0.5$

pH	Buffer	Buffer concn (M)	Average isoAsp/Asp
12.0	NaOH	0.01	4.2
11.1	Phosphate	0.01	4.3
		0.05	4.3
		0.08	4.3
10.0	Bicarbonate	0.005	4.0
		0.05	4.1
		0.1	4.6
9.0	Tris	0.05	4.1
		0.2	4.0
		0.4	4.0
8	Tris	0.05	4.1
		0.2	4.0
		0.4	4.2
7.5	Phosphate	0.005	3.8
		0.05	3.8
		0.1	3.8
7.0	Phosphate	0.005	3.8
		0.05	4.1
		0.1	3.6
6.5	Phosphate	0.005	3.8
		0.05	3.6
		0.1	3.6
6.0	Phosphate	0.005	3.7
		0.05	3.5
		0.1	3.2

The appearance of the isoAsp-peptide suggests that the mechanism of deamidation at neutral and basic pHs involves the formation of the cyclic imide intermediate (Fig. 2). The fact that the ratio of isoAsp-peptide to Asp-peptide is essentially constant over the pH range 6–12 suggests that the major pathway of deamidation involves the formation of the cyclic imide, not direct hydrolysis of the Asn residue. It was also assumed that the rate-limiting step (slow step) in this pathway was the formation of the cyclic imide and the fast step was its hydrolysis to the Asp-peptide and isoAsp-peptide. To confirm this conclusion, hydrolysis of the cyclic imide Val-Tyr-Pro-Asu*-Gly-Ala was studied at various pHs and the results are shown in Table IV

* Asu is used as an abbreviation of the cyclic imide form of Asn.

Table IV
Rates of Hydrolysis of the Cyclic Imide Hexapeptide and the Distribution of the Deamidated Products at 37°C[a]

pH	Buffer ($I = 0.5$)	k_{obs} (hr^{-1})	IsoAsp/Asp
10.0	0.05 M borate	3.85×10^{1}	3.9
		$(4.15 \times 10^{-1})^{b}$	$(4.1)^{b}$
7.5	0.1 M phosphate	3.94×10^{-1}	3.6
		$(3.48 \times 10^{-2})^{b}$	$(3.8)^{b}$
6.0	0.05 M phosphate	2.65×10^{-2}	3.2
		$(1.67 \times 10^{-3})^{b}$	$(3.5)^{b}$
5.0	0.025 M acetate	3.9×10^{-3}	3.1
		$(2.6 \times 10^{-4})^{b}$	
1.0	0.1 N HCl	NR[c]	—
		$(2.33 \times 10^{-3})^{b}$	(100% Asp)[b]

[a] $N = 2$.
[b] Data obtained for hydrolysis of Asn-hexapeptide under identical experimental conditions.
[c] No loss up to 38 days.

(Patel and Borchardt, 1990a). At all pHs studied, the rate of hydrolysis of the cyclic imide was substantially more rapid than the rate of hydrolysis of the Asn-Gly hexapeptide. In addition, the ratios of isoAsp-peptide to Asp-peptide obtained from hydrolysis of the cyclic imide were essentially identical to those observed upon hydrolysis of the Asn-Gly hexapeptide. Since the ratio of Asp- and isoAsp-peptides in the deamidation reaction at neutral to alkaline pH coincides with the values obtained in the hydrolysis of the cyclic imide, it is proposed that in the pH range 6–12, the deamidation reaction occurs exclusively through the cyclic imide rather than direct hydrolysis of the amide. These results are consistent with those reported by Geiger and Clarke (1987) at a single pH (pH 7.4).

At acidic pHs (pH 1–2), deamidation of the Asn-Gly hexapeptide was observed to be much slower than deamidation at neutral or alkaline pHs and only the Asp-peptide was observed (Fig. 5b). No isoAsp-peptide was detected upon hydrolysis of the Asn-Gly hexapeptide at acid pHs. Interestingly, the cyclic imide was also detected, but the maximum concentration was only 10% of the total. At least one major new degradation product was observed by HPLC (Fig. 5b). This product was identified as a tetrapeptide by coinjecting the authentic sample. Since peptides with Asp residues are prone to undergo peptide backbone cleavage in acidic pH at Asp-X linkage (Manning *et al.*, 1989), this tetrapeptide could have formed from the degradation of Asp-peptide. The reaction at pH 3 and 4 was found to be negligible, e.g., degradation of the Asn-Gly hexapeptide was not detected at 37°C up to 60 days. The observation that only the Asp-peptide is generated from the Asn-Gly hexapeptide under acidic conditions is consistent with the hypothesis that at acidic pH the deamidation of the Asn-Gly hexapeptide involves the direct hydrolysis of the Asn residue and not the intermediate formation of the cyclic imide. Evidence in support of this hypothesis is the observation that no loss of the cyclic imide was found at pH 1.0 up to 38 days, whereas the Asn-Gly hexapeptide under identical conditions has a $t_{1/2}$ = 301.3 hr (Tables II, IV).

It is interesting to note that the rates for deamidation of ACTH and $ACTH^{22-27}$ (Asn-Gly hexapeptide) at a specific pH and buffer concentration are quite similar (Table I). For both peptides only the Asp-peptide was produced at acidic pH while both the Asp- and isoAsp-peptides were produced at neutral and alkaline pHs.

2.2.2. pH Rate Profile

The pH dependence of the overall rate of deamidation of the Asn-Gly hexapeptide at 37°C is shown in Fig. 6. The rate constants used in the construction of the graph were obtained from the intercepts of the graphs of k_{obs} versus total buffer concentrations at various pH values. Results from runs

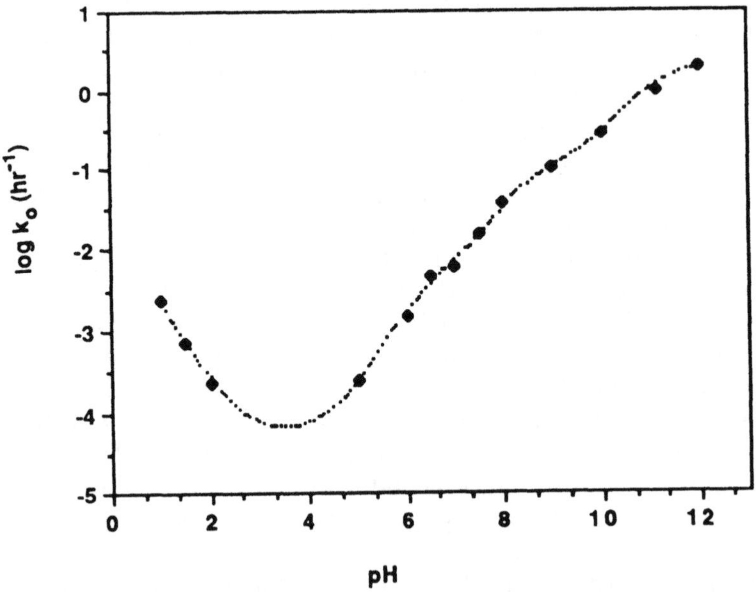

Figure 6. pH–rate profile for the deamidation of $ACTH^{22-27}$ in aqueous solution at 37°C.

performed in dilute HCl (pH 1–2) and sodium hydroxide (pH 12) are also included. At 37°C, deamidation shows strong dependence in the pH range 5–12. The pH of maximum stability is found to be about pH 3 to 4, since the rate of deamidation again increases at pH 1. The pH–rate profile for the deamidation of Asn in the Asn-Gly hexapeptide has a unit negative slope on the acid side (pH 1–2), indicating specific hydronium catalysis. From pH 5 to 12, the rate constant increases with hydroxide ion concentration, but is not first order in hydroxide ion (approximate slope = 0.54). This may be due to the complexity of the mechanism involving the cyclic imide. Several kinetic models could be proposed to describe this pH–rate profile. One particular kinetic model, which adequately describes the rate profile and is kinetically possible, has been presented. Since in the pH range studied the Asn-Gly hexapeptide has three ionizable groups (α-carboxyl group of COOH-terminus, α-amino group of NH_2-terminus, and Tyr-OH), there are numerous ionic species, all of which may or may not have different propensities to deamidation (Fig. 7). The shape of the pH–rate profile suggests that the overall velocity of degradation of the Asn-Gly hexapeptide at a given pH can be expressed as

$$k_0 = k_H[H_3O^+][AH_4/A_T] + k_1[AH_4/A_T] + k_2[AH_3/A_T] + k_3[AH_2/A_T] + k_4][AH/A_T] \qquad (1)$$

where $[A_T] = [AH_4] + [AH_3] + [AH_2] + [AH]$, k_0 is the pH-dependent first-order rate constant, k_H is the second-order specific acid catalysis constant, and k_1, k_2, k_3, and k_4 are first-order constants for the reaction of AH_4, AH_3, AH_2, and AH species, respectively. Equation (1) can be rewritten as follows:

$$k_0 = \frac{k_H[H_3O^+]^4 + k_1[H_3O^+]^3 + k_2[H_3O^+]^2K_1 + k_3[H_3O^+]K_1K_2 + k_4K_1K_2K_3}{[H_3O^+]^3 + [H_3O^+]^2K_1 + [H_3O^+]K_1K_2 + K_1K_2K_3} \qquad (2)$$

where K_1, K_2, and K_3 are the ionization constants for the Asn-Gly hexapeptide. The experimental data were fitted to Eq. (2) using the RS1 nonlinear

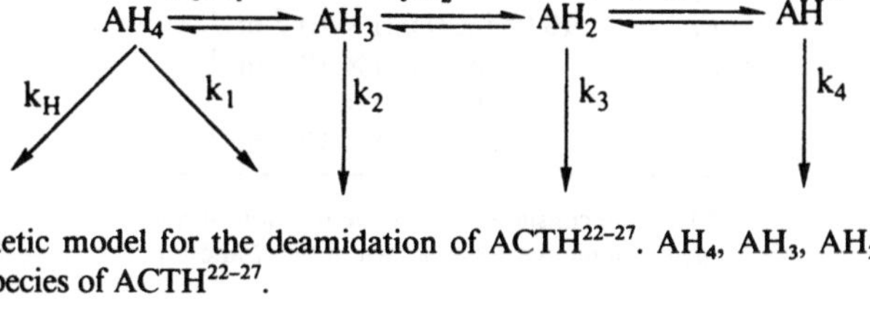

Figure 7. Kinetic model for the deamidation of ACTH[22-27]. AH_4, AH_3, AH_2, and AH are various ionic species of ACTH[22-27].

curve-fitting program. The data were weighted using reciprocal variance as the weight. The best fit was obtained by using $K_1 = 3.16 \times 10^{-7}$ (pK_1 = 6.5), $K_2 = 3.16 \times 10^{-9}$ (pK_2 = 8.5), and $K_3 = 1 \times 10^{-11}$ (pK_3 = 11.0). The individual values of the rate constants are given in Table V. In Fig. 6 the full line represents the theoretical curve calculated by substituting the rate constants from Table V into Eq. (2), while the points show the experimental results. Since no data are available between pH 2 and 5, it is assumed that Eq. (2) also holds in this pH range. The good agreement observed shows that Eq. (1) can adequately describe the kinetics of the total degradation. Even though individual ionization constants were not determined using standard techniques, it is not unreasonable to assign pK_2 = 8.5 and pK_3 = 11 to NH_2-terminus (α-NH_2) and Tyr-OH, respectively (Tanford, 1962). While pK_1 = 6.5 seems reasonably high for COOH-terminus (α-COOH), it is not unusual for many amine bases and carboxylic acids in proteins to have anomalously high or low pK_a because of their microenvironment (Fersht, 1985). It is speculated that either the pK_a for α-COOH is unusually high due to some structural effects such as hydrogen bonding or this is a kinetic pK_a representing change in the rate-determining step.

2.2.3. Effect of Temperature

Deamidation of the Asn-Gly hexapeptide was studied at pH 7.5 (0.1 M phosphate buffer, $I = 0.5$) from 25 to 70°C. Similar experiments were carried out at pH 5.0 (0.05 M acetate buffer, $I = 0.5$) from 50 to 90°C. For pH 7.5, at all temperatures, Asp- and isoAsp-peptides were the only products formed upon deamidation, with no indication of side reactions. However, at pH 5.0, the tetrapeptide Val-Tyr-Pro-Asp was also observed after one half-life, probably because of further degradation of deamidated products. For both pHs, the rates of deamidation were found to increase with the temperature, obey-

Table V
Calculated Rate Constants for Deamidation of ACTH[22–27a]

k_H	$(2.2 \pm 0.1) \times 10^{-2}$ M^{-1} hr^{-1}
k_1	$(0.054 \pm 0.02) \times 10^{-3}$ hr^{-1}
k_2	$(6.6 \pm 0.5) \times 10^{-3}$ hr^{-1}
k_3	$(11.6 \pm 0.9) \times 10^{-2}$ hr^{-1}
k_4	$(20.1 \pm 1.1) \times 10^{-1}$ hr^{-1}

[a] The rate constants are for the reaction of numerous ionic species as shown in Fig. 7 and Eq. (1).

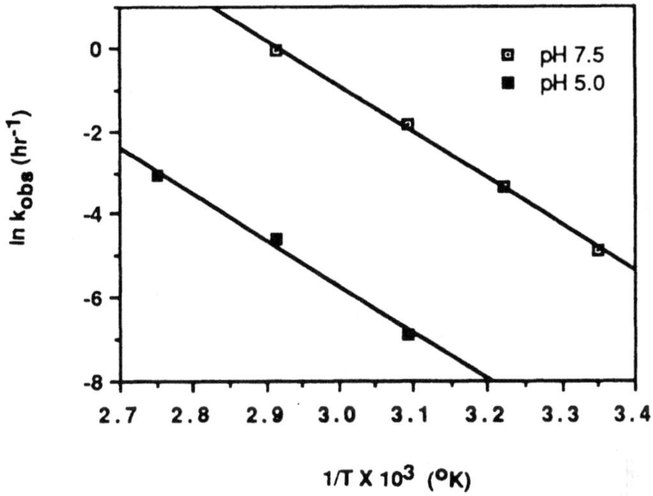

Figure 8. Arrhenius plot for the deamidation of $ACTH^{22-27}$ at pH 7.5, 0.1 M phosphate (I = 0.5), and pH 5.0, 0.05 M acetate buffer (I = 0.5).

ing the Arrhenius equation (Fig. 8). From the Arrhenius plot, it was possible to calculate observed activation energies of 21.7 and 22.2 kcal/mole for pH 7.5 and 5.0, respectively. The large value of the activation energy indicates that the rate of deamidation is temperature sensitive and that peptides exposed to high temperatures will be prone to this type of degradation.

2.2.4. Effect of Primary Structure

The effects of primary sequence on the pathways of deamidation of Asn in $ACTH^{22-27}$ were studied using Val-Tyr-X-Asn-Y-Ala hexapeptides with substitution on the COOH-terminal side (Y) and on the NH_2-terminal side (X) of the Asn residue (Patel and Borchardt, 1990b). The peptides used were: (1) Val-Tyr-Pro-Asn-Ser-Ala (Asn-Ser hexapeptide), (2) Val-Tyr-Pro-Asn-Val-Ala (Asn-Val hexapeptide), and (3) Val-Tyr-Gly-Asn-Gly-Ala (Gly-Asn hexapeptide). The deamidation reactions were carried out at pH 1.0 (0.1 M HCl), pH 7.5 (0.1 M phosphate), and pH 10.0 (0.05 M borate), I = 0.5, at 37°C. At a given pH, the deamidation followed pseudo-first-order kinetics for all peptides. Table VI lists the observed rate constants for deamidation of the Asn residue in these hexapeptides. The reaction for the Asn-Val hexapeptide was very slow at pH 7.5 and 37°C. Therefore, experiments were carried out at higher temperatures (50–90°C) and the rate constant at 37°C was calculated using an Arrhenius equation and an activation energy of 20.7

Table VI
Pseudo-First-Order Rate Constants for the Deamidation Reaction of Hexapeptides at 37°C and $I = 0.5$

Peptide	k_{obs} (hr^{-1}) × 10^2[a] pH 1.0 (0.1 M HCl)	pH 7.5 (0.1 M phosphate)	pH 10.0 (0.05 M borate)
Val-Tyr-Pro-Asn-Gly-Ala	0.233 ± 0.006	3.47 ± 0.13	41.5 ± 4.3
Val-Tyr-Pro-Asn-Ser-Ala	0.196 ± 0.006	0.52 ± 0.01	2.31 ± 0.04
Val-Tyr-Pro-Asn-Val-Ala	0.199 ± 0.003	0.027 ± 0.002[b]	0.23 ± 0.06
Val-Tyr-Gly-Asn-Gly-Ala	0.245 ± 0.007	5.59 ± 0.11	48.7 ± 5.6

[a] $N = 3$.
[b] Calculated using an Arrhenius equation and an activation energy of 20.7 kcal/mole.

kcal/mole. Like the parent peptide (Asn-Gly hexapeptide), both Asp- and isoAsp-peptides were detected at pH 7.5 and 10.0 for all peptides studied. Interestingly, a tetrapeptide with a composition of Val, Tyr, Pro, Asx (after amino acid analysis) was also detected for Asn-Ser and Asn-Val hexapeptides. This peptide may represent the product of the attack of the side chain amide nitrogen on the peptide bond carbonyl to release a dipeptide (Ser-Ala or Val-Ala, respectively) and form a COOH-terminal cyclic imide (Fig. 2). This cyclic imide can then be further hydrolyzed to form a tetrapeptide with a COOH-terminal Asn or Asp amide residue (Fig. 2). Generally, the preferential hydrolysis of this cyclic imide occurs to form a product with a COOH-terminal Asn (Voorter *et al.*, 1988). Similar cleavage products were also observed with Val-Tyr-Pro-Asn-Leu-Ala and Val-Tyr-Pro-Asn-Pro-Ala hexapeptides by Geiger and Clarke at pH 7.4 and 37°C (Geiger and Clarke, 1987).

Table VII
The Ratio of isoAsp- and Asp-peptides at 37°C and $I = 0.5$

Peptide	pH 1.0 (0.1 M HCl)	pH 7.5 (0.1 M phosphate)	pH 10.0 (0.05 M borate)
Val-Tyr-Pro-Asn-Gly-Ala	100% Asp	3.8	4.1
Val-Tyr-Pro-Asn-Ser-Ala	100% Asp	4.3	4.2
Val-Tyr-Pro-Asn-Val-Ala	100% Asp	2.0–2.6[a]	2.1
Val-Tyr-Gly-Asn-Gly-Ala	100% Asp	4.0	4.2

[a] Reactions were carried out at 50 to 90°C and the ratio varied between 2.0 and 2.6 in the temperature range studied.

For all of the peptides, at both pH 7.5 and 10.0, the ratio of isoAsp- and Asp-peptides (Table VII) remained constant during the entire reaction. The ratio of isoAsp- and Asp-peptides for Asn-Ser and Gly-Asn hexapeptides is similar to that for the parent peptide (Asn-Gly hexapeptide) at pH 7.5 and 10.0, while for the Asn-Val hexapeptide the ratio is smaller than for the Asn-Gly hexapeptide.

Similar to the results obtained with the Asn-Gly hexapeptide, at pH 1.0 all three peptides studied here produce only Asp-peptides upon deamidation. Formation of isoAsp-peptides was not observed. However, the Asp-peptides were found to degrade further at pH 1 to form tetra- and dipeptides due to cleavage of the Asp-X peptide bonds. Interestingly, hydrolysis of the Asp-peptide also formed the tripeptides Val-Tyr-Gly and Asp-Gly-Ala following the deamidation of the Gly-Asn hexapeptide. These fragments probably formed due to the peptide bond cleavage on the NH_2-terminal side of the Asp residue (Gly-Asp) as described by Inglis (1983). Such cleavage on the NH_2-terminal side of the Asp residue was not observed for any of the other peptides.

Geiger and Clarke (1987) and Stephenson and Clarke (1989) have reported deamidation rates for Val-Tyr-Pro-Asn-Ala-Ala (Asn-Ala hexapeptide), Val-Tyr-Pro-Asn-Leu-Ala (Asn-Leu hexapeptide), and Val-Tyr-Pro-Asn-Pro-Ala (Asn-Pro hexapeptide) at pH 7.4, 37°C (Table VIII). Interestingly, the difference in the rate of degradation between the Asn-Val, Asn-Pro, and Asn-Leu hexapeptides is very small. In fact, the Asn-Val hexapeptide is slightly more stable than the Asn-Leu hexapeptide even though

Table VIII
Effect of Amino Acid Sequence on Rates of Deamidation of Hexapeptides

Peptide	$t_{1/2}$ (days)
Val-Tyr-Pro-Asn-Gly-Ala	1.89[a]
Val-Tyr-Pro-Asn-Ser-Ala	5.55[a]
Val-Tyr-Pro-Asn-Ala-Ala	20.2[b]
Val-Tyr-Pro-Asn-Val-Ala	106[a]
Val-Tyr-Pro-Asn-Leu-Ala	70[b]
Val-Tyr-Pro-Asn-Pro-Ala	106[b]

[a] pH 7.5 (0.1 M phosphate buffer, $I = 0.5$), 37°C (Patel and Borchardt, 1990b).
[b] pH 7.4 (0.1 M phosphate buffer), 37°C (Geiger and Clarke, 1987; Stephenson and Clarke, 1989).

Leu is bulkier than Val (R = isobutyl versus isopropyl). It is not clear why the Asn-Val hexapeptide is more stable than the Asn-Leu hexapeptide, but similar results were also obtained in a study with an Asp-β ester peptide in alkaline media by Bodanszky and Kwei (1978).

These data suggest that the primary sequence does not influence significantly the rates of hydrolysis of the Asn residues under acidic conditions. Based on the mechanism of hydrolysis under acidic conditions, i.e., direct hydrolysis of the Asn residue, this observation was not unexpected. However, the primary sequence does have a dramatic effect on the rate of deamidation of the Asn residues under neutral and alkaline conditions. Particularly, modifications of the amino acid residue on the COOH-terminal side of the Asn influences the rate of deamidation. Again these results are not unexpected considering the mechanism, which involves the formation of the cyclic imide. The rate of cyclic imide formation would be highly dependent on the size (steric factors) of the adjacent amino acid residue. Modifications on the NH_2-terminal side of the Asn residue does not appear to have as dramatic an effect on the rate of deamidation as substitution on the COOH-terminal end. Since this amino acid is not directly involved in the formation of cyclic imide, these results are not surprising.

3. CONCLUSIONS

Depending on the pH, different deamidation products are generated from ACTH. Under neutral to alkaline conditions, a mixture of Asp- and isoAsp-containing peptides is produced, whereas under acidic conditions, only Asp-containing peptide is produced. Maximum stability of ACTH to deamidation can be achieved by avoiding strongly acidic or basic conditions, high buffer concentrations, and high temperatures. Using model peptides, it has been shown that primary structure, particularly the amino acid on the COOH-terminal side of the Asn residue, can influence the rates of deamidation of Asn at neutral and alkaline pHs but with no apparent effect at acidic pHs.

REFERENCES

Aswad, D. W., 1984, Stoichiometric methylation of porcine adrenocorticotropin by porcine carboxyl methyltransferase requires deamidation of asparagine 25, *J. Biol. Chem.* **259:**10714–10721.

Bhatt, N. P., Patel, K., and Borchardt, R. T., 1990, Chemical pathways of peptide degradation. I. Deamidation of adrenocorticotropic hormone, *Pharm. Res.* **7**:593–599.

Bodanszky, M., and Kwei, J. Z., 1978, Side reactions in peptide synthesis. VII. Sequence dependence in the formation of aminosuccinyl derivatives from beta-benzyl-aspartyl peptides, *Int. J. Pep. Protein Res.* **12**:69–74.

Dixon, H. B. F., and Stack-Dunne, M. P., 1955, Chromatographic studies on corticotropin, *Biochem. J.* **61**:483–487.

Fersht, A., 1985, The pH dependence of enzyme catalysis, in: *Enzyme Structure and Mechanism,* 2nd ed. (A. Fersht, ed.), Freeman, San Francisco, pp. 155–175.

Friedman, M., 1969, The duration of action of the synthetic pentapeptide D-serine^1-norleucine^4-valinamide^{25}-beta-1–25-corticotropin (DW-75) in man, *Experientia* **25**:416–419.

Geiger, T., and Clarke, S., 1987, Deamidation, isomerization, and racemization at asparaginyl and aspartyl residues in peptides, *J. Biol. Chem.* **262**:785–794.

Graf, L., Bajusz, S., Patthy, A., Barat, E., and Cseh, G., 1971, Revised amide location for porcine and human adrenocorticotropic hormone, *Acta Biochim. Biophys. Acad. Sci. Hung.* **6**:415–418.

Graf, L., Hajos, G., and Cseh, G., 1973, The influence of deamidation on the biological activity of porcine adrenocorticotropic hormone, *Horm. Metab. Res.* **5**:142–143.

Haynes, R. C., and Larner, J., 1975, Adrenocorticotropic hormone; adrenocortical steroids and their synthetic analogs; inhibitors of adrenocortical steroid biosynthesis, in: *The Pharmacological Basis of Therapeutics* (L. S. Goodman and A. Gilman, eds.), Macmillan Co., New York, pp. 1451–1471.

Hofmann, K., Yajima, H., Yanaihara, N., Liu, T. Y., and Lande, S., 1961, Studies on polypeptides. XIII. The synthesis of a tricosapeptide possessing essentially the full biological activity of natural ACTH, *J. Am. Chem. Soc.* **83**:487–489.

Hofmann, K., Liu, T. Y., Yajima, H., Yanaihara, N., Yanaihara, C., and Humes, J. L., 1962, Studies on polypeptides. XXII. High adrenocorticotropic activity in the rat and in man of a synthetic eicosapeptide amide, *J. Am. Chem. Soc.* **84**:1054–1056.

Inglis, A. S., 1983, Cleavage at aspartic acid, *Methods Enzymol.* **91**:324–332.

Johnson, B. A., and Aswad, D. A., 1985, Enzymatic protein carboxyl methylation at physiological pH: Cyclic imide formation explains rapid methyl turnover, *Biochemistry* **24**:2581–2586.

Jubiz, W., West, C. D., and Tayler, F. H., 1968, Lack of prolonged activity of a new ACTH preparation (DW-75) in human subjects, *J. Clin. Endocrinol. Metab.* **28**:1377–1379.

Li, C. H., 1972, Adrenocorticotropin 45. Revised amino acid sequences for sheep and bovine hormones, *Biochem. Biophys. Res. Commun.* **49**:835–839.

Li, C. H., Geschwind, I. I., Dixon, J. S., Levy, A. L., and Harris, J. I., 1955, Corticotropins (ACTH) I. Isolation of alpha-corticotropin from sheep pituitary glands, *J. Biol. Chem.* **213**:171–185.

Li, C. H., Meienhofer, J., Schnabel, E., Chung, D., Lo, T. B., and Ramachandran, J., 1960, The synthesis of a nonadecapeptide possessing adrenocorticotropic and melanotropic activities, *J. Am. Chem. Soc.* **82**:5760–5762.

Low, M., Kisfaludy, L., and Fermandjian, S., 1975, Proposed preferred conformation of ACTH, *Acta Biochim. Biophys. Acad. Sci. Hung.* **10**:229–231.

Manning, M. C., Patel, K., and Borchardt, R. T., 1989, Stability of protein pharmaceuticals, *Pharm. Res.* **6**:903–918.

Merck Index, 1989, *An Encyclopedia of Chemicals, Drugs, and Biologicals* (S. Budavri, ed.), 11th ed., Merck & Co., Rahway, N.J., pp. 21–22.

Patel, K., and Borchardt, R. T., 1990a, Chemical pathways of peptide degradation. II. Kinetics of deamidation of an asparaginyl residue in a model hexapeptide, *Pharm. Res.* **7**:703–711.

Patel, K., and Borchardt, R. T., 1990b, Chemical pathways of peptide degradation. III. Effect of primary sequence on the pathways of deamidation of asparaginyl residues in hexapeptides, *Pharm. Res.* **7**:787–793.

Pickering, B., Anderson, T. R. N., Lohmar, P., Birk, Y., and Li, C. H., 1963, Adrenocorticotropin. XXVII. On the presence of pig type adrenocorticotropin in sheep pituitaries and a simple method for the isolation of alpha-adrenocorticotropin, *Biochim. Biophys. Acta* **74**:763–773.

Riniker, B., Seiber, P., Rittel, W., and Zuber, H., 1972, Revised amino-acid sequences for porcine and human adrenocorticotropin hormone, *Nature New Biol.* **235**:114–115.

Schiller, P., 1972, Study of adrenocorticotropic hormone conformation by evaluation of intramolecular resonance energy transfer in N-dansyllysine^{21}-ACTH-(1-24)-tetrakosipeptide, *Proc. Natl. Acad. Sci. USA* **69**:975–979.

Schwyzer, R., 1964, Chemistry and metabolic action of nonsteroid hormones, *Annu. Rev. Biochem.* **33**:259–286.

Schwyzer, R., 1977, ACTH: An introductory review, *Ann. N.Y. Acad. Sci.* **297**:3–26.

Schwyzer, R., Rittel, W., Kappeler, H., and Iselin, B., 1960, Synthese eines nonadekapeptides mit hoher corticotroper wirksamkeit, *Angew. Chem.* **72**:915–917.

Shepherd, R. G., Howard, K. S., Bell, P. H., Cacciola, A. R., Child, R. G., Davies, M. C., English, J. P., Finn, B. M., Meisenhelder, J. H., Moyer, A. W., and Van der Scheer, J., 1956, Studies with corticotropin. I. Isolation, purification and properties of beta-corticotropin, *J. Am. Chem. Soc.* **78**:5051–5059.

Squire, P. G., and Bewley, T., 1965, Adrenocorticotropins. XXXV. The optical rotatory dispersion of sheep adrenocorticotropic hormone in acidic and basic solutions, *Biochim. Biophys. Acta* **109**:234–240.

Stephenson, R. C., and Clarke, S., 1989, Succinimide formation from aspartyl and asparaginyl peptides as a model for the spontaneous degradation of proteins, *J. Biol. Chem.* **264**:6164–6170.

Tanford, E., 1962, The interpretation of hydrogen ion titration curves of proteins, *Adv. Prot. Chem.* **17**:69–165.

Voorter, C. E. M., de Haard-Hoekman, W. A., Van den Oetelaar, P. J. M., Bloemendal, H., and de Jong, W. W., 1988, Spontaneous peptide bond cleavage in aging alpha-crystalline through a succinimide intermediate, *J. Biol. Chem.* **263**: 19020–19023.

Chapter 7

Stability and Characterization of Human Interleukin-1β

Leo Gu and Jodi Fausnaugh

1. INTRODUCTION

Human interleukin-1 (hIL-1) is a peptide hormone produced by activated macrophages and other cell types which is involved in the regulation of immune and inflammatory responses. Many activities have been attributed to hIL-1, which include stimulation of murine thymocyte proliferation *in vitro,* production of other cytokines (e.g., IL-2, IL-6, interferons, colony-stimulating factors), activation of lymphocytes, endogenous pyrogenicity, fibroblast proliferation, stimulation of prostaglandin E_2 production, and induction of acute-phase protein synthesis by hepatocytes (Kampschmidt, 1984; Dinarello, 1988). Some of the systemic effects of hIL-1 are summarized in Table I.

hIL-1 has been purified from stimulated macrophages (Blyden and Handschumacher, 1977; Togawa *et al.,* 1979; Mizel and Mizel, 1981; Schmidt *et al.,* 1982; Schmidt, 1984; Kronheim *et al.,* 1985) and also from leukemic monocytes (Lachman, 1983). The activity was associated with proteins of 15 and 30 kDa and isoelectric points (pI) of approximately 5 and 7.

Leo Gu • Institute of Pharmaceutical Sciences, Syntex Research, Palo Alto, California 94304. *Jodi Fausnaugh* • Institute of Analytical Research, Syntex Research, Palo Alto, California 94304.

Stability and Characterization of Protein and Peptide Drugs: Case Histories, edited by Y. John Wang and Rodney Pearlman. Plenum Press, New York, 1993.

Table I
Systemic Effects of hIL-1[a]

Central nervous system	Metabolic	Hematologic	Vascular wall
Fever	↑ hepatic proteins	Neutrophilia	↑ leukocyte adherence
Sleep	↑ Na excretion	Lymphopenia	↑ PGI_2/PGE_2
↑ ACTH	↓ plasma Zn/Fe	↑ tumor killing	↑ platelet activating factor
↑ neuropeptides	↑↓ insulin	↑ bone marrow growth factors	Hypotension
↓ appetite	↓ cytochrome P450	↑ nonspecific resistance	Capillary leak syndrome
	Lactic acidosis		↓ vessel resistance

[a] Reproduced from Dinarello (1988).

Two forms of human IL-1, α and β, have been identified and cloned by several groups (Auron *et al.*, 1984; Furutani *et al.*, 1985; March *et al.*, 1985; Gubler *et al.*, 1986; Wingfield *et al.*, 1986; Kikumoto *et al.*, 1987; Meyers *et al.*, 1987). Both molecules are synthesized as larger precursor polypeptides that are processed to smaller forms. hIL-1α has an acid pI, approximately 5.0, and hIL-1β has a more basic pI, approximately 7.0. There is only 26% homology between hIL-1α and hIL-1β, but both molecules have the same receptors and possess the same spectrum of biological activities (Dinarello, 1988).

1.1. Preparation of Recombinant hIL-1β

cDNA encoding a precursor for IL-1β was first cloned by Auron *et al.* (1984) by reverse transcription of poly(A)-RNA isolated from adherent endotoxin-stimulated human monocytes. The cDNA possessed an open reading frame encoding a polypeptide of 269 amino acids (Fig. 1). mRNA isolated by hybridization to the cDNA clone directed the synthesis and secretion of a polypeptide of 35 kDa. Analysis of the nucleotide sequence of the cDNA and *in vitro* translation products suggested that hIL-1β is initially synthesized as a much larger precursor which is subsequently processed to yield a new NH_2-terminus at Ala-117 (Fig. 1). Mature hIL-1β is a 153-amino-acid protein of 17.4 kDa. Examination of the nucleotide sequence of the precursor polypeptide showed that the NH_2-terminus did not possess a

1 10	20	30	40
MAEVPKLASE	MMAYYSGNED	DLFFEADGPK	QMKCSFEDLD
50	60	70	80
LCPLDGGIQL	RISDHHYSKG	FRQAASVVVA	MDKLRKMLVP
90	100	110	120
CPQTFQENDL	STFFPFIFEE	EPIFFDTWDN	EAYVHDAPVR
			1
130	140	150	160
SLNCTLRDSQ	QKSLVMSGPY	ELKALHLQGQ	DMEQQVVFSM
10	20	30	40
170	180	190	200
SFVQGEESND	KIPVALGLKE	KNLYLSCVLK	DDKPTLQLES
50	60	70	80
210	220	230	240
VDPKNYPKKK	MEKRFVFNKI	EINNKLEFES	AQFPNWYIST
90	100	110	120
250	260	269	
SQAENMPVFL	GGTKGGQDIT	DFTMQFVSS	
130	140	150 153	

Figure 1. Amino acid sequence of hIL-1β precursor. The amino acids are designated by the one-letter code. The numbers above the sequence are for the precursor polypeptide, while those below are for the residues in the mature hIL-1β. The NH_2-terminal residue of the mature hIL-1β (Ala-117) and a potential glycosylation site (Asn-123) are underlined. Reprinted from Auron *et al.* (1984).

signal peptide typical of most secreted proteins. This may explain why monocytes or macrophages synthesize hIL-1 in response to endotoxin stimulation but do not release the protein efficiently (Gery *et al.*, 1981; Gery and Lepe-Zuniga, 1984). Stimulants known to cause cellular damage, however, have been found to release a greater amount of hIL-1 than endotoxin (Gery *et al.*, 1981). This conclusion is supported by the inefficient release of hIL-1 activity by *Xenopus* oocytes injected with hIL-1 mRNA or by COS-7 monkey kidney cells transfected with hIL-1 expression vectors (Furutani *et al.*, 1985).

Recombinant human IL-1β (rhIL-1β) has been expressed and purified from *E. coli* by several groups (Kronheim *et al.*, 1986; Wingfield *et al.*, 1986; Kikumoto *et al.*, 1987; Meyers *et al.*, 1987). Kronheim *et al.* (1986) produced IL-1β in an insoluble form in *E. coli* RR1 harboring the plasmid pRK248cI_{ts}. The first step in the purification involved the acid extraction (pH 3.9) of the *E. coli* cell suspension. This step solubilized the rhIL-1β while precipitating the bulk of the *E. coli* proteins. The crude extract was applied to a sulfopropyl Sephadex column followed by DEAE Sephacel and procion red agarose chromatography. These steps resulted in purification to apparent homogeneity as demonstrated by SDS-PAGE.

1.2. Pharmacology Evaluation and Clinical Applications

One of the major indications under active clinical investigation for rhIL-1β is protection against the effects of radiation and chemotherapeutic agents in cancer patients (Gauldie *et al.*, 1987). However, despite the extensive knowledge on the biological activities and systemic effects of hIL-1β, the mechanism for the apparent efficacy is not clear. Neta *et al.* (1986) demonstrated that a single injection of 0.2 μg rhIL-1β (or 0.1 μg rhIL-1α) at 20 hr before lethal radiation could rescue up to 80% of the irradiated C57BL/6 mice. The dose reduction factor measured by $LD_{50/30}$ was 1.25. The timing of the administration was critical as the apparent protective effect was greatly reduced with administration at 45 or 4 hr before or 1 hr after irradiation. Castelli *et al.* (1988) in a separate study found, in addition to protection against myelosuppression induced by radiation, that rhIL-1 was able to provide protection and recovery from an otherwise lethal dose (400 mg/kg) of cyclophosphamide (an alkylating agent) in C57BL/6 mice. Again, the timing of the administration was important with maximum protection observed when rhIL-1β was given 20 hr before or 48 hr after the cyclophosphamide administration.

Gasparetto *et al.* (1989) investigated the effect of rhIL-1β on the hematopoietic recovery in cynomolgus monkeys following 5-fluorouracil treatment (75 mg/kg for 2 days). It was found that administration of 1 μg/kg rhIL-1β for 2 or 7 days after 5-fluorouracil treatment resulted in a significantly shortened time to achieve hematopoietic recovery (as measured by absolute neutrophil count) when compared with controls or a 14-day rhIL-1β prolonged treatment. Although other mechanisms cannot be ruled out, the authors suggested that a physiologic balance between the positive (or myeloprotective) and negative (or myelosuppressive) effects must exist to explain the delayed recovery associated with the prolonged administration of rhIL-1β. Among the positive effects are synergism with colony-stimulating factors (CSFs) for proliferation of progenitors and promotion of the production of CSFs. On the other hand, rhIL-1β can induce the production of hematopoietic inhibitors such as tumor necrosis factor-α or other cytokine inhibitors. This would indicate that prolonged administration of rhIL-1β (i.e., 14 days' treatment) may not be better than the short-term courses (i.e., 2 or 7 days' treatment).

Another indication under clinical evaluation for rhIL-1β is wound healing (F-D-C reports, 1989). However, no pharmacology reports for this indication were available in the literature for IL-1β. Sauder *et al.* (1990) found that topically applied 4–40 μg of rhIL-1α in a water-miscible vehicle significantly enhanced the healing rate of the wound (made in $10 \times 7 \times 0.3$ mm in the paravertebral and thoracic areas) in adult pigs.

Preliminary clinical testing of rhIL-1β in humans has indicated potential myeloprotection (Crown *et al.*, 1989) and antitumor activity in metastatic malignant melanoma (Starnes *et al.*, 1991).

2. STRUCTURAL CHARACTERIZATION AND ANALYTICAL METHODS

Once large quantities of hIL-1β were available by recombinant techniques, extensive analytical characterization including sequence and structural studies could be performed. The rhIL-1β used in the authors' studies was obtained from Immunex Corporation and was purified as described in Kronheim *et al.* (1986).

2.1. Spectroscopy

The ultraviolet absorption spectrum of rhIL-1β in 100 mM Tris/HCl, pH 7.4, and 1 mM sodium azide exhibits an absorption maximum at 277 nm and a minimum at 246 nm (Wingfield *et al.*, 1986). The molar extinction coefficient was determined to be 10.1×10^3 cm^{-1} M^{-1} at 280 nm based on the concentration as determined by quantitative amino acid analysis and the sequence-predicted molecular mass of 17.4 kDa (Meyers *et al.*, 1987). This value for the extinction coefficient is in good agreement with the value of 10.3×10^3 cm^{-1} M^{-1}, which was calculated using the predicted spectrophotometric contributions of Trp and Tyr according to Edelhoch (1967).

Craig *et al.* (1987) have used absorption spectroscopy to examine the environment of the single tryptophan residue of rhIL-1β and found it to be 40% exposed to solvent. Epps *et al.* (1989) have confirmed these results by fluorescence spectroscopy.

2.2. Amino Acid Analysis

The amino acid composition of rhIL-1β is shown in Table II. The experimental values are in good agreement with the expected composition as predicted by the cDNA sequence. From the expected composition, the pre-

Table II
Amino Acid Composition of rhIL-1β

	Expected[a]	Obtained
Asx	17	19.7
Ser	14	17.6
Gly	8	7.7
Glx	23	22.9
Thr	6	7.6
Ala	5	5.2
Val	11	10.3
Met	6	5.9
Tyr	4	3.1
Ile	5	4.2
Leu	15	13.6
Phe	9	8.0
His	1	0.6
Lys	15	13.4
Trp	1	ND[b]
Arg	3	2.2
Pro	8	5.5
Cys	2	ND

[a] Calculated from cDNA sequence.
[b] ND, not determined.

dicted isoelectric point of this protein would be 6.9. The amino acid composition of both natural and rhIL-1β has been determined by several different groups (Matsushima *et al.*, 1985; Kronheim *et al.*, 1986; Wingfield *et al.*, 1986; Meyers *et al.*, 1987) with similar results to those shown here.

2.3. Sequence Analysis

2.3.1. Amino-Terminal Sequence

March *et al.* (1985) purified hIL-1β from the supernatant of activated macrophages. Automated NH_2-terminal sequencing was performed on approximately 680 pmole. Nineteen of the first twenty residues were identified. The NH_2-terminus of the hIL-1β began at residue 117 (Ala) of the open reading frame of the cDNA clone (Fig. 1). Only a single NH_2-terminal residue was observed.

NH_2-terminal sequence analysis of rhIL-1β produced in *E. coli* yielded two sequences, one for the mature form, rhIL-1β, and one for the form missing the first amino acid, [des-Ala[1]] rhIL-1β. The mature form was 67% of the initial yield from the sequencer and the [des-Ala[1]] rhIL-1β form 33%. The NH_2-terminal sequence for rhIL-1β has been reported by several groups (Kronheim *et al.*, 1986; Wingfield *et al.*, 1986; Meyers *et al.*, 1987). Kronheim *et al.* (1986) and Meyers *et al.* (1987) detected two sequences, one for rhIL-1β and one for [des-Ala[1]] rhIL-1β. Neither group detected the presence of the NH_2-terminal initiator methionine, [Met[0]] rhIL-1β. Meyers *et al.* (1987) reported that the ratio of rhIL-1β to [des-Ala[1]] rhIL-1β was the same in the *E. coli* lysate indicating that the truncation did not occur during purification. Wingfield *et al.* (1986), expressing rhIL-1β in *E. coli* W3110cI, reported the presence of three NH_2-termini for rhIL-1β: rhIL-1β (67%), [Met[0]] rhIL-1β (20%), and [des-Ala[1]] rhIL-1β (13%).

2.3.2. Carboxy-Terminal Sequence

The COOH-terminal sequence of rhIL-1β has been determined by time course release of amino acids with carboxypeptidase Y. A twofold amount of serine (relative to subsequent amino acids) was observed at early time points which is consistent with the presence of the sequence Ser-Ser at the COOH-terminus.

2.3.3. Glycosylation

From the amino acid sequence of hIL-1β, as predicted by the cDNA sequence, Asn-123 (Fig. 1) is part of an *N*-linked glycosylation consensus sequence and is, therefore, a potential glycosylation site. However, direct protein sequencing of hIL-1β detected asparagine at this position, confirming the lack of glycosylation (Cameron *et al.*, 1985; Kronheim *et al.*, 1986). Krakauer (1984) found that tunicamycin (an inhibitor of lipid carrier-dependent glycosylation) had no effect on the activity or isoelectric point of hIL-1β, nor did hIL-1β bind to a concanavalin A–Sepharose column. These results indicate that hIL-1β does not have carbohydrate groups attached through an *N*-glycosidic bond to asparagine.

2.3.4. Peptide Mapping

The entire amino acid sequence of rhIL-1β has been determined by peptide mapping (Kikumoto *et al.*, 1987) with lysyl-endopeptidase. The digest was subjected to reversed-phase chromatography (Fig. 2). The purified peptides were identified by amino acid analysis and amino acid sequence analysis.

2.4. Secondary and Tertiary Structural Analysis

The amino acid composition of rhIL-1β shows the presence of two cysteine residues. The cysteine may be present as free sulfhydryl groups or disulfide bonds. Meyers *et al.* (1987) examined rhIL-1β for the presence of

A.

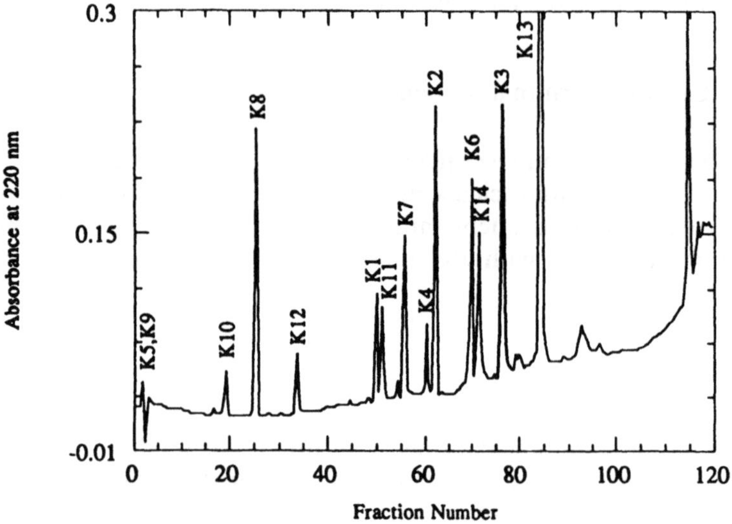

B.

```
1        10        20        30        40        50
APVRSLNCTLRDSQQKSLVMSGPYELKALHLQGQDMEQQVVFSMSFVQGE
--------------K1------------ --------K2--------- -------------------------K3---------------

          60        70        80        90        100
ESNDKIPVALGLKEKNLYLSCVLKDDKPTLQLESVDPKNYPKKKMEKRFV
--------- -----K4----- K5 -------K6------ ------------K7---------- ---K8--K9 K10 -K11

          110       120       130       140       150
FNKIEINNKLEFESAQFPNWYISTSQAENMPVFLGGTKGGQDITDFTMQFVSS
----- --K12--- -----------------------K13------------------------ -----------K14-----------
```

Figure 2. Peptide mapping of rhIL-1β with lysyl-endopeptidase. (A) The hydrolysate was applied to a Cosmosil 5C18 column. Elution was performed with a trifluoroacetic acid/acetonitrile buffer system. (B) Assignment of peptide fragments obtained by the peptide mapping. Reprinted from Kikumoto *et al.* (1987).

free sulfhydryls with Ellman's reagent (Ellman, 1959). Under nonreducing conditions in 6 M guanidine hydrochloride (Gdn-HCl), 1.94 moles SH/mole protein were found, indicating that the cysteines do not form a disulfide bridge. The cysteines were unreactive in the absence of 6 M Gdn-HCl, indicating that the residues are buried and inaccessible in the protein interior. The requirement of free sulfhydryl groups in the activity of rhIL-1β was examined by reaction with iodoacetamide (Meyers *et al.*, 1987). Carboxyamidomethylated protein was found to have approximately 50% of the activity of control samples of rhIL-1β in the EL-4 bioassay. Bioassays can exhibit as much as 50% variation in results. Therefore, the retention of 50% activity indicates that sulfhydryl groups are not required for the biological activity, at least in the IL-2 induction assay.

Craig *et al.* (1987) and Meyers *et al.* (1987) have examined the circular dichroism (CD) spectrum of native rhIL-1β. The spectrum was typical of a protein with strong β-sheet structure (Fig. 3). The lack of a positive rotational signal at 195 nm indicated the presence of little α-helical structure (Meyers *et al.*, 1987). Analysis of the CD spectrum with the CONTIN program for secondary structure yielded estimates of 0% helix and 59% sheet (Craig *et al.*, 1987).

The three-dimensional crystal structure of rhIL-1β has been determined at 3.0 Å (Priestle *et al.*, 1988) and 2.0 Å (Finzel *et al.*, 1989) resolution. The structure is composed of 12 antiparallel β-strands (Fig. 4) which are folded into a six-stranded β-barrel similar to soybean trypsin inhibitor. The core of the structure is a tetrahedron with each edge formed by two antiparallel β-strands. There is an approximate threefold axis of symmetry about the axis of the β-barrel. Sedimentation velocity analysis showed rhIL-1β to be approx-

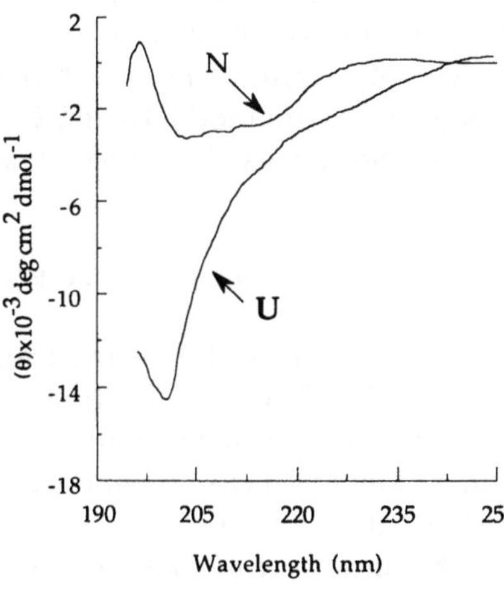

Figure 3. Far-UV circular dichroism spectra of native and unfolded rhIL-1β. Spectra are an average of ten scans with the baseline subtracted. Reprinted from Craig *et al.* (1987).

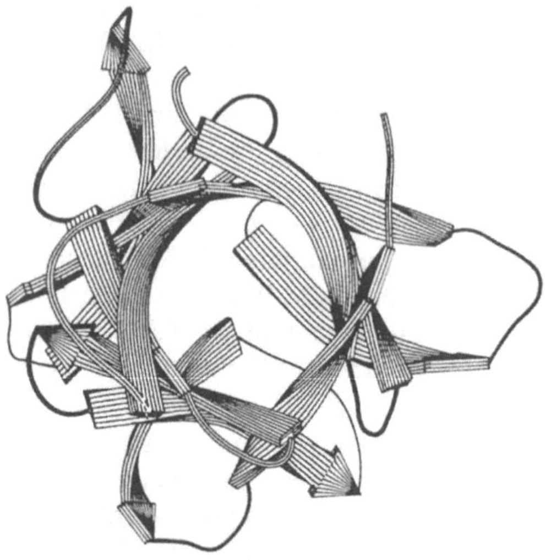

Figure 4. The cartoon of three-dimensional structure of rhIL-1β. The twisted arrows represent β-strands. The view is down the axis of the barrel formed by six of the β-strands. Reprinted from Priestle *et al.* (1988) by permission of Oxford University Press.

imately a spherical molecule, since the frictional ratio (f/f_0) was close to 1 (Wingfield *et al.*, 1986).

2.5. Chromatographic and Electrophoretic Methods

2.5.1. Reversed-Phase Chromatography (RP-HPLC)

rhIL-1β was analyzed by RP-HPLC on either a Vydac or a Rainin Dynamax C4 column. The gradient (method I) was 20% to 80% acetonitrile in 0.1% trifluoroacetic acid. rhIL-1β eluted as a symmetrical peak at approximately 44% acetonitrile. For stability studies of rhIL-1β, another gradient mobile phase (method II) using 20% to 80% acetonitrile containing 1% phosphoric acid was used (Kontor *et al.*, 1990).

2.5.2. Ion Exchange Chromatography (IEC)

Two major peaks for the Immunex rhIL-1β can be seen by IEC on a TSK gel SP-5PW strong cation exchange column. Elution was performed isocratically at 30 mM sodium chloride in 50 mM sodium acetate, pH 5.5 buffer. The first peak to elute was rhIL-1β at 17.8 min and then [des-Ala^1] rhIL-1β at 21.8 min.

Kikumoto *et al.* (1987) and Yem *et al.* (1988) have purified the three forms of rhIL-1β—rhIL-1β, [Met[0]] rhIL-1β, and [des-Ala[1]] rhIL-1β—by cation exchange chromatography on a TSK gel SP-5PW column (Fig. 5). Elution was performed with a sodium chloride gradient in 50 mM sodium acetate, pH 5.5 buffer.

2.5.3. Size Exclusion Chromatography and SDS-PAGE

rhIL-1β chromatographs as an approximately 17.4-kDa species on a size exclusion column in 0.1 M sodium sulfate/0.1 M sodium phosphate buffer, pH 6.8. Therefore, the protein is a monomer. SDS-PAGE of purified rhIL-1β shows a single band of 17.5 kDa. In addition, the molecular mass has been determined by sedimentation equilibrium analysis to be 17.7 kDa (Wingfield *et al.*, 1986).

2.5.4. Isoelectric Focusing (IEF)

The pI of rhIL-1β was determined to be 6.87 on Servalyt IEF gels with a pH range of 3–10. This is in good agreement with the predicted pI of 6.9. The three forms of rhIL-1β are indistinguishable by SDS-PAGE, but differ in their pI (Kikumoto *et al.*, 1987). rhIL-1β and [des-Ala[1]] rhIL-1β have a pI of 6.9, while [Met[0]] rhIL-1β has a lower pI of 6.8.

In our laboratory, rhIL-1β and [des-Ala[1]] rhIL-1β can be seen as two separate bands on IEF gels under certain conditions (Fausnaugh *et al.*, man-

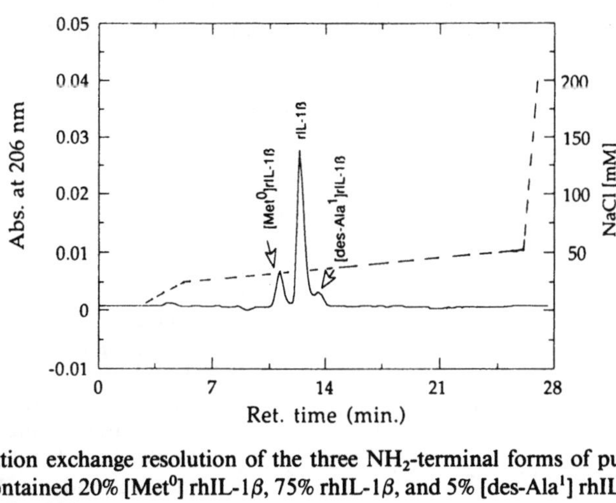

Figure 5. Cation exchange resolution of the three NH_2-terminal forms of purified rhIL-1β. The sample contained 20% [Met[0]] rhIL-1β, 75% rhIL-1β, and 5% [des-Ala[1]] rhIL-1β. Reprinted from Yem *et al.* (1988).

uscript in preparation). When the sample is loaded at the basic end of the slab gel, two separate diffuse bands can be seen, one with a pI of 6.79 and the other 6.73 (Fig. 6). However, if electrophoresis is allowed to continue, the two bands will focus into one. The same effect is seen if electrophoresis is performed at lowered temperatures. The two bands were isolated and subjected to amino acid sequencing. The pI 6.79 band corresponded to pure mature rhIL-1β and the pI 6.73 band corresponded to pure [des-Ala[1]] rhIL-1β.

2.6. Bioassays

2.6.1. *In Vitro* Assays

2.6.1a. Murine Thymocyte Proliferation Assay. The murine thymocyte proliferation assay is the most widely used assay for routine analysis of hIL-1 activity (Gery *et al.,* 1972). This assay is based on the ability of hIL-1 to stimulate the proliferation of thymocytes. The proliferation is due to the IL-2 released by IL-1-stimulated T cells, since monoclonal antibodies specific for hIL-2 can block this response (Oppenheim and Gery, 1982). The assay is sensitive to levels of hIL-1 of ~10–50 pg/ml.

In this assay, the hIL-1 preparation is added in dilutions to microplate wells containing 1–5 × 10^5 cells in 100–200 μl of culture medium. A co-mitogen such as phytohemagglutinin or concanavalin A is then added at a suboptimal concentration. Proliferation is assessed by [^{3}H]thymidine incorporation after 72 hr of culture (Lachman *et al.,* 1980).

A major problem with this assay is its lack of specificity. Proliferation

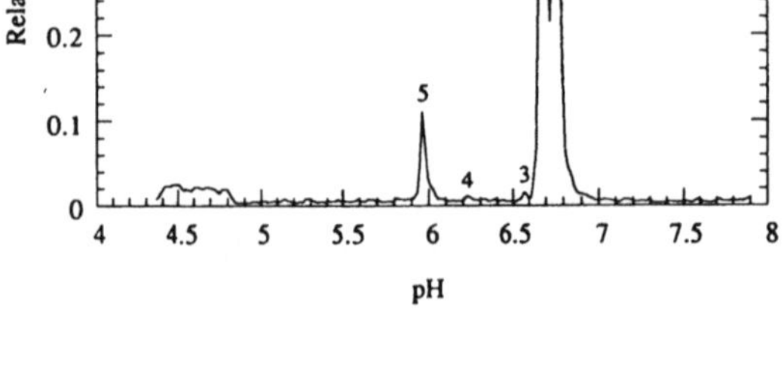

Figure 6. Densitometer trace of IEF separation of rhIL-1β and [des-Ala[1]] rhIL-1β. The sample was loaded at the basic end of the slab gel and electrophoresed for 2800 V-hr. Band 1 is rhIL-1β and band 2 is [des-Ala[1]] rhIL-1β. Bands 3, 4, and 5 are rhIL-1β related.

can be stimulated by hIL-2 itself and by compounds used to induce hIL-1 production (e.g., lipopolysaccharide and phorbol myristate acetate). Mouse strain C3H/HeJ is often used because it is hyporesponsive to endotoxin and can eliminate this problem. Different mouse strains also show different response levels to hIL-1. hIL-1 activity in tissue fluids and also from culture supernatants can be obscured by inhibitory material of which the most common in culture is prostaglandin E_2.

2.6.1b. D10.G4.1 Proliferation Assay. There are several bioassays involving T cell clones. One of these is the D10.G4.1 proliferation assay. D10.G4.1 is a cloned murine helper T cell line specific for hen's egg conalbumin (Kaye *et al.*, 1984). This cell line is 100–1000 times more sensitive to hIL-1 than to hIL-2. As with the thymocyte proliferation assay, the hIL-1 activity is measured by incorporation of [^{3}H]thymidine. This assay has several advantages over the thymocyte proliferation assay: increased sensitivity to hIL-1 and decreased sensitivity to hIL-2. The maintenance of this cloned cell line, however, is time-consuming.

2.6.1c. LBRM-33-1A5 Conversion Assay. The LBRM-33-1A5 conversion assay developed by Conlon (1983) uses LBRM-33 spleen cells metabolically inactivated with mitomycin C. The cells, stimulated with phytohemagglutinin, are mixed with an hIL-2-dependent cell line, CTLL-2. Addition of hIL-1 to this mixture causes release of hIL-2 from the LBRM-33 cells which stimulates proliferation of the CTLL-2 cells. This is called a conversion assay because it converts the LBRM-33 cell line from being an hIL-2 nonproducer to an hIL-2 producer. Proliferation is assayed by [^{3}H]thymidine incorporation in the CTLL-2 cells after 24 hr of culture. This assay produces results 48 hr faster and can be up to 1000-fold more sensitive than the murine thymocyte assay. It has an added advantage in that it is insensitive to endotoxin. hIL-2, however, is active in this assay so that hIL-1 preparations need to be checked for hIL-2 activity.

2.6.1d. EL-4 Assay. Another bioassay using a T cell clone is the EL-4 cell line assay. EL-4 is a murine T cell thymoma. As with the LBRM-33-1A5 conversion assay, the proliferation of the IL-2-dependent cell line CTLL-2 is measured by [^{3}H]thymidine incorporation as an index of hIL-1 activity (Simon *et al.*, 1985). The EL-4 cell line assay has approximately the same sensitivity as the LBRM-33-1A5 assay and is insensitive to endotoxin. It has the advantage, however, that the EL-4 cell line is phenotypically stable. hIL-2, however, is active in this assay.

An assay involving a subclone of the EL-4 cell line, NOB-1, has been developed by Gearing *et al.* (1987). In this assay, the NOB-1 cells do not have to be metabolically inactivated as the cells do not incorporate significant amounts of [^{3}H]thymidine during the assay. A preincubation step is included which makes the assay unresponsive to hIL-2.

2.6.2. *In Vivo* Assay

The leukocyte pyrogen test is an *in vivo* assay which can be used to detect hIL-1 by measuring the increase in body temperature following administration of an hIL-1 preparation (Duff and Durum, 1983). In this test, rabbits or mice are injected with the preparation and their temperatures are monitored. Fever should develop within 60 min, producing increases in temperature of 0.6–0.9°C. This assay is very insensitive, requiring up to 1000-fold more hIL-1 than the murine thymocyte assay to give a significant rise in body temperature in rabbits.

2.6.3. Bioassay of rhIL-1β's

The three NH_2-terminal variants of rhIL-1β have been shown to have differing activities in the bioassays. Kikumoto *et al.* (1987) have shown that [Met^0] rhIL-1β has one-tenth the activity of mature rhIL-1β, while [des-Ala^1] rhIL-1β has one-half the activity in a growth inhibitory assay with A375 human melanoma cells. Bioactivity measurements, using the murine C3H/HeJ thymocyte proliferation assay, showed that mature rhIL-1β was 300–400% more active than [des-Ala^1] rhIL-1β, and was 400–600% more active than [Met^0] rhIL-1β (Yem *et al.*, 1988).

2.7. Immunoassays

Immunoassays have several advantages over bioassays. They can measure concentrations of hIL-1 in samples containing interfering factors which would mask the activity in a bioassay. They can also distinguish hIL-1α from hIL-1β (Kenney *et al.*, 1987). One disadvantage is that the detection of hIL-1 by the immunoassay may not be directly correlated to the activity, unless the antibody is specific for the active site.

Kenney *et al.* (1987) have developed a two-site ELISA to detect hIL-1β. The two monoclonal antibodies, H6 and H67, recognize different epitopes on the protein surface. The sensitivity of this assay was 15 pg/ml, which is comparable to the murine thymocyte proliferation bioassay.

A radioimmunoassay (RIA) for hIL-1β is commercially available from Cistron Biotechnology. This assay has a sensitivity limit of 250 pg/ml.

3. PHYSICOCHEMICAL STABILITY

3.1. Unfolding and Refolding of rhIL-1β

Craig *et al.* (1987) have studied the unfolding and refolding kinetics of rhIL-1β (80% rhIL-1β and 20% [Met^0] rhIL-1β) in the presence of Gdn-HCl and urea by CD and fluorescence spectroscopy. The near-UV (for tertiary structure) and the far-UV (for secondary structure) CD spectrum of the native and unfolded rhIL-1β are shown in Figs. 7 and 3, respectively.

The CD spectrum did not change with pH from pH 6.5 to 7.8 and the unfolding kinetics of rhIL-1β followed pseudo-first-order behavior. The unfolding rate was faster at higher denaturant concentrations and at pH 7.8 than at pH 6.5. The fraction of unfolded rhIL-1β was studied as a function of Gdn-HCl concentrations by both near-UV and far-UV CD (Fig. 8). The two experiments agreed well with each other suggesting that there were no structured equilibrium intermediates and the unfolding process followed a simple two-state model.

At pH 6.5 [0.01 M 4-morpholinoethane sulfonic acid (MES), 0.09 M NaCl, and 0.001 M Na_2EDTA], 90–100% of the unfolded rIL-1β could be consistently refolded and the refolding kinetics appeared to be biphasic with a fast reaction ($t_{1/2} < 1$ sec) followed by a slow and measurable reaction. The biphasic kinetics are consistent with a mechanism that involves a rapid formation of secondary structure followed by a slow rearrangement of the compact globule to form the native protein. At pH 7.8, the unfolded protein was not stable and aggregated with time (Craig *et al.*, 1987). Thus, the yield of refolding at pH 7.8 decreased sharply (50%–0%). Removal of the denaturant by dialysis also resulted in aggregation. Finally, the yield of refolded protein at pH 7.8 was improved to 85% in the presence of 1% β-mercaptoethanol. It was concluded that intermolecular disulfide bond formation is responsible for the low yield of refolding at pH 7.8 (Craig *et al.*, 1987).

3.2. Freeze–Thaw Effect

The effect of freeze–thaw cycles on the stability of rhIL-1β solutions (pH 7.0, 1 ml) in polypropylene containers (Sarstedt, 3 ml) was investigated. The

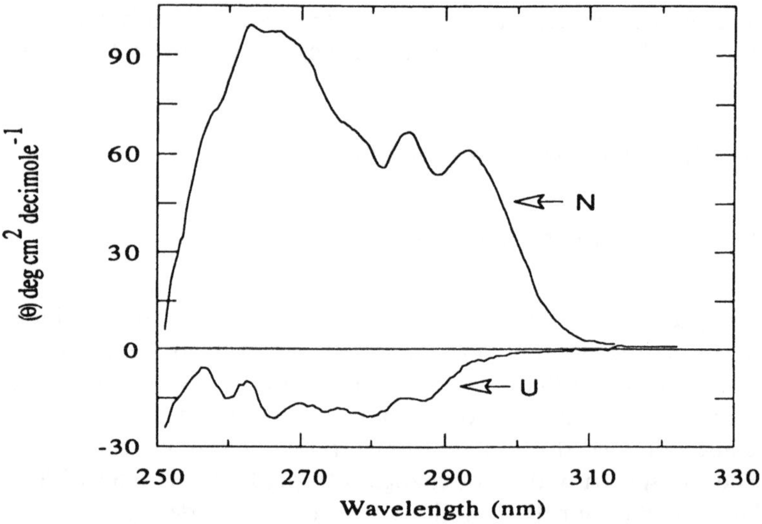

Figure 7. Near-UV circular dichroism spectra of native and unfolded rhIL-1β. Spectra are an average of ten scans with the baseline subtracted. Reprinted from Craig *et al.* (1987).

cycle involved storage of freshly prepared rhIL-1β solutions in a −80°C freezer for 24 hr and thawing in a 5°C ice bath. The cycle was repeated daily on three consecutive days and control samples were stored in a refrigerator. Table III summarizes the results. Except for solutions containing ≥0.025% human serum albumin (HSA), all other control samples showed various degrees of loss (measured by RP-HPLC) due to apparent adsorption. Com-

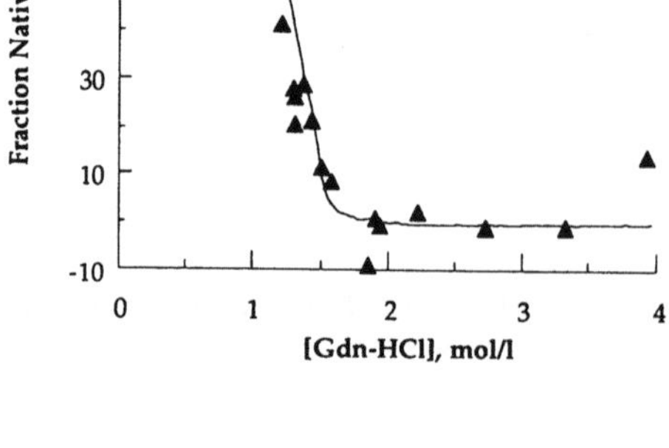

Figure 8. Reversible unfolding transition of rhIL-1β at pH 6.5 measured by far-UV CD. CD at 205 nm with a 0.005-cm path length was measured after overnight incubation at 25°C to reach equilibrium. The solid line is the theoretical curve calculated from the near-UV unfolding transition (data not shown). Reprinted from Craig *et al.* (1987).

pared with control samples, freeze–thaw cycles appeared to have little or no adverse effects on rhIL-1β in the presence of sucrose (3%), HSA (0.005%–0.025%), or 0.01 M Tris buffer, but significant loss of rhIL-1β at 15 μg/ml concentration occurred in mannitol (3%) and in NaCl (2%)/citrate (2%) solutions. Interestingly, even though Tris (0.01 M), sucrose (3%), and mannitol (3%) alone had no protective effect for rhIL-1β at 15 μg/ml, the combination of these reagents showed enhanced protection against freeze–thaw cycles and adsorption.

To substantiate the HPLC results, selected sample solutions used in the freeze–thaw studies were also evaluated by ELISA and murine thymocyte proliferation assay and in each case the ELISA and bioassay results are in good agreement with those obtained by RP-HPLC (Table III). This indicates that, with proper protection, rhIL-1β is conformationally stable to unfolding or aggregation in freeze–thaw cycles.

3.3. Aggregation and Precipitation

The stability of rhIL-1β in aqueous solution was studied as a function of pH (2–9) and temperature (5–60°C), and in the presence of several cryoprotectants (sugars, HSA) (Gu *et al.*, 1991). It was found that at or above 39°C, all solutions (100–500 μg/ml) became cloudy upon storage. When the supernatants of the stressed sample solutions were analyzed by IEF, less than 7% of reacted rhIL-1β could be accounted for by newly formed protein bands. RP-HPLC (method I) analysis of the supernatants also showed insignificant amounts of degradation products. The amount of protein remaining in solution as measured by RP-HPLC agreed well with data from ELISA and murine thymocyte proliferation assay, when the standard errors involved in these assays were considered, ~4, ~13, and ~26%, respectively (Fig. 9). These results demonstrated that the RP-HPLC method I was stability indicating for processes leading to aggregation (i.e., storage at or above 39°C) and the subsequent precipitate formation.

The precipitates were found to be soluble in a 2.5% SDS solution and SDS-PAGE analysis of the dissolved precipitates revealed, in addition to monomers (≤30%), oligomers (≥70%) under nonreducing conditions (Fig. 10). All oligomers were converted to the monomer form in the presence of β-mercaptoethanol. This suggests that most of the precipitates were covalent aggregates resulting from disulfide bond formation.

At 60°C, the aggregation/precipitation followed apparent first-order kinetics. The rate data at an rhIL-1β concentration of 100 μg/ml are summarized in Table IV. The reaction rate was relatively independent of pH from

Table III
Effect of Freeze–Thaw on Stability of rhIL-1β

Solution	Conc. (μg/ml)	Freeze–thaw cycle[a] (°C)	HPLC (ELISA, murine thymocyte proliferation assay) results (% initial)		
			Day 1	Day 2	Day 3
3% sucrose	15	5	76	70	65
		−80/5	71	69	73
2% NaCl	60	5	93	85	85
2% citrate		−80/5	94	90	90
	15	5	90	82	80
		−80/5	76	62	71
0.05% HSA	60	5	100	101	103
		−80	100	101	100
	15	5	100 (100, 1.6 E+5)	100	94 (91, 1.4 E+5)
		−80/5	97 (100, 1.4 E+5)	98 (96, 1.3 E+5)	98 (92, 1.6 E+5)
0.025% HSA	5	5	92	102	100
		−80/5	102	97	92
0.005% HSA	5	5	82	86	70
		−80/5	89	92	90
3% mannitol	15	5	64	60	59
		−80/5	63	55	43
0.01 M Tris	15	5	66	61	65
		−80/5	65	68	67
0.01 M Tris 3% sucrose 3% mannitol	15	5	83 (87, 1.6 E+5)	—	79 (87, 1.6 E+5)
		−80/5	89 (106, 1.5 E+5)	76 (100, 1.3 E+5)	76 (99, 1.6 E+5)

[a] One cycle was performed on each day.

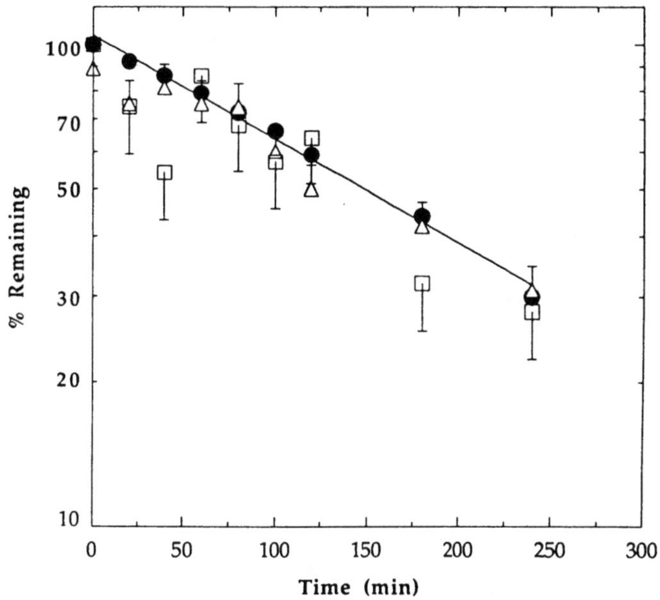

Figure 9. Semilog percent drug remaining versus time profile for the degradation of rhIL-1β at pH 7 and 60°C analyzed using HPLC method I (●), ELISA (△), and murine thymocyte proliferation assay (□). Standard error bars were each drawn in one direction for the ELISA (up) and murine thymocyte proliferation assay (down) results for clarity. The line indicates a linear fit of the HPLC data.

pH 3.0 to 6.2 ($k_{obs} = 5.7\text{–}9.4 \times 10^{-5}\ sec^{-1}$), but was faster at pH ~7 ($2.5\text{–}3.6 \times 10^{-4}\ sec^{-1}$). Excipients such as HSA, Tween 80, dextran, EDTA, or reducing agents were found to have no stabilizing effect under these extreme conditions.

At or below 55°C, the aggregation/precipitation kinetics at 100 μg/ml became biphasic, showing an initial lag phase followed by a more rapid one (Fig. 11). The lag phase was more pronounced at lower pH values and at lower temperatures (39–45°C) (data not shown), but was shortened considerably at an rhIL-1β concentration of 500 μg/ml.

The kinetics and product analyses described above for the aggregation/precipitation of rhIL-1β are consistent with a mechanism involving autooxidation of the cysteine residues. Since both cysteine residues of rhIL-1β are buried inside its native conformation (Meyers *et al.*, 1987), it is likely that only the denatured form of rhIL-1β is oxidized. Consistent with this suggestion is the observation by Craig *et al.* (1987) who showed that in the presence of denaturants, rhIL-1β could form covalent aggregates at alkaline pH values.

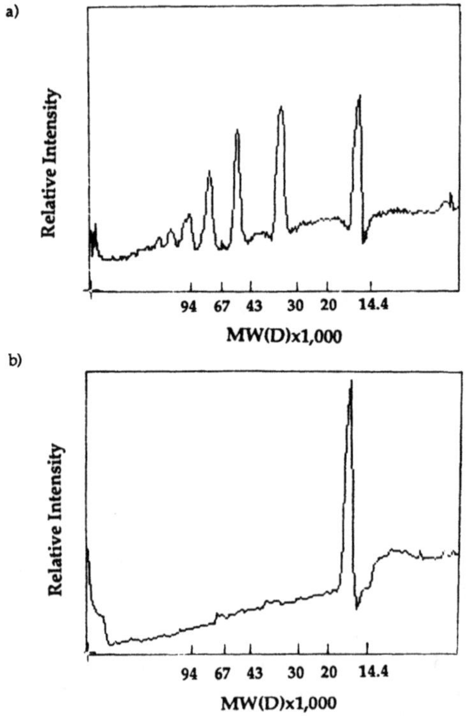

Figure 10. Densitometry analysis of the SDS-PAGE gels of the dissolved precipitates under (a) nonreducing and (b) reducing conditions.

Auto-oxidation of the sulfhydryl group has been shown to be catalyzed by metal ions. EDTA, an effective metal chelator, was evaluated as a possible stabilizer. At 60°C, where the lag phase was not apparent, EDTA had no effect on the reaction rate. At 50°C, EDTA appeared to prolong the lag phase. Finally, when the sample was degassed at 50°C, negligible amounts of degradation occurred (Fig. 11).

3.4. Stability at or below 30°C

Although less well defined, the mechanism of the degradation of rhIL-1β at or below 30°C was found to differ from that above 39°C. Three products were apparent by RP-HPLC (method II) (Fig. 12) and IEF showed two major products with pI values of 5.7 and 5.2, respectively. SDS-PAGE analysis of the 90-day aged samples, on the other hand, revealed no apparent protein bands with molecular weights different from that of rhIL-1β upon Coomassie brilliant blue staining. Although other mechanisms cannot be ruled out, these results indicate deamidation is a possible degradation path-

Table IV
Apparent First-Order Rate Constants of the Degradation of 100 μg/ml rhIL-1β Solutions at 60°C[a,b]

Solvent[c]	pH[d]	k ($sec^{-1} \times 10^5$)
0.001 N HCl	3.0	6.4
Acetate buffer	4.0	9.1
	5.2	9.4
Tris buffer	5.2	5.7
	6.2	8.4
	7.2	25.
Phosphate buffer	7.0	36.
Acetate buffer/isopropyl sulfide (1%)	5.2	16.
Acetate buffer/ethyl methyl sulfide (1%)	5.2	16.
Acetate buffer/dithiothreitol (0.02%)	5.2	14.
Tris buffer/dextran (3%)	6.2	12.
Tris buffer/Tween 80 (0.1%)	6.2	8.3
Tris buffer/HSA (0.1%)	6.2	8.9
Phosphate buffer/EDTA (5 mM)	7.0	43.

[a] Reproduced from Gu *et al.* (1991).
[b] Based on HPLC results of percentage protein remaining in solution.
[c] All solutions contained 0.01 M buffer and 0.15 M NaCl, unless noted.
[d] Measured at reaction temperature.

way. In very acid or basic solutions, the degradation at 30 or 5°C followed apparent first-order kinetics (Fig. 13). From pH 4 to 8, no kinetic order could be assigned because of the limited degradation observed at the 5-month time point. The addition of 0.05% HSA at a given pH and the use of citrate buffer at pH 7 did not affect the stability significantly.

The 5-month stability samples were also analyzed by ELISA and murine thymocyte proliferation assay. Figure 14 shows the results obtained for the 5°C samples at 154 days. The percent remaining data by RP-HPLC are in good agreement with those by ELISA and murine thymocyte proliferation assay when the errors involved in the various assay methods are taken into consideration. From Fig. 14, it is clear that several aqueous formulations of rhIL-1β exhibit at least 6 months' stability (>90% remaining).

4. rhIL-1β FORMULATION AND ITS STABILITY

The development and characterization of a lyophilized formulation for clinical evaluation of rhIL-1β has recently been reported (Visor *et al.*,

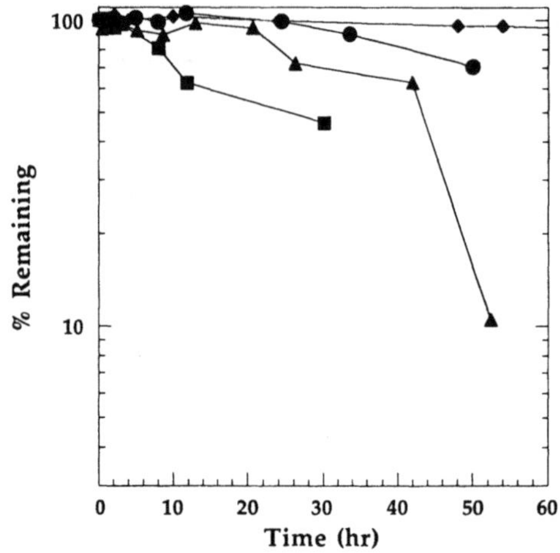

Figure 11. Semilog percent drug remaining versus time profile at pH 7 (phosphate buffer) and 50°C for 100 μg/ml rhIL-1β (▲), 500 μg/ml rhIL-1β (■), 100 μg/ml rhIL-1β in the presence of 5 mM EDTA (●), and degassed 100 μg/ml rhIL-1β (♦).

1990a). The formulation contains, before lyophilization, 3% mannitol, 3% sucrose, and 100 μg/ml of rhIL-1β in 0.01 M Tris buffer (pH 7.5). At pH 7.5, rhIL-1β is negatively charged and the interaction between rhIL-1β and the negatively charged surface of glass containers is potentially reduced. The mannitol/sucrose matrix could replace water in stabilizing the protein structure and thus minimize the potential for denaturation and aggregation during lyophilization. The lyophilized product was found to be amorphous by a powder x-ray diffraction study and was reconstituted in normal saline without visible precipitation. The moisture content of the lyophilized product was typically in the range of 1–2%.

The suitability of intravenous delivery of rhIL-1β in a programmable syringe-pump-based infusion system and in intravenous admixture bags has

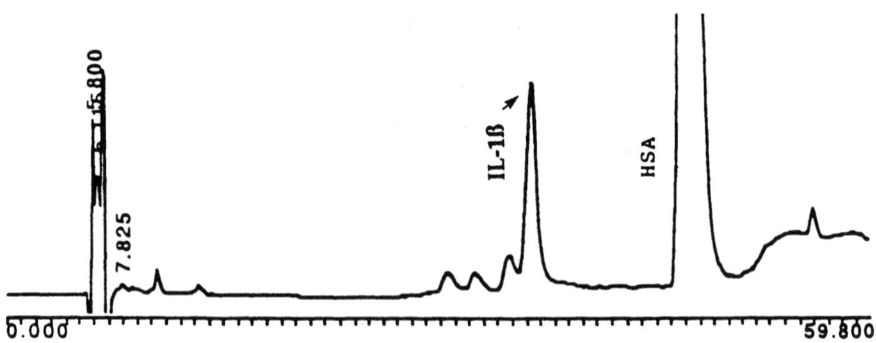

Figure 12. A RP-HPLC (method II) chromatogram of a 100 μg/ml rhIL-1β solution (67% remaining) aged for 154 days at pH 7 and 30°C.

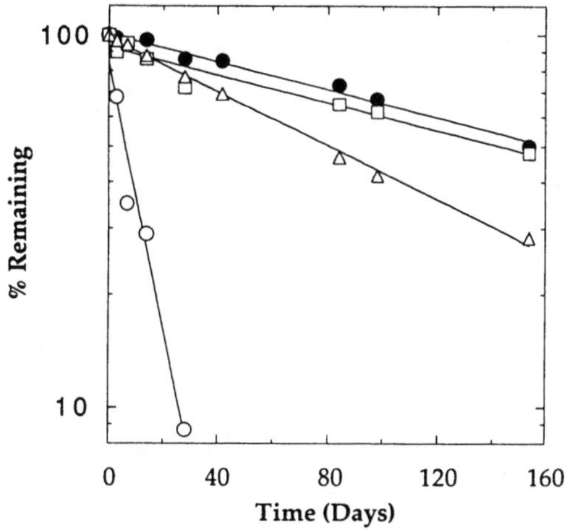

Figure 13. Semilog percent drug remaining versus time plots of the degradation of 100 μg/ml rhIL-1β in aqueous solution at 30°C and pH 2.1 (○), pH 4 (●), pH 7 (□), or pH 8 (△).

been evaluated (Visor *et al.*, 1990b). Intravenous delivery of rhIL-1β is expected to have wide applications because of the broad pharmacological effects of rhIL-1β and the clinical experiences obtained with two other cytokines, IL-2 and granulocyte–macrophage CSF, both of which were found to have fewer adverse effects via infusion relative to bolus injection (West *et al.*, 1987; Brandt *et al.*, 1988). The lyophilized formulation of rhIL-1β was initially reconstituted to 1 μg/ml with normal saline or saline containing 1% polysorbate-80 or 1% HSA and was found to be physically and chemically stable (by HPLC, ELISA, and murine thymocyte proliferation assay) following a 1-hr simulated infusion through a syringe-pump system (Harvard Mini Infuser System, Bard Microsystem). The 1 μg/ml rhIL-1β formulation in

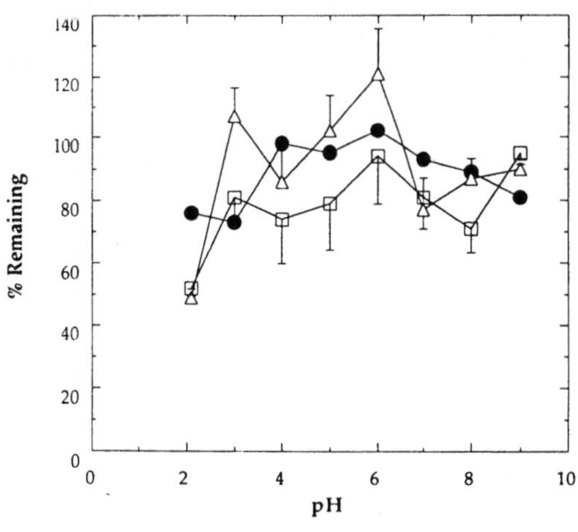

Figure 14. Stability of 100 μg/ml rhIL-1β solutions after storage at 5°C for 154 days and analyzed using HPLC method II (●), ELISA (△), and murine thymocyte proliferation assay (□).

normal saline or in 5% dextrose solutions was also stable (by HPLC) in PVC admixture bags for at least 6 hr although up to 8% adsorption was evident after storage for 24 hr. Reconstitution of rhIL-1β formulation with normal saline to 100 ng/ml, on the other hand, was found to be inappropriate for syringe-pump delivery as over 80% (by ELISA) of the drug was immediately lost by adsorption to the polypropylene syringes. The apparent adsorption was effectively suppressed by the addition of 1% HSA.

The long-term stability of the lyophilized 100 μg/ml rhIL-1β formulation was evaluated at 5–40°C (Visor *et al.,* 1990a) by HPLC, ELISA, and murine thymocyte proliferation assay (Table V). At 5 or 25°C, no significant loss ($\leq$4%) was observed by HPLC (method I) and no apparent differences could be identified by IEF or SDS-PAGE analysis between the control (or initial) and the 1-year aged samples. The aged samples also retained full biological activity as demonstrated by the ELISA and murine thymocyte proliferation assay results. However, after storage for 1 year at 40°C, up to 30% loss of rhIL-1β was detected by HPLC, ELISA, and murine thymocyte proliferation assay. Several degradation products on IEF gels were detected in the aged 40°C, lyophilized samples and these products appeared to have similar or identical molecular weight as that of rhIL-1β.

ACKNOWLEDGMENTS

The authors acknowledge the contribution of the following people: Lisa Erdös for the work on the freeze–thaw and stability studies; Linda Foster for

Table V
Stability of Lyophilized rhIL-1β over a 12-Month Period at 5, 25, and 40°C (Mean ± S.D., N = 3)[a]

Storage condition	Time (months)	HPLC[b] (%)	ELISA (%)	Bioassay[c] (units/μg, $\times 10^5$)
	0	100.5 ± 3.2	96.0 ± 5.0	1.56 ± 0.26
5°C	3	98.0 ± 0.0	93.4 ± 6.0	1.78 ± 0.35
	6	98.0 ± 7.1	92.4 ± 4.0	2.14 ± 0.29
	12	97.5 ± 0.71	92.0 ± 3.1	1.41 ± 0.32
25°C	3	98.5 ± 0.71	93.5 ± 3.0	1.74 ± 0.33
	6	96.5 ± 0.71	95.3 ± 6.0	1.89 ± 0.19
	12	96.0 ± 2.83	88.5 ± 4.9	1.07 ± 0.11
40°C	3	92.5 ± 0.71	84.1 ± 4.0	1.51 ± 0.81
	6	79.0 ± 0.0	79.3 ± 5.0	1.51 ± 0.16
	12	70.5 ± 0.71	58.2 ± 12	0.73 ± 0.02

[a] Reproduced from Visor *et al.* (1990a).
[b] Method I.
[c] Murine thymocyte proliferation assay.

the reaction of degassed samples; J. Duffy and W. C. Hsu for the bioassays and ELISA; and K. Hsia for participation in the analytical characterization. The authors also thank Dr. David Lokensgard, Linda Foster, Hi-Shi Chiang, and Lisa Erdös for their helpful discussions.

REFERENCES

Auron, P. E., Webb, A. C., Rosenwasser, L. J., Mucci, S. F., Rich, A., Wolff, S. M., and Dinarello, C. A., 1984, Nucleotide sequence of human monocyte interleukin 1 precursor cDNA, *Proc. Natl. Acad. Sci. USA* **81**:7907–7911.

Blyden, G., and Handschumacher, R. E., 1977, Purification and properties of human lymphocyte activating factor (LAF), *J. Immunol.* **118**:1631–1637.

Brandt, S. J., Peters, W. P., Atwater, S. K., Kurtzberg, J., Borowitz, M. J., Johes, R. B., Shpall, E. J., Bast, R. C., Gilbert, C. J., and Dette, D. H., 1988, Effect of recombinant human granulocyte–macrophage colony-stimulating factor on hematopoietic reconstitution after high-dose chemotherapy and autologous bone marrow transplantation, *N. Engl. J. Med.* **318**:869–876.

Cameron, P., Limjuco, G., Rodkey, J., Bennett, C., and Schmidt, J. A., 1985, Amino acid sequence analysis of human interleukin 1 (IL-1): Evidence for biochemically distinct form of IL-1, *J. Exp. Med.* **162**:790–801.

Castelli, M. P., Black, P. L., Schneider, M., Pennington, R., Abe, F., and Talmadge, J. E., 1988, Protective, restorative and therapeutic properties of recombinant human IL-1 in rodent models, *J. Immunol.* **140**:3830–3838.

Conlon, P. J., 1983, A rapid biologic assay for the detection of interleukin 1, *J. Immunol.* **131**:1280–1282.

Craig, S., Schmeissner, U., Wingfield, P., and Pain, R. H., 1987, Conformation, stability, and folding of interleukin 1β, *Biochemistry* **26**:3570–3576.

Crown, J., Kemeny, N., Jakubowski, A., Sheridan, C., Gasparetto, C., Toner, G., Meisenberg, B., Abboud, M., Sinha, S., Gordon, M., Houston, C., Oettgen, H., Buhles, W. C., Cheney, T., Moore, M. A. S., Kelsen, D., and Gabrilove, J., 1989, Phase I/II trial of recombinant human interleukin-1β (IL-1) in patients (PTS) with metastatic colorectal cancer (MCC) receiving 5-fluorouracil (5-FU), *Blood* **74**(Suppl.):15a.

Dinarello, C. A., 1988, Biology of interleukin 1, *FASEB J.* **2**:108–115.

Duff, G. W., and Durum, S. K., 1983, The pyrogenic and mitogenic actions of interleukin-1 are related, *Nature* **304**:449–451.

Edelhoch, H., 1967, Spectroscopic determination of tryptophan and tyrosine in proteins, *Biochemistry* **6**:1948–1954.

Ellman, G. L., 1959, Tissue sulfhydryl groups, *Arch. Biochem. Biophys.* **82**:70–77.

Epps, D. E., Yem, A. T., and Deibel, M. R., Jr., 1989, Characterization of the tryptophan environments of interleukins 1α and 1β by fluorescence quenching and lifetime measurements, *Arch. Biochem. Biophys.* **275**:82–91.

F-D-C reports, 1989, Immunex IL-1 topical wound-healing gel Phase I scheduled for August, F-D-C Reports 51:T&G-9.

Finzel, B. C., Clancy, L. L., Holland, D. R., Muchmore, S. W., Watenpaugh, K. D., and Einspahr, H. M., 1989, Crystal structure of recombinant human interleukin-1β at 2.0 Å resolution, *J. Mol. Biol.* **209:**779–791.

Furutani, Y., Notake, M., Yamayoshi, M., Yamagishi, J., Nomura, H., Ohue, M., Furuta, R., Fukui, T., Yamada, M., and Nakamura, S., 1985, Cloning and characterization of the cDNAs for human and rabbit interleukin-1 precursor, *Nucleic Acid Res.* **13:**5869–5882.

Gasparetto, C., Laver, J., Abboud, M., Gillo, A., Smith, C., O'Reilly, R. J., and Moore, M. A. S., 1989, Effects of interleukin-1 on hematopoietic progenitors: Evidence of stimulatory and inhibitory activities in a primary model, *Blood* **74:**547–550.

Gauldie, J., Sauder, D. N., McAdam, K. P., and Dinarello, C. A., 1987, Purified interleukin-1 (IL-1) from human monocytes stimulates acute-phase protein synthesis by rodent hepatocytes *in vitro, Immunology* **60:**203–207.

Gearing, A. J. H., Bird, C. R., Bristow, A., Poole, S., and Thorpe, R., 1987, A simple sensitive bioassay for interleukin-1 which is unresponsive to 10^3 U/mL of interleukin-2, *J. Immunol. Methods* **99:**7–11.

Gery, I., and Lepe-Zuniga, J. L., 1984, Interleukin 1: Uniqueness of its production and spectrum of activities, *Lymphokines* **9:**109–125.

Gery, I., Gershon, R. K., and Waksmann, B. H., 1972, Potentiation of the T-lymphocyte response to mitogens, *J. Exp. Med.* **136:**128–142.

Gery, I., Davies, P., Derr, J., Krett, N., and Barranger, J. A., 1981, Relationship between production and release of lymphocyte-activating factor (interleukin 1) by murine macrophages, *Cell. Immunol.* **64:**293–303.

Gu, L. C., Erdös, E. A., Chiang, H.-S., Calderwood, T., Tsai, K., Visor, G. C., Duffey, J., Hsu, W. C., and Foster, C. L., 1991, Stability of interleukin 1β (IL-1β) in aqueous solution—Analytical methods, kinetics, products and solution formulation implications, *Pharm. Res.* **8:**485–490.

Gubler, U., Chua, A. O., Stern, A. S., Hellmann, C. P., Vitek, M. P., Dechiara, T. M., Benjamin, W. R., Collier, K. J., Dukovich, M., Familletti, P. C., Fiedler-Nagy, C., Jenson, J., Kaffka, K., Kilian, P. L., Stremlo, D., Wittreich, B. H., Woehle, D., Mizel, S. B., and Lomedico, P. T., 1986, Recombinant human interleukin 1α: Purification and biological characterization, *J. Immunol.* **136:**2492–2497.

Kampschmidt, R. F., 1984, The numerous postulated biological manifestations of interleukin-1, *J. Leuk. Biol.* **36:**341–355.

Kaye, J., Gillis, S., Mizel, S. B., Shevach, E. M., Malek, T. R., Dinarello, C. A., Lachmann, L. B., and Janeway, C. A., 1984, Growth of a cloned helper T cell line induced by a monoclonal antibody specific for the antigen receptor: Interleukin 1 is required for the expression of receptors for interleukin 2, *J. Immunol.* **133:**1339–1345.

Kenney, J. S., Masada, M. P., Eugui, E. M., Delustro, B. M., Mulkins, M. A., and Allison, A. C., 1987, Monoclonal antibodies to human recombinant interleukin

1 (IL-1)β: Quantitation of IL-1β and inhibition of biological activity, *J. Immunol.* **138**:4236–4242.

Kikumoto, Y., Hong, Y.-H., Nishida, T., Nakai, S., Masui, Y., and Hirai, Y., 1987, Purification and characterization of recombinant human interleukin-β produced in *Escherichia coli, Biochem. Biophys. Res. Commun.* **147**:315–321.

Kontor, K., Calderwood, T. S., Chan, R., and Malefyt, T. R., 1990, A reversed-phase HPLC method for the separation and quantitation of interleukin-1, *Pharm. Res.* **7**:S23.

Krakauer, T., 1984, Structural analysis of the charge heterogeneity of human interleukin 1, *Arch. Biochem. Biophys.* **234**:371–376.

Kronheim, S. R., March, C. J., Erb, S. K., Conlon, P. J., Mochizuki, D. Y., and Hopp, T. P., 1985, Human interleukin 1: Purification to homogeneity, *J. Exp. Med.* **161**:490–502.

Kronheim, S. R., Cantrell, M. A., Deeley, M. C., March, C. J., Glackin, P. J., Anderson, D. M., Hemenway, T., Merriam, J. E., Cosman, D., and Hopp, T. P., 1986, Purification and characterization of human interleukin-1 expressed in *Escherichia coli, Bio/Technology* **4**:1078–1082.

Lachman, L. B., 1983, Human interleukin 1: Purification and properties, *Fed. Proc.* **42**:2639–2645.

Lachman, L. B., Page, S. O., and Metzgar, R. S., 1980, Purification of human interleukin 1, *J. Supramol. Struct.* **13**:457–462.

March, C. J., Mosley, B., Larsen, A., Cerretti, D. P., Braedt, G., Price, V., Gillis, S., Henney, C. S., Kronheim, S. R., Grabstein, K., Conlon, P. J., Hopp, T. P. and Gosman, D., 1985, Cloning, sequence and expression of two distinct human interleukin-1 complementary DNAs, *Nature* **315**:641–647.

Matsushima, K., Durum, S. K., Kimball, E. S., and Oppenheim, J. J., 1985, Purification of human interleukin 1 from human monocyte culture supernatants and identity of thymocyte comitogenic factor, fibroblast-proliferation factor, acute-phase protein-inducing factor, and endogenous pyrogen, *Cell. Immunol.* **92**:290–301.

Meyers, C. A., Johanson, K. O., Miles, L. M., McDevitt, P. J., Simon, P. L., Webb, R. L., Chen, M. J., Holskin, B. P., Lillquist, J. S., and Young, P. R., 1987, Purification and characterization of human recombinant interleukin-1β, *J. Biol. Chem.* **262**:11176–11181.

Mizel, S. B., and Mizel, D., 1981, Purification to apparent homogeneity of murine interleukin 1, *J. Immunol.* **126**:834–837.

Neta, R., Vogel, S. N., Oppenheim, J. J., and Douches, S. D., 1986, Cytokines in radioprotection. Comparison of the radioprotective effects of IL-1 to IL-2, GM-CSF and IFNα, *Lymphokine Res.* **5**:S105–S110.

Oppenheim, J. J., and Gery, I., 1982, Interleukin 1 is more than an interleukin, *Immunol. Today* **3**:113–119.

Priestle, J. P., Schar, H.-P., and Grutter, M. G., 1988, Crystal structure of the cytokine interleukin-1β, *EMBO J.* **7**:339–343.

Sauder, D. N., Kilian, P. L., Mclane, J. A., Quick, W. T., Jakubovic, H., Davis, S. C.,

Eaglstein, W. H., and Mertz, P. M., 1990, Interleukin-1 enhances epidermal wound healing, *Lymphokine Res.*, **9**:465–473.

Schmidt, J. A., 1984, Purification and partial biochemical characterization of normal human interleukin 1, *J. Exp. Med.* **160**:772–787.

Schmidt, J. A., Mizel, S. B., Cohen, D., and Green, I., 1982, Interleukin 1, a potential regulator of fibroblast proliferation, *J. Immunol.* **128**:2177–2182.

Simon, P. L., Laydon, J. T., and Lee, J. C., 1985, A modified assay for interleukin-1 (IL-1), *J. Immunol. Methods* **84**:85–94.

Starnes, H. F., Jr., Martman, G., Torti, F., Carson, R. W., Twari, A., and Buhles, W. C., Jr., 1991, Recombinant human interleukin-1β (IL-1β) has anti-tumor activity and acceptable toxicity in metastatic malignant melanoma, *Proc. ASCO* **10**:292 (Abstract 1023).

Togawa, A., Oppenheim, J. J., and Mizel, S., 1979, Characterization of lymphocyte-activating factor (LAF) produced by human mononuclear cells: Biochemical relationship of high and low molecular weight forms of LAF, *J. Immunol.* **122**:2112–2118.

Visor, G. C., Tsai, K. P., Miller, M. D., Duffy, J., Calderwood, T., Lokensgard, D., Killian, J., Malefyt, T., Gu, L., Massey, I., and Knepp, V. M., 1990a, Development and characterization of a lyophilized dosage form of IL-1β, *Lymphokine Res.* **9**:425–434.

Visor, G. C., Tsai, K. P., Duffy, J., Miller, M. D., Calderwood, T., Lokensgard, D., and Knepp, V. M., 1990b, Quantitative evaluation of the stability and delivery of interleukin-1β by infusion, *J. Parenteral Sci. Technol.* **44**:130–132.

West, W. H., Tauer, K. W., Yannelli, J. R., Marshall, G. D., Orr, D. W., Thurman, G. B., and Oldham, R. K., 1987, Constant-infusion recombinant interleukin-2 in adoptive immunotherapy of advanced cancer, *N. Engl. J. Med.* **316**:898–905.

Wingfield, P., Payton, M., Tavernier, J., Barnes, M., Shaw, A., Rose, K., Simona, M. G., Demczuk, S., Williamson, K., and Dayer, J.-M., 1986, Purification and characterization of human interleukin-1β expressed in recombinant *Escherichia coli*, *Eur. J. Biochem.* **160**:491–497.

Yem, A. W., Richard, K. A., Staite, N. D., and Deibel, M. R., Jr., 1988, Resolution and biological properties of three N-terminal analogues of recombinant human interleukin-1β, *Lymphokine Res.* **7**:85–92.

Chapter 8

Development and Shelf-Life Determination of Recombinant Interleukin-2 (Proleukin)

John Geigert, Nicholas Solli, Peggy Woehleke, and Sriram Vemuri

1. THE IL-2 MOLECULE

Human interleukin-2 (IL-2), formerly known as T cell growth factor (TCGF), is a growth and/or differentiation factor for a variety of immunologically active cells including T cells, B cells, natural killer (NK) cells, lymphokine-activated killer cells, and macrophages (Doyle *et al.*, 1985; Ralph *et al.*, 1987).

Molecular and functional characterization of IL-2 had been hampered for years by a problem common to all lymphokines, namely that they are produced in extremely small quantities. The problem of quantity was successfully addressed by the discovery that certain phorbol esters potentiate IL-2 synthesis and by the identification of high-producing cell lines. In addi-

John Geigert, Nicholas Solli, and Peggy Woehleke • Quality Control Department, Chiron Corporation, Emeryville, California 94608. *Sriram Vemuri* • Scios Nova Inc., Mountain View, California 94043. *Present address of J.G.:* Immunex Corporation, Seattle, Washington 98101.

Stability and Characterization of Protein and Peptide Drugs: Case Histories, edited by Y. John Wang and Rodney Pearlman. Plenum Press, New York, 1993.

tion, cloning of cDNA encoding the human IL-2 protein sequence has made possible the generation of virtually unlimited amounts of this protein.

Human IL-2 is a protein of approximately 15.4 kDa and has the primary amino acid sequence shown in Fig. 1. Native IL-2 isolated from human JURKAT cells and human peripheral blood lymphocytes (Robb, 1985; Grote *et al.,* 1987), as well as recombinant IL-2 produced by Chinese hamster ovary cells (Marchese *et al.,* 1990), are *O*-linked glycosylated at threonine-3. Recombinant IL-2 produced in *E. coli* is not glycosylated; however, glycosylation is not required for full biological potency of many of the human lymphokines such as IL-2. Certain strains of *E. coli* yield IL-2 with methionine at the NH_2-terminus (Lahm and Stein, 1985; Yamaka *et al.,* 1986). An intramolecular disulfide linkage between cysteine-58 and cysteine-105 is necessary for full biological activity (Wang *et al.,* 1984; Yamada *et al.,* 1987). Methionine-104 is especially sensitive to oxidation resulting in the sulfoxide derivative (Sasaoki *et al.,* 1989).

1 10

ala pro thr ser ser ser thr lys lys thr gln leu gln leu glu his leu leu leu

20 30

asp leu gln met ile leu asn gly ile asn asn tyr lys asn pro lys leu thr arg met

40 50

leu thr phe lys phe tyr met pro lys lys ala thr glu leu lys his leu gln cys leu

60 70

glu glu glu leu lys pro leu glu glu val leu asn leu ala gln ser lys asn phe his

80 90

leu arg pro arg asp leu ile ser asn ile asn val ile val leu glu leu lys gly ser

100 110

glu thr thr phe met cys glu tyr ala asp glu thr ala thr ile val glu phe leu asn

120 130

arg trp ile thr phe cys gln ser ile ile ser thr leu thr

Figure 1. Primary amino acid sequence of human IL-2.

IL-2 has found a clinically useful role in the management of metastatic renal cell carcinoma, for which one company, Chiron Corporation, has received marketed authorization from the FDA. The drug product, Proleukin, a lyophilized formulation of IL-2, is administered intravenously at doses of up to 2 mg IL-2 per patient.

The human IL-2 discussed in this chapter as manufactured by Cetus Corporation is produced in *E. coli,* and has the generic name Aldesleukin. The IL-2 is not glycosylated and it does not contain methionine on the NH_2-terminus. It does contain the required intramolecular disulfide linkage for full biological activity. By site-specific mutagenesis, the IL-2 gene was altered to produce the following changes in the primary amino acid sequence: alanine-1 is deleted and cysteine-125 is replaced by serine. Proleukin is a lyophilized formulation of recombinant IL-2 admixed with mannitol to provide bulk and a sufficient amount of sodium dodecyl sulfate (SDS) to ensure the solubility of IL-2 upon reconstitution with water (Fernandes and Taforo, 1986).

2. THE IL-2 MANUFACTURING PROCESS

2.1. Synthesis

Recombinant IL-2, Aldesleukin, is synthesized by *E. coli* containing the gene for modified human IL-2. The recombinant bacteria are grown in a fermenter having a 1500-liter capacity and equipped with controls that regulate temperature, dissolved oxygen, pH, and antifoam. IL-2 expression occurs after cell growth has consumed extracellular tryptophan.

2.2. Recovery

At high expression levels, the IL-2 protein is found as aggregated inclusion bodies in the cells, and this property is exploited in the recovery strategy.

The harvest is concentrated using crossflow filters in order to decrease the working volume. Cells are disrupted by homogenization to facilitate the isolation of the IL-2 inclusion bodies. Fermentation medium salts are removed by diafiltration and the remaining host bacteria in the dialfiltered solution are killed by the addition of alcohol. Sucrose is added to obtain a mixture of higher density and thus permit separation of the IL-2 inclusion bodies from the cell debris during centrifugation. The IL-2 fermentation paste is collected, aliquoted, and stored at −70°C until further processing.

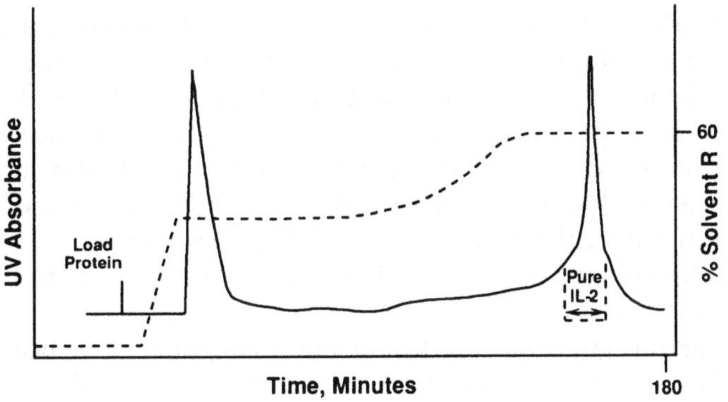

Figure 3. The second chromatographic step (a preparative RP-HPLC column) of the IL-2 manufacturing process.

2.4. Formulating and Finishing

The formulation of Proleukin consists of the purified recombinant IL-2 bulk, SDS (which is added as a surfactant), phosphate buffer, and mannitol (which is added as a bulking agent). The formulated mixture is then filled into 5-cm glass vials with rubber stoppers and lyophilized. The vials are sealed under vacuum with nitrogen.

3. STABILITY-INDICATING TEST METHODS

3.1. Methods for IL-2

Various analytical methods have been employed to assess the stability of IL-2: bioassay, SDS-PAGE, RP-HPLC, high-performance size exclusion chromatography, peptide mapping, and circular dichroism.

The bioassay of IL-2 using T-cell proliferation has been found to be an insensitive measure of IL-2 stability. Solutions of recombinant IL-2 held for over 2 months at 4°C showed significant change in physicochemical structure but no apparent change in potency. The SDS-PAGE analysis of these solutions also showed no change (Kenny *et al.*, 1986), even though this method has been found to be a sensitive stability-indicating method on other solutions of IL-2.

RP-HPLC has been used to monitor the oxidation of methionine residues to the corresponding methionine sulfoxides (Sasaoki *et al.*, 1989). Methionine residue 104 in the IL-2 molecule is especially sensitive to oxidation (Kunitani *et al.*, 1986). There are three cysteines in IL-2 at positions 58, 105, and 125. The native protein forms a disulfide linkage between the two cysteines at 58 and 105. Cleavage of this disulfide in IL-2 and the subsequent formation of two less active isomers with disulfide bonds at incorrect positions are promoted by high pH and copper ion. RP-HPLC has also been used to detect intramolecular disulfide scrambling (Browning *et al.*, 1986). Non-native disulfide-linked isomers exhibited less hydrophobicity to the C_4 column than the native form.

High-performance size exclusion chromatography has been used to monitor IL-2 aggregate formation over time at elevated temperatures (Kenny *et al.*, 1986; Watson and Kenny, 1988). First-order kinetics was observed.

Tryptic digest peptide mapping was capable of detecting deamidation of glutamine/asparagine side chains between amino acid residues 9 and 35 (Blackstock *et al.*, 1988).

Circular dichroism has been used to demonstrate the thermal stability of IL-2 solutions at 50 and 80°C (Landgraf *et al.*, 1989), and also to demonstrate that cleavage of the intramolecular disulfide bond results in a significant decrease of the secondary structure (Yamada *et al.*, 1987).

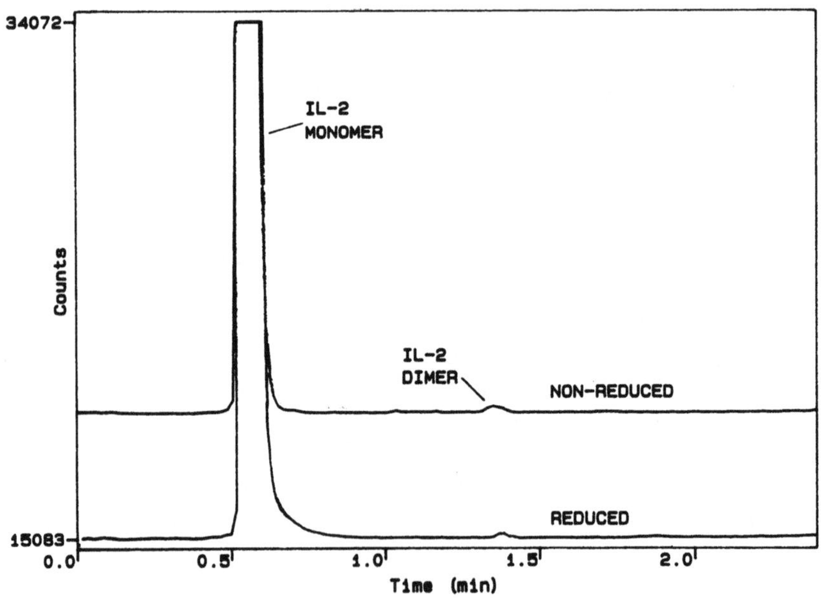

Figure 4. SDS-PAGE densitometer scan of Proleukin.

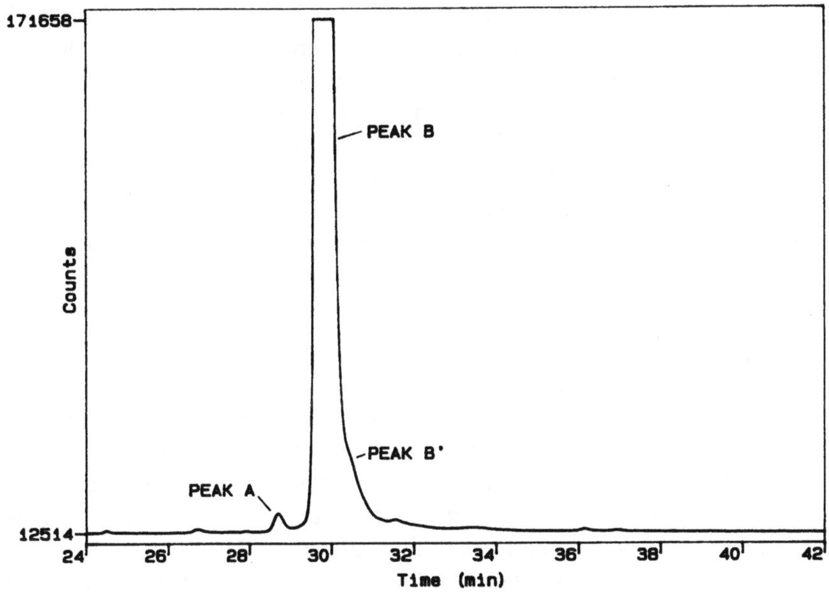

Figure 5. RP-HPLC of Proleukin.

3.2. Methods for Proleukin

To assess the stability of the lyophilized Proleukin final product, the following analytical methods were employed.

3.2.1. Potency by T-Cell Proliferation Bioassay

The cell line-derived bioassay is based on the ability of IL-2 to induce proliferation of the HT-2 mouse cell line which proliferates in the presence of IL-2 in a dose-dependent manner. Proliferation is measured by a colorimetric method based on the ability of viable cells to cleave 3-(4,5-dimethyl-thiazol-2-yl)-2,5-diphenyltetrazolium bromide (MTT) to form a dark blue formazan dye (Tada *et al.*, 1986). The assay is calibrated using the First International Standard for IL-2 and the results expressed as international units (IU) per milligram IL-2 (Gearing and Thorpe, 1988). Proleukin has a nominal specific activity of 18×10^6 IU/mg IL-2.

3.2.2. Protein Purity by SDS-PAGE

This assay separates proteins on the basis of molecular size. The assay is run under reducing and nonreducing conditions using a 12 to 20% linear gradient gel. The reducing conditions employ approximately 5% reducing agent, and reduced and nonreduced samples are heated at 90 to 100°C for about 3 min. Quantitative staining of the protein bands with the acid dye Fast Green FCF, followed by scanning densitometry in the 600-nm wavelength region, quantitates the amount of protein within each band. Figure 4 presents the SDS-PAGE densitometer scan of Proleukin. Comparison of the area of the IL-2 monomeric (main) band with the total band area establishes the percentage purity. The method is capable of detecting IL-2 monomer, oligomer, and fragments.

3.2.3. Protein Purity by RP-HPLC

This technique separates proteins onthe basis of differences in hydrophobicity. Separation is performed on a C_4 column using a nonlinear gradient of 10 to 100% acetonitrile and 0.1% trifluoroacetic acid. Elution is monitored by absorbance at 214 nm. The usual chromatogram of IL-2 consists of three main peaks which are biologically active IL-2 as shown in Fig. 5. Peak B is the predominate species and represents IL-2 monomer. Peak B′ is a shoulder on Peak B and is also IL-2 monomer. Peak A represents a modified form of IL-2 monomer resulting from the oxidation of the methionine resi-

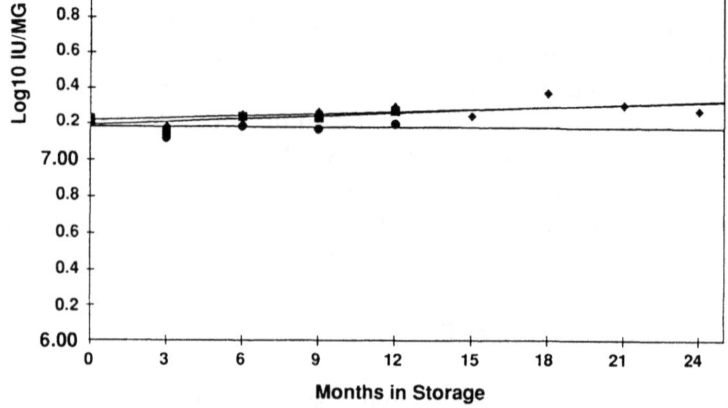

Figure 6. Summary of potency results for Proleukin stability lots (♦, 4°C; ■, 25°C; ●, 37°C).

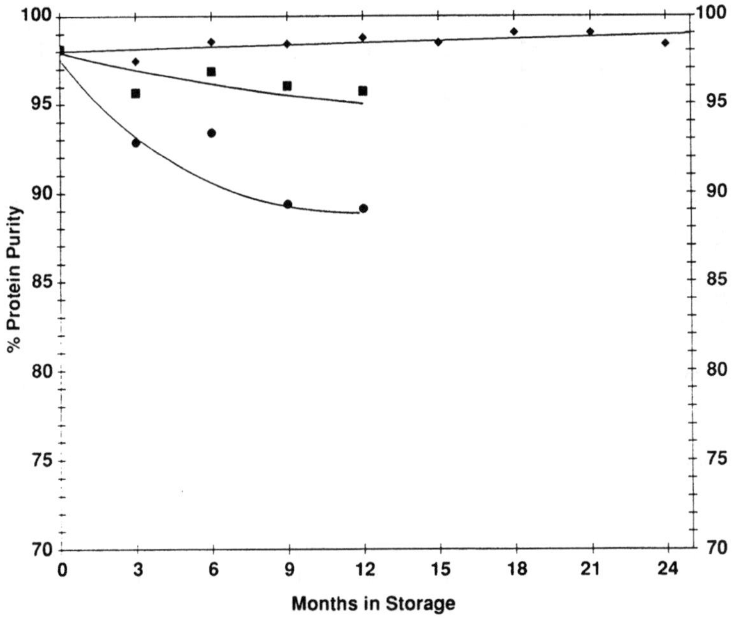

Figure 7. Protein purity by SDS-PAGE reducing gel (♦, 4°C; ■, 25°C; ●, 37°C).

due at amino acid position 104. "Other" peaks, which include IL-2 oligomers, show little or no activity and are treated as impurities.

3.2.4. Product Quality

Product quality is assessed by measurement of pH, visual appearance, residual moisture (coulometric Karl Fischer titration), and retained vacuum.

4. PROLEUKIN STABILITY

Samples from manufactured lots of Proleukin were placed on controlled temperature storage at 4°C for 24 months and at 25 and 37°C for 12 months.

Figure 6 presents the potency results over time. As found previously with another recombinant IL-2 product (Kenny *et al.*, 1986), the bioassay method was an insensitive measure of stability for Proleukin. There was no significant change in potency even after 12 months at 37°C.

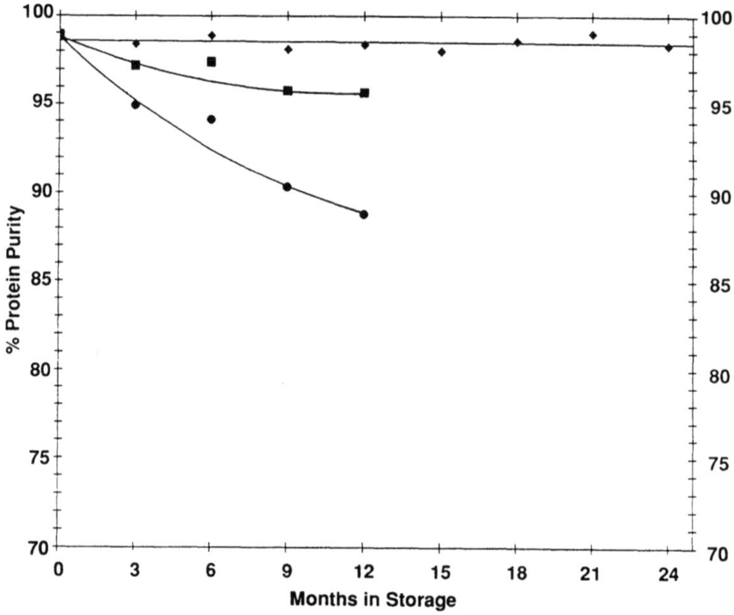

Figure 8. Protein purity by SDS-PAGE nonreducing gel (♦, 4°C; ■, 25°C; ●, 37°C).

Figures 7 (reducing conditions) and 8 (nonreducing conditions) present the SDS-PAGE protein purity results over time. There was only a slight decrease in protein purity over 24 months at 4°C. Significant decreases in protein purity, which were temperature dependent, were observed by 3 months at 25 and 37°C storage. The decrease in protein purity was due to formation of "nonreducible" oligomer (as evidenced by its presence on reduced SDS-PAGE gels). Nonreducible oligomers are formed through covalent bonding between IL-2 protein molecules, but this bonding does not occur through disulfide bridges. Such oligomers have been reported in natural proteins (Fisher and Porter, 1981; Fernandes and Lundblad, 1980; Ross *et al.*, 1984). There was no evidence for fragmentation of the IL-2 protein, which would have been noted by new bands migrating below IL-2 monomer on the reduced SDS-PAGE gels.

Figure 9 presents the RP-HPLC protein purity results over time. There was only a slight decrease in protein purity over 24 months at 4°C. Significant decreases in protein purity, which were temperature dependent, were readily observed even by 3 months at 25 and 37°C storage. The decrease in protein purity was due to the formation of the "other" peaks, which include IL-2 oligomers, that elute after peak B′. There was no evidence for oxidation of the methionine residue at amino acid position 104, which would have

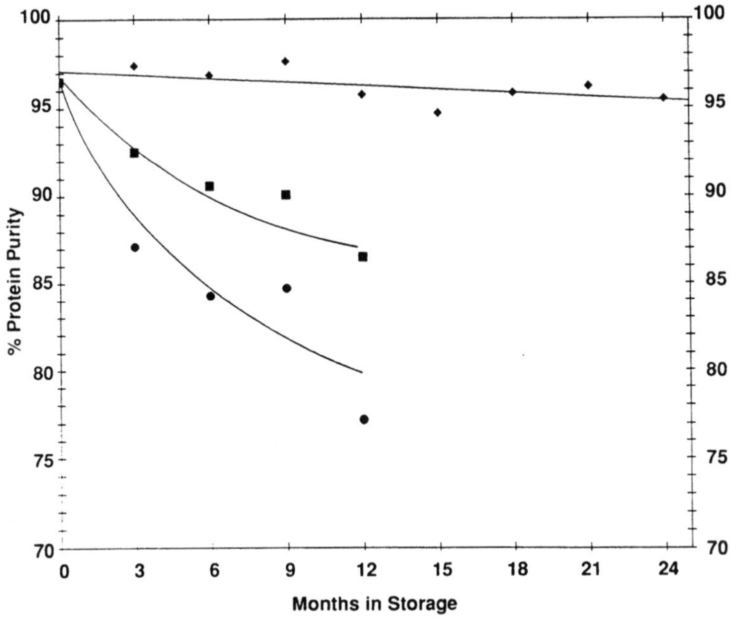

Figure 9. Protein purity by RP-HPLC (♦, 4°C; ■, 25°C; ●, 37°C).

been noted by an increase in peak A percentage. Because Proleukin is genetically engineered IL-2 (i.e., IL-2 containing not three but only two cysteine residues), it cannot undergo intramolecular disulfide scrambling.

Visual appearance, pH, residual moisture, and retained vacuum were not significantly changed over 24 months' storage at 4°C. However, at 25 and 37°C storage, Proleukin upon reconstitution yielded a yellow solution whose color intensity was dependent on both time in storage and temperature of storage.

As these data demonstrate, Proleukin is a stable product, retaining full biological potency, purity, and quality, when held at 4°C storage. Temperature stressing of the product shows that Proleukin degrades through formation of "nonreducible" oligomers and through chemical reactions that form yellow components.

5. STABILITY UPON RECONSTITUTION

In addition to determining the stability of the lyophilized product, the stability of the product upon reconstitution was established. Samples from

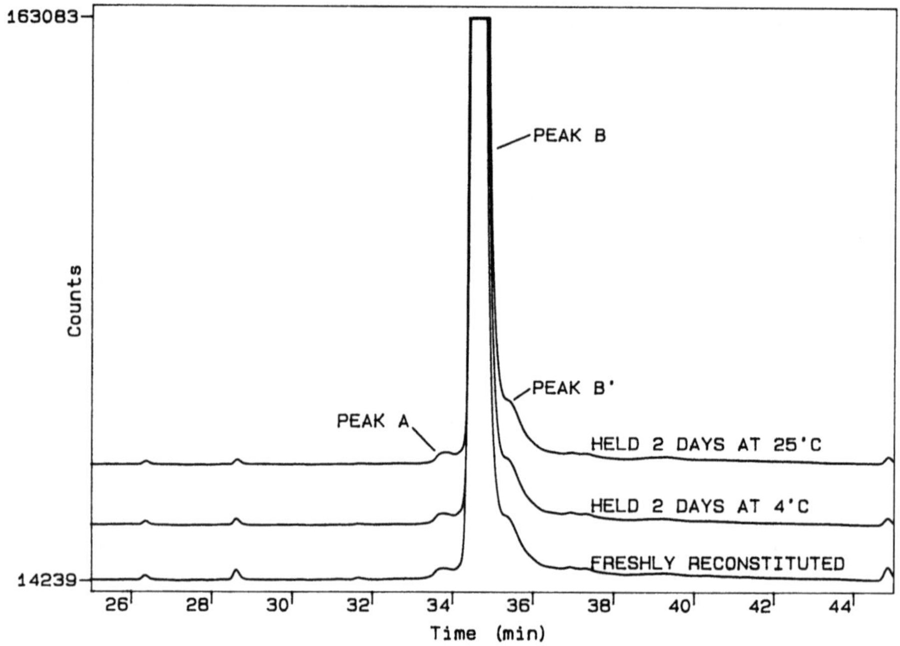

Figure 10. RP-HPLC results.

manufactured lots of Proleukin were reconstituted with Sterile Water for Injection and placed on controlled temperature storage at 4 and 25°C for 48 h. Samples were held both inverted and right-side up to evaluate a worst case scenario, sample/stopper interactions.

By the following test parameters, reconstituted Proleukin was found to be chemically and bioactively stable for up to 48 h at both 4 and 25°C: potency, protein purity by SDS-PAGE and RP-HPLC, visual appearance, and pH. Figure 10 demonstrates the RP-HPLC protein purity results of the reconstituted solutions.

Acknowledgments

We would like to thank individually the staff from the Quality Control Department whose work over several years forms the basis of this chapter: Gary Baker, Anita Beshore (presently at Synergen), Sharon Burgos, Elaine Chan, Christine Despues, Stanley Fong, Paul Huie (presently at Syva), Holly Huston, Vida Kong, Beverly McCalla, Sarah Powell, Eva Santos, Rowena Smith, Dana Wong, David Wong, and James Wong.

REFERENCES

Blackstock, W. P., Dennis, R. J., Lane, S. J., Sparks, J. I., and Weir, M. P., 1988, The analysis of recombinant interleukin-2 by thermospray liquid chromatography–mass spectrometry, *Anal. Biochem.* **175:**319–326.

Browning, J. L., Mattaliano, R. J., Chow, E. P., Liang, S. M., Allet, B., Rosa, J., and Smart, J. E., 1986, Disulfide scrambling of interleukin-2: HPLC resolution of the three possible isomers, *Anal. Biochem.* **155:**123–128.

Doyle, M. V., Lee, M. T., and Fong, S., 1985, Comparison of the biological activities of human recombinant interleukin-2_{125} and native interleukin-2, *J. Biol. Response Modifiers* **4:**96–109.

Fernandes, P. M., and Lundblad, J. L., 1980, Preparation of a stable intravenous gamma-globulin: Process design and scale up, *Vox Sang.* **39:**101–112.

Fernandes, P., and Taforo, T. A., 1986, Pharmaceutical compositions of microbially produced interleukin-2, U.S. Patent No. 4,604,377.

Ferrara, P., Pecceu, F., Marchese, E., Vita, N., Roskam, W., and Lupker, J., 1987, Characterization of recombinant glycosylated human interleukin-2 produced by a recombinant plasmid transformed CHO cell line, *FEBS* **226:**47–52.

Fisher, B. V., and Porter, P. B., 1981, Stability of bovine insulin, *J. Pharm. Pharmacol.* **33:**203–206.

Gearing, A. J. H., and Thorpe, R., 1988, The international standard for human IL-2: Calibration by international collaborative study, *J. Immunol. Methods* **114:**3–9.

Grote, W., Klaar, J., Muhlradt, P. F., and Monner, D. A., 1987, Large scale production and purification of human IL-2 from buffy coat lymphocytes stimulated with 12-O-tetradecanoyl-phorbol 13-acetate and calcium ionophore A23187, *J. Immunol. Methods* **103:**15–25.

Kato, K., Yamada, T., Kawahara, K., Onda, H., Asano, T., Sugino, H., and Kakinuma, A., 1985, Purification and characterization of recombinant human interleukin-2 produced in *Escherichia coli, Biochem. Biophys. Res. Commun.* **130:**692–699.

Kenny, W. C., Watson, E., Bartley, T., Boone, T., and Altrock, B. W., 1986, Parameters for the evaluation of IL-2 stability, *Lymphokine Res.* **5:**S23–S27.

Kunitani, M., Hirtzer, P., Johnson, D., Halenbeck, R., Boosman, A., and Koths, K., 1986, Reversed-phase chromatography of interleukin-2 muteins, *J. Chromatogr.* **359:**391–402.

Lahm, H. W., and Stein, S., 1985, Characterization of recombinant human interleukin-2 with micromethods, *J. Chromatogr.* **326:**357–361.

Landgraf, B., Cohen, F. E., Smith, K. A., Gadski, R., and Ciardelli, T. L., 1989, Structural significance of the C-terminal amphophilic helix of interleukin-2, *J. Biol. Chem.* **264:**816–822.

Marchese, E., Vita, N., Maureaud, T., and Ferrara, P., 1990, Separation by cation-exchange high-performance liquid chromatography of three forms of Chinese hamster ovary cell-derived recombinant interleukin-2, *J. Chromatogr.* **504:** 351–358.

Ralph, P., Nakoinz, I., Doyle, M., Lee, M. T., Jeong, G., Halenbeck, R., Mark, D. F., and Koths, K., 1987, Human B and T lymphocyte stimulating properties of interleukin-2 (IL-2) muteins, in: *Immune Regulation by Characterized Polypeptides,* Liss, New York, pp. 453–462.

Robb, R. J., 1985, Human interleukin-2, *Methods Enzymol.* **116:**493–525.

Ross, P. D., Finlayson, J. S., and Shrake, A., 1984, Thermal stability of human albumin measured by differential scanning colorimetry, *Vox Sang.* **47:**19–27.

Sasaoki, K., Hiroshima, T., Kusumoto, S., and Nishi, K., 1989, Oxidation of methionine residues of recombinant human interleukin-2 in aqueous solutions, *Chem. Pharm. Bull.* **37:**2160–2164.

Tada, H., Shiho, O., Kurshima, K. I., Koyama, M., and Tsukamoto, K., 1986, An improved colorimetric assay for interleukin-2, *J. Immunol. Methods* **93:** 157–165.

Wang, A., Lu, S. D., and Mark, D. F., 1984, Site-specific mutagenesis of the human interleukin-2 gene: Structure–function analysis of the cysteine residues, *Science* **224:**1431–1433.

Watson, E., and Kenny, W. C., 1988, High-performance size-exclusion chromatography of recombinant derived proteins and aggregated species, *J. Chromatogr.* **436:**289–298.

Weber, D. V., and Bailon, P., 1990, Application of receptor-affinity chromatography to bioaffinity purification, *J. Chromatogr.* **510:**59–69.

Yamada, T., Fujishima, A., Kawahara, K., Kato, K., and Nishimura, O., 1987, Importance of disulfide linkage for constructing the biologically active human interleukin-2, *Arch. Biochem. Biophys.* **257:**194–199.

Yamaka, T., Kato, K. Kawahara, K., and Nishimura, O., 1986, Separation of recombinant human interleukin-2 and methionyl interleukin-2 produced in *Escherichia coli, Biochem. Biophys. Res. Commun.* **135:**837–843.

Chapter 9

Formulation and Stability of Recombinant α_1-Antitrypsin

Sriram Vemuri, C. Tony Yu, and Niek Roosdorp

1. INTRODUCTION

Congenital α_1-antitrypsin (AAT) deficiency is one of the major causes of chronic emphysema (Laurell and Eriksson, 1963; Eriksson, 1965; Gadek and Crystal, 1983). Cigarette smoking is also known to cause chronic emphysema. The abnormal degradation of the connective tissue of the lung is believed to be due to uncontrolled proteolysis by an elastolytic enzyme released from neutrophils (Mittman, 1971). This enzyme, neutrophil elastase, degrades elastin and collagen (Barrett, 1975) and, unless controlled, will ultimately cause extensive lung damage. Normally, plasma AAT (pAAT) rapidly inactivates neutrophil elastase (Beatty *et al.*, 1980). However, when levels of pAAT are reduced, either due to congenital deficiency or by oxidative inactivation of pAAT (Johnson and Travis, 1979), enzymatic degradation of lung tissue may occur which ultimately leads to pulmonary emphysema (Travis and Salvesen, 1983). pAAT has become commercially available (Prolastin, Cutter Biological, Miles Inc., West Haven, Conn.), but the supply is limited. Development of a recombinant DNA-derived AAT

Sriram Vemuri • Scios Nova Inc., Mountain View, California 94043. *C. Tony Yu* • Bristol–Myers Squibb, Pharmaceutical Research Institute, Buffalo, New York 14213. *Niek Roosdorp* • Chiron Corporation, Emeryville, California 94608.

Stability and Characterization of Protein and Peptide Drugs: Case Histories, edited by Y. John Wang and Rodney Pearlman. Plenum Press, New York, 1993.

(rAAT) is highly desirable. Recently, cDNA clones of AAT have been constructed in plasmids and the protein expressed by yeast cells (Rosenberg *et al.*, 1984).

Human pAAT is a glycoprotein of 52 kDa. The rAAT described in this chapter is an unglycosylated protein of 44 kDa which consists of 394 amino acids. The amino acid sequence (Fig. 1) is identical to that of the native protein with the exception of an added *N*-acetylmethionine at the NH_2-terminus. The protein is heat-labile and susceptible to oxidation (Travis *et al.*, 1985; Vemuri *et al.*, 1989; Yu *et al.*, 1988). It loses biological activity rapidly upon storage at elevated temperatures or in the absence of an antioxidant (Travis *et al.*, 1985). The loss of activity is partially attributable to the oxidation of methionine-358 which is near the active binding site of the protein. The stability problem presents a great challenge to scientists who are involved with the isolation, purification, and formulation of AAT.

```
     1                                                   20
Ac-M E  D  P  Q  G  D  A  A  Q  K  T  D  T  S  H  H  D  Q  D  H
                                                         40
     P  T  F  N  K  I  T  P  N  L  A  E  F  A  F  S  L  Y  R  Q
                                                         60
     L  A  H  Q  S [N] S  T  N  I  F  F  S  P  V  S  I  A  T  A
                                                         80
     F  A  M  L  S  L  G  T  K  A  D  T  H  D  E  I  L  E  G  L
                                                         100
     N  F [N] L  T  E  I  P  E  A  Q  I  H  E  G  F  Q  E  L  L
                                                         120
     R  T  L  N  Q  P  D  S  Q  L  Q  L  T  T  G  N  G  L  F  L
                                                         140
     S  E  G  L  K  L  V  D  K  F  L  E  D  V  K  K  L  Y  H  S
                                                         160
     E  A  F  T  V  N  F  G  D  T  E  E  A  K  K  Q  I  N  D  Y
                                                         180
     V  E  K  G  T  Q  G  K  I  V  D  L  V  K  E  L  D  R  D  T
                                                         200
     V  F  A  L  V  N  Y  I  F  F  K  G  K  W  E  R  P  F  E  V
                                                         220
     K  D  T  E  E  E  D  F  H  V  D  Q  V  T  T  V  K  V  P  M
                                                         240
     M  K  R  L  G  M  F  N  I  Q  H  C  K  K  L  S  S  W  V  L
                                                         260
     L  M  K  Y  L  G [N] A  T  A  I  F  F  L  P  D  E  G  K  L
                                                         280
     Q  H  L  E  N  E  L  T  H  D  I  I  T  K  F  L  E  N  E  D
                                                         300
     R  R  S  A  S  L  H  L  P  K  L  S  I  T  G  T  Y  D  L  K
                                                         320
     S  V  L  G  Q  L  G  I  T  K  V  F  S  N  G  A  D  L  S  G
                                                         340
     V  T  E  E  A  P  L  K  L  S  K  A  V  H  K  A  V  L  T  I
                                                         360
     D  E  K  G  T  E  A  A  G  A  M  F  L  E  A  I  P (M) S  I
                                                         380
     P  P  E  V  K  F  N  K  P  F  V  F  L  M  I  E  Q  N  T  K
                                           394
     S  P  L  F  M  G  K  V  V  N  P  T  Q  K
```

Figure 1. Protein sequence of recombinant α_1-antitrypsin ((M), active site; [N], glycosylation attachment site).

2. OVERVIEW OF THE MANUFACTURING PROCESS

The manufacture of rAAT consists of the following steps: yeast fermentation, rAAT purification, concentration, formulation, and packaging. The fermentation process is comparable to that used in the food and beverage industries, e.g., the brewing of beer. Conditions such as nutrient feed, pH, agitation, partial pressure of oxygen, and air flow in the reactor are optimized to obtain the maximum yield of rAAT. Extraction of rAAT protein from the yeast cells is accomplished by disruption of the cells and separation of the cell debris from rAAT protein. Separation of rAAT protein from the yeast proteins is achieved by gel filtration, a preliminary coarse filtration. Further purification is accomplished by methods such as ion-exchange chromatography, affinity chromatography, hydrophobic interaction chromatography, and size-exclusion chromatography. The various steps of the chromatographic purification yield a dilute aqueous solution of rAAT which is then concentrated by using an ultrafiltration device equipped with an appropriate molecular weight cutoff membrane. The concentrated rAAT solution is combined with other pharmaceutical excipients such as sodium citrate and lactose. The final formulation is then filtered aseptically by bacterial retentive filter, filled into small vials, and lyophilized.

3. ANALYTICAL METHODS

The monomeric content of rAAT is determined by size-exclusion HPLC. The biological activity of rAAT is determined by measuring its ability to inhibit the enzymatic activity of human sputum elastase. The total rAAT protein is measured by ELISA. The total nitrogen content is determined by a chemiluminescence method.

3.1. Determination of Monomeric rAAT Content

The monomeric content of rAAT is determined by size-exclusion HPLC for the monomer peak as the percent of total peak area. The column separates proteins according to their molecular weight. The higher the molecular weight, the shorter the retention time is. A typical chromatogram is shown as Fig. 2. The HPLC system (Nicolet Analytical Instruments, Madison, Wisc.) consists of a liquid chromatograph (LC/9560), a UV detector (LC/9563), and an autosampler (LC/9505). A Zorbax GF 250 analytical column (25 cm × 9.4 mm) is used. The eluent is 0.2 M sodium phosphate buffer (pH 7.0). The flow rate is 1.0 ml/min. The wavelength of detection is set at 278 nm. The percent area of each peak with respect to total peak area is calculated.

3.2. Determination of rAAT Activity

AAT is a protease inhibitor which binds to human sputum elastase, among other proteases, and blocks the reaction of the elastase to its substrate. Methoxysuccinyl-alanyl-alanyl-prolyl-valyl-*p*-nitroanilide is chosen as the substrate because it is converted to *p*-nitroanilide, a yellow substance which can be quantitated spectrophotometrically at 405 nm. The binding of rAAT to elastase occurs stoichiometrically and irreversibly. Human sputum elastase is obtained from Elastin Products Co. and the substrate is obtained from Sigma.

3.3. Determination of Total rAAT Protein

The total rAAT protein is determined by an enzyme-linked immunosorbant assay (ELISA). The assay is based on competition between an un-

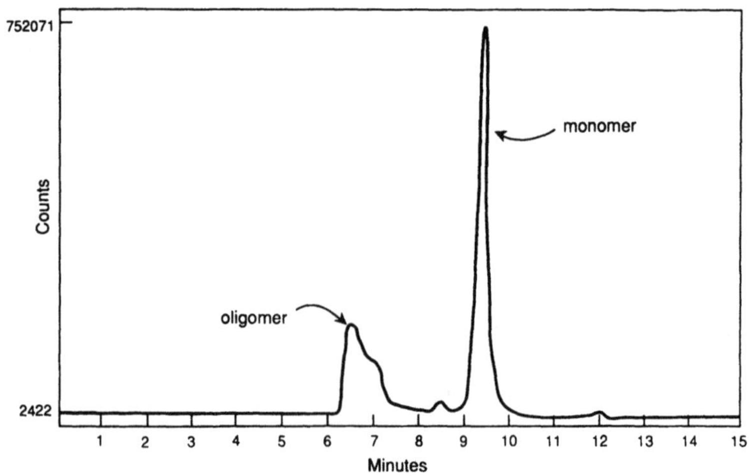

Figure 2. Typical size-exclusion chromatogram of rAAT. The major peak with a retention time of approximately 9.5 min is the rAAT monomer.

known amount of rAAT in the sample and a fixed amount of horseradish peroxidase-conjugated rAAT (HRP-rAAT), for the rAAT-specific antibody bound to a plastic surface. The more rAAT in the test sample, the less HRP-rAAT will be bound to the limited amount of anti-rAAT antibody. The amount of HRP-rAAT bound is measured spectrophotometrically by the addition of chromogenic substrate to the bound HRP-rAAT. The reaction of the chromogenic substrate and the horseradish peroxidase is followed at 414 nm. The substrate, 2,2-azinobis(3-ethylbenzthiazoline sulfonic acid), is obtained from Sigma.

3.4. Determination of Total Nitrogen

The rAAT sample is subjected to high-temperature oxidation. The nitrous oxide generated is exposed to ozone, creating a chemiluminescence in which the light emission is proportional to the amount of nitrogen in the sample. The chemiluminescent nitrogen detection system (Model 703C, Antec Instruments) consists of a pyroreactor and a nitrogen detector. The assay is performed by injecting a known amount of bovine serum albumin (BSA) as a standard, and the unknown sample. The total nitrogen in the test sample is calculated with respect to the known value of BSA.

4. PREFORMULATION STUDIES

AAT is labile in solution (Lieberman, 1973; Travis *et al.*, 1985). Earlier studies in our laboratory revealed that a solution of rAAT in sodium phosphate buffer (pH 7.0) had a shelf life of 2½ weeks at 4°C. The shelf life was improved to 6 weeks at 4°C by formulating the rAAT in sodium citrate/sodium phosphate buffer (pH 7.0). Effects of pH, potassium chloride, sodium citrate, and ionic strength on the rAAT stability are discussed in this section.

4.1. Effect of pH on rAAT Solution Stability

The effect of pH on the stability of rAAT was investigated. The rAAT was formulated in a buffer solution containing 20 mM citrate and 10 mM phosphate. The solution was divided into several aliquots and each aliquot was adjusted to a different pH with 1 N HCl or 1 N NaOH. The initial concentration of rAAT was 50 mg/ml. Each aliquot was kept on stability at room temperature for 2 months. Sample vials were taken and analyzed for monomeric rAAT by HPLC, and activity by the elastase inhibition assay.

Results showed that monomeric rAAT decreased rapidly at both low and high pH. Oligomeric and polymeric rAATs were formed accordingly (Fig. 2). Two other peaks (retention time 8.5 and 12 min) are seen in addition to rAAT monomeric peak, in degraded samples. Degradation kinetics of rAAT and its aggregation products was followed at various pHs for 8 weeks. Degradation rate constant (k) was calculated at various pH values. A plot of log k versus pH revealed a V-shaped stability profile (Fig. 3). The best stability results were obtained from the aliquot at pH 7.5. At that pH, rAAT aggregates at a rate of 1.4% per week. At pH 6.5 and 9.0, the aggregation rates are about one and a half times higher than the rate at pH 7.5. The stability of the enzymatic activity of rAAT, however, does not seem to be affected by the pH in the range studied, as revealed in Table I, although the variability of the assay is relatively high.

4.2. Effect of Potassium Chloride on rAAT Solution Stability

Potassium chloride has been reported to prevent freezing damage to catalase when stored at −40°C (Fishbein and Winkert, 1979). In order to study this effect on rAAT, formulations of rAAT in a 0.02 M KH_2PO_4 buffer were spiked with various amounts of KCl. Samples were stored at 4°C for 8 weeks. Product stability was assessed by monitoring percent monomeric content of rAAT using size-exclusion HPLC and enzymatic activity by the elastase inhibition assay. No filtration prior to analysis was performed. Results show that the higher the KCl content, the greater the rAAT stability is (Fig. 4). The zero-order aggregation rates at 4°C are 1.44% per week for 0.1

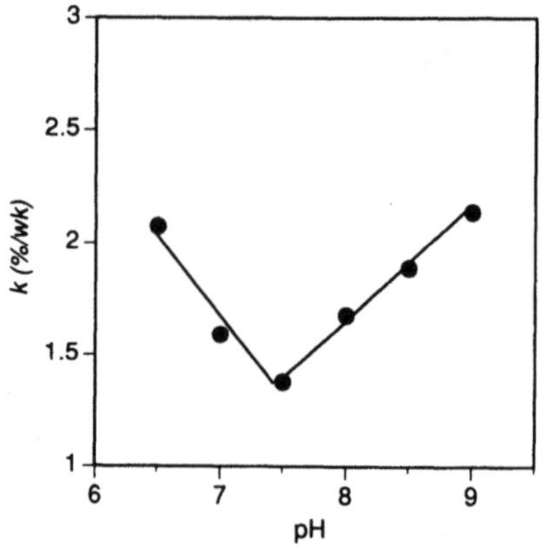

Figure 3. pH stability profile (at 25°C) of monomeric rAAT by SE-HPLC assay. Monomeric rAAT decreased rapidly at both low and high pHs. A plot of log k versus pH revealed a V-shaped stability profile with optimal stability at pH 7.5.

Table I
Effect of pH on Enzymatic Activity of rAAT by Elastase Inhibitor Assay

	mg/ml active rAAT (% of initial)		
pH	0	2 weeks	8 weeks
6.5	25.0 (100)	24.1 (96.4)	28.6 (114.4)
7.0	24.1 (100)	25.5 (105.8)	33.0 (136.9)
7.5	25.2 (100)	26.7 (106.0)	33.1 (131.3)
8.0	23.7 (100)	29.7 (125.3)	32.6 (137.6)
8.5	27.1 (100)	30.1 (111.1)	32.1 (118.5)
9.0	25.4 (100)	29.3 (115.4)	33.3 (131.1)

M KCl-, 0.61% per week for 0.5 M KCl-, and 0.30% per week for 1.0 M KCl-containing formulations. The enzymatic activity of rAAT, however, does not seem to be affected by KCl, as is evident from Fig. 5. No apparent loss of enzymatic activity was observed from the three KCl-containing formulations.

4.3. Effect of Ionic Strength on rAAT Solution Stability

Ionic strength is widely known to affect the stability of a protein to a large extent. The current study investigates the effect of 0.1, 0.5, and 1.0 M NaCl on the stability of rAAT in a phosphate/citrate-buffered solution (0.02 M each of sodium phosphate and sodium citrate, pH 7.5). The study was conducted at 4°C in a similar fashion to the one described for KCl in the

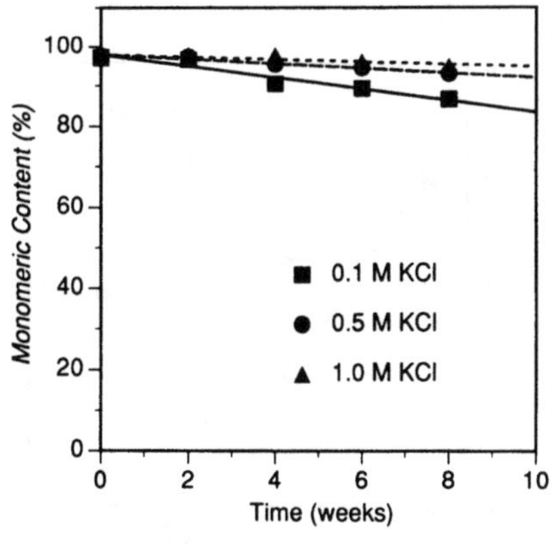

Figure 4. Effect of KCl content on stability of rAAT in solution at 4°C. The product stability was assessed by SE-HPLC.

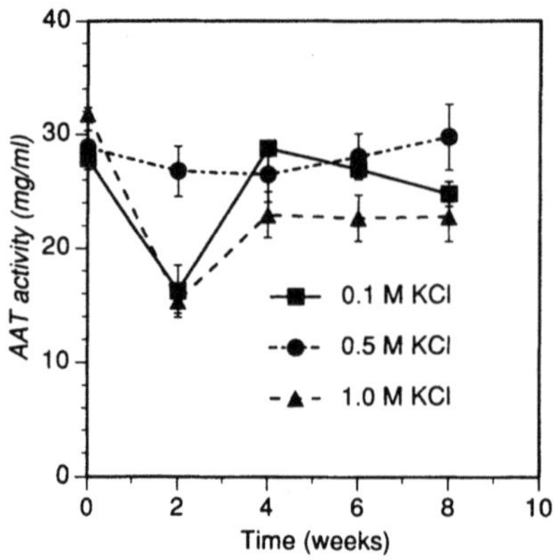

Figure 5. Effect of KCl concentration in formulation on enzymatic activity of rAAT solution at 4°C. Enzymatic activity was assessed by the elastase inhibition assay.

preceding section. Results of the disappearance rate constants of monomeric rAAT are shown in Fig. 6. The rate constants from the KCl study are also plotted in Fig. 6 for comparison. As can be seen, the higher the salt concentration, the more stable the rAAT solution is. The 1.0 M NaCl-containing rAAT solution aggregates at two-thirds the rate of the 0.1 M NaCl-containing rAAT solution, while the 1.0 M KCl-containing rAAT solution aggregates at one-fifth the rate of the 0.1 M KCl-containing rAAT solution.

4.4. Effect of Citrate on rAAT Solution Stability

Two buffer solutions of rAAT were examined: a phosphate-buffered (18 mM phosphate, pH 7.0) solution and a citrate-buffered (18 mM citrate + 18 mM phosphate, pH 7.0) solution. The concentrations studied were 10 and 50 mg/ml. The solutions were divided into several aliquots and packaged in small vials. The sample vials were stored at 4°C and at room temperature for

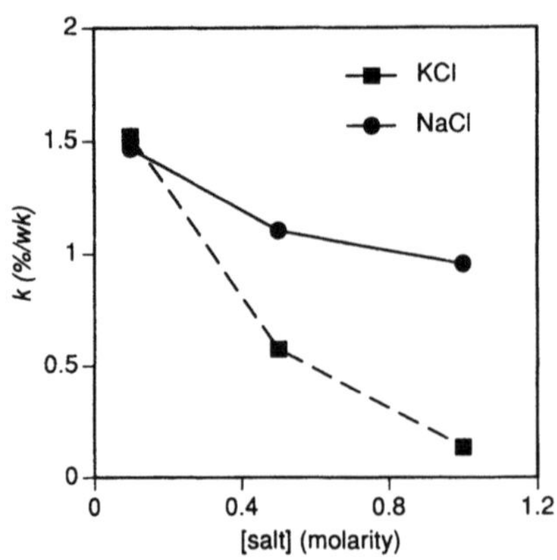

Figure 6. Effect of salt concentration on rAAT solution stability. The study was carried out in the presence of either NaCl or KCl. Results of the disappearance rate constants of the monomeric rAAT at various salt concentrations are shown.

up to 9 weeks. Sample vials were taken from storage weekly and analyzed for monomeric rAAT by HPLC, and activity by the elastase inhibition assay.

Figure 7 shows the stability of rAAT by HPLC assay. The results are expressed as percent of the initial monomeric content. It is evident that the citrate solution has a smaller loss of the monomeric rAAT over time than the phosphate solution. Linear regression parameters of the curves are summarized in Table II. The outlier at one-week time point for the 10 mg/ml rAAT in the citrate solution is excluded from the linear regression. As shown in Table II, the shelf life of the citrate solution (50 mg/ml) at 4°C is about five times longer than that of the phosphate solution.

The stability of the elastase inhibitory capacity of rAAT, expressed as percent of the initial activity (Fig. 8), is difficult to interpret because the data show no consistent trend. The apparent loss, or increase of the activity, may be due to the large assay uncertainty.

4.5. Physical Stability of rAAT Solution

AAT is known to be very labile in solution (Lieberman, 1973; Travis *et al.,* 1985; Vemuri *et al.,* 1989; Yu *et al.,* 1988). For example, Lieberman (1973) reported an *in vitro* half-life of 60 min for normal plasma-derived AAT at 55°C as measured by its trypsin-inhibitory capacity. In the same study, he reported that the half-lives for other phenotype AATs were about 30 min. AATs (Met 358 and Val 358) were reported (Travis *et al.,* 1985) to have an *in vitro* half-life of 40 min at 56°C and an *in vivo* (rabbits) half-life of 8.5 hr. The rapid loss of AAT activity was attributed to oxidation of the

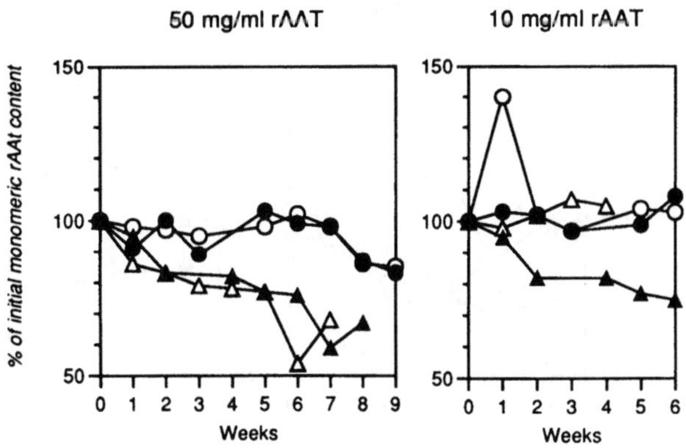

Figure 7. Stability of rAAT solution with regard to its monomeric purity assessed by SE-HPLC. ○, citrate formulation at 4°C; △, phosphate formulation at 4°C; ●, citrate formulation at room temperature; ▲, phosphate formulation at room temperature.

Table II
Linear Regression Parameters and Estimated Shelf Life (t_{90}) for Monomeric rAAT

	Citrate formulation				Phosphate formulation			
	10 mg/ml		50 mg/ml		10 mg/ml		50 mg/ml	
	4°C	RT	4°C	RT	4°C	RT	4°C	RT
Correlation coefficient *r*	0.29	0.32	0.62	0.38	0.85	0.52	0.90	0.95
k (%/week)	−1.99	0.49	−1.17	−0.83	1.80	−1.90	−5.43	−4.40
Shelf life, t_{90} (weeks)	5.0		8.5	12.0		5.3	1.8	23.0

exposed methionine residue (Travis *et al.*, 1985) and the general thermal lability (Lieberman, 1973) of the AAT. Oxidation and/or thermal inactivation of protein are often accompanied by a conformational change of the protein molecule. The conformational change frequently results in aggregation, particulate formation, or precipitation. An AAT solution may, therefore, encounter physical stability problems in addition to the problem of activity loss. Efforts are warranted to ensure an adequate physical shelf life in addition to the required potency shelf life for an injectable rAAT formulation.

4.5.1. Particulate Formation

A solution of rAAT (1.0 mg/ml) was made by dissolving the lyophilized rAAT powder which consisted of 95.2% monomers and 4.8% dimers in a

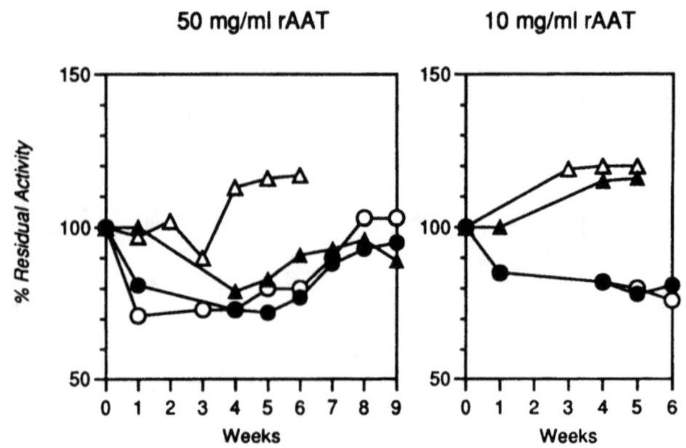

Figure 8. Stability of rAAT solution with regard to its elastase inhibitory activity assay. ○, citrate formulation at 4°C; △, phosphate formulation at 4°C; ●, citrate formulation at room temperature; ▲, phosphate formulation at room temperature.

triple buffer (Tris, 0.075 M; glycine, 0.075 M; sodium phosphate, 0.075 M; pH 7.5). The solution was passed through a 0.2-μm filter and subdivided into 0.5-ml aliquots. The filter membrane used was a low-protein-binding polyvinylidene difluoride membrane (Millex GV, Millipore, Bedford, Mass.). No protein binding problem was encountered. Each aliquot was filled into a small vial and capped with a rubber stopper and an aluminum seal. A given number of vials were placed in ovens at 25, 35, 45, 60, 80, or 90°C. A vial was removed from each oven at a predetermined time interval. The content of the vial was filtered through a 0.2-μm filter to remove the aggregates. The filtrate was then assayed for total soluble protein by the biuret–phenol method (Layne, 1957).

Figure 9 shows semilog plots of the percent rAAT remaining versus time for given storage temperatures. A typical first-order loss of rAAT was observed. Data were fitted to a first-order kinetics and a linear regression line was drawn through the data points. The half-life (t_{50}) and the shelf life (t_{90}) were calculated from the resultant rate constants, as shown in Table III. Also shown are the coefficients of correlation (r) for the linear regression. The half lives and the shelf lives were then plotted against the reciprocals of storage temperatures (a modified Arrhenius plot) as shown in Fig. 10. For data interpretation, temperatures as degrees centigrade corresponding to reciprocals of temperatures expressed as degrees Kelvin are also indicated in Fig. 10. A linear regression line ($r = 0.9831$) was drawn through the data points and extrapolated to 4°C. At 4°C, the predicted shelf life of the rAAT solution is 5.1 months. In real time, a batch of the rAAT solution showed that 61% of the rAAT remained in the filtrate after 18 months of storage at 4°C. The observed stability compares fairly well with the predicted value of 69% for the same duration of storage at 4°C.

An Arrhenius plot of rate constants versus reciprocals of storage temperatures is shown in Fig. 11. From the slope of the linear regression line, the activation energy for the physical changes of the rAAT injection was calculated to be 19 kcal/mole. This value is comparable to the activation energy of 20 kcal/mole for ovalbumin denaturation reaction reported by Simpson and Kauzmann (1953). It is possible that the physical changes observed in this study are due predominantly to protein denaturation.

4.5.2. Formation of Soluble Aggregate

Hydrophobic interactions between protein molecules can lead to noncovalent bonding or aggregation. Perturbation of the tertiary structure of the protein leads to denaturation which may not be reversible. The partially unfolded intermediates have been shown to exist for many proteins (Arakawa *et al.*, 1987; Brems *et al.*, 1988).

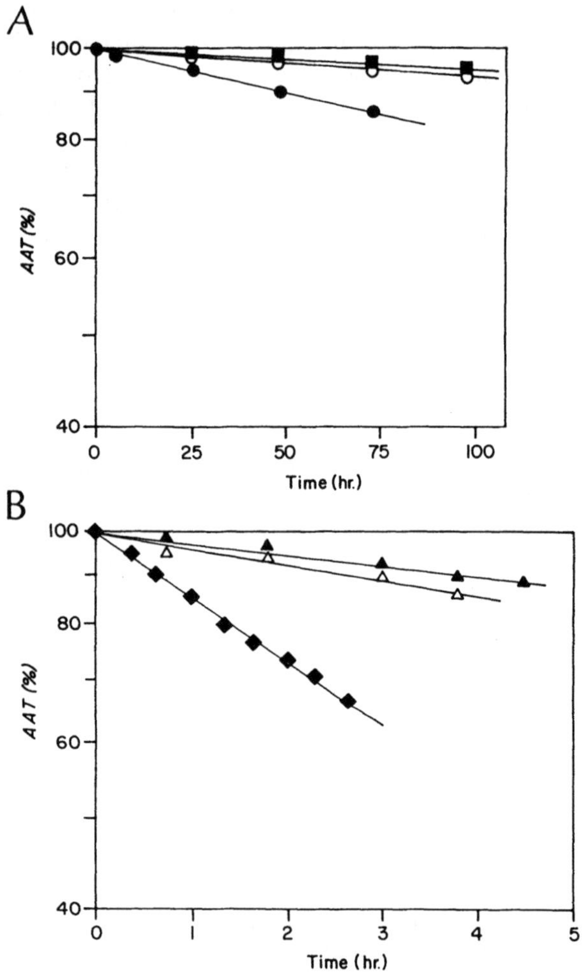

Figure 9. First-order loss of rAAT to precipitation. The data were fitted to a first-order kinetics and a linear regression line was drawn through the data points. (A) The semilog plot of percentage of rAAT remaining versus time at 45°C (●), 35°C (○), and 25°C (■). (B) The semilog plot of percentage of rAAT remaining versus time at 90°C (◆), 80°C (△), and 60°C (▲).

The current study separates the insoluble particulates from the soluble aggregates by way of filtration through a 0.2-μm filter. The filtrate was then analyzed by SE-HPLC. The results were expressed as percent remaining of the monomeric rAAT. Three concentrations of rAAT were studied: 10, 20, and 25 mg/ml. The solutions were made in the phosphate–citrate buffer (pH 7.4) described in an earlier section. The solutions were filtered through a

Table III
Estimated Shelf Life and Half-Life for the Formation of Particulate Matter in a rAAT Formulation

	Storage temperature					
	90°C	80°C	60°C	45°C	35°C	25°C
Correlation coefficient (r)	0.986	0.970	0.960	0.944	0.969	0.983
Rate constant × 10^3 (hr^{-1})	138.0	34.3	15.0	1.59	0.68	0.50
Half-life (t_{50}, hr)	5.02	20.2	46.2	436.0	1017.0	1383.0
Shelf life (t_{90}, hr)	0.76	3.07	7.02	66.3	154.7	210.3

0.2-μm filter and subdivided into 0.5-ml aliquots. Each aliquot was placed in a small vial and capped with a rubber stopper and an aluminum seal. A given number of the vials were placed at 50, 37, or 25°C or in a refrigerator at 5°C. A vial was removed from each storage condition at a predetermined time interval. The content of the vial was passed through a 0.2-μm filter to remove the particulates. The filtrate was then assayed by SE-HPLC. A typical chromatogram is shown in Fig. 2. The rate constant for the disappearance of the monomeric rAAT at a given storage temperature is then calculated. Figure 12 shows the Arrhenius plot of the rate constant versus the reciprocal of the

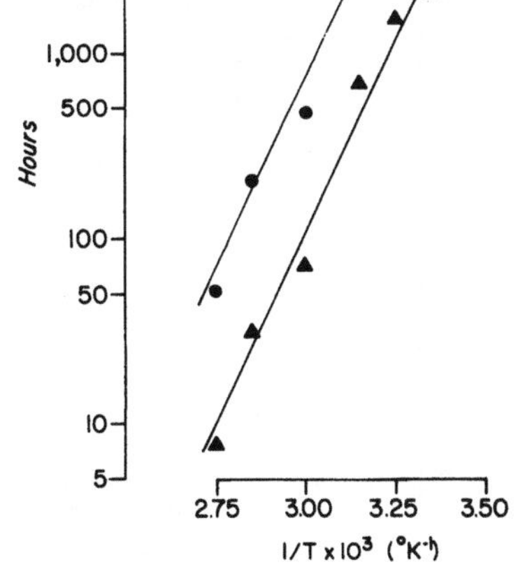

Figure 10. A modified Arrhenius plot is shown for loss of rAAT monomer. The half lives and shelf lives are plotted against the reciprocals of storage temperature. The half-life and shelf life were calculated from the resultant rate constants, as shown in Table III.

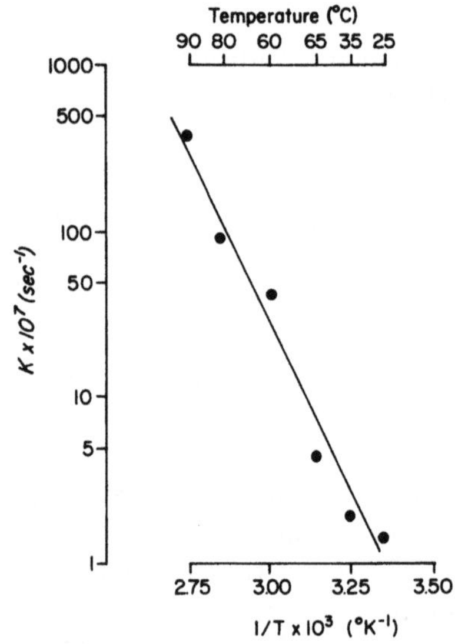

Figure 11. An Arrhenius plot of thermal denaturation of rAAT in solution. The rate constants versus reciprocals of storage temperatures are shown. From the slope of the line, the activation energy for the physical changes of rAAT was calculated.

storage temperature. From the slope of the linear regression line, the activation energy for the protein aggregation can be calculated. Table IV lists the activation energy for the formation of the soluble aggregates. The activation energy ranges from 13.0 to 17.5 kcal/mole for various concentrations of rAAT. These values are comparable to the 19.0 kcal/mole observed for the formation of insoluble particulates (Yu *et al.*, 1988). The mechanism for

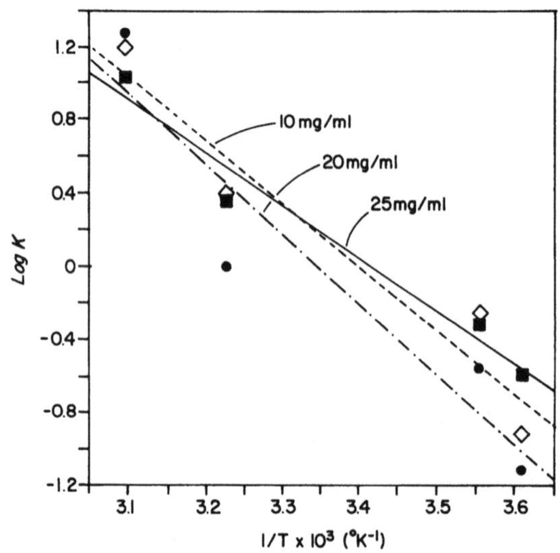

Figure 12. The rate constants for the disappearance of monomeric rAAT at a given temperature are plotted. In this study, soluble aggregates are separated from nonsoluble aggregates by 0.2-μm membrane filtration. The rate constants versus reciprocals of storage temperatures are shown.

Table IV
Activation Energy (E_a) for rAAT Particulate Formation and rAAT Aggregate Formation

Physical parameter tested	rAAT concentration	Test method	E_a (kcal/mole)
Particulate matter (remaining protein in solution)	1 mg/ml	Lowry	19.1
Aggregates in solution (remaining monomers in solution)	10 mg/ml	SE-HPLC	17.5
	20 mg/ml	SE-HPLC	16.0
	25 mg/ml	SE-HPLC	13.0

formation of the soluble aggregates appears to be similar to that of the insoluble particulates, despite the fact that the rate of the former is almost 1000 times that of the latter at 25°C (Table V).

Preformulation studies of rAAT identified several optimal conditions to stabilize protein in solution. These conditions will be considered further in the formulation development work.

5. FORMULATION DEVELOPMENT

5.1. Solution Stability Screening

Results (Section 4.5.1) showed a shelf life (t_{90}) of 5.1 months at 4°C. The current study represents an initial attempt to formulate rAAT in a stable solution. Adjuvants examined were polyethylene glycol 400, sucrose, zinc chloride, Tween 20, sodium citrate, EDTA, thioglycerol, ascorbic acid, and sodium formaldehyde sulfoxylate. Polyethylene glycol and sucrose were known to protect proteins from freeze–thaw damage (Fennema, 1979; Tsuchiya *et al.*, 1975; Fishbein and Winkert, 1979). Zinc chloride was chosen to investigate the effect of complexation on rAAT stability, because the Zn^{2+} ion is known to form complexes with proteins. Tween 20 was chosen to investigate the effect of surfactant on rAAT stability. Sodium citrate and EDTA were chosen to study the effect of chelating agents on rAAT stability. Thioglycerol, ascorbic acid, and sodium formaldehyde sulfoxylate were studied for their antioxidation effects.

Twenty-four solution formulations were screened for their stability upon storage at 4°C or room temperature. Each formulation consists of 10 or 50 mg/ml rAAT in 0.02 M sodium phosphate buffer (pH 7.0) and one of the following adjuvants at the specified concentration:

1. Polyethylene glycol (PEG) 400, 2%
2. PEG 400, 4%
3. Zinc chloride, 0.1%
4. Tween 20, 0.01%
5. Sodium citrate, 0.018 M
6. Sodium citrate, 0.018 M, and disodium-EDTA, 0.01%
7. Sucrose, 5%
8. Sucrose, 10%
9. Thioglycerol, 0.5%
10. Ascorbic acid, 0.1%
11. Sodium formaldehyde sulfoxylate, 0.1%
12. Control formulation (i.e., no adjuvant)

The solutions were sterile-filtered and aseptically packaged into small vials. The vials were stored at 4°C or room temperature (about 22°C). Sample vials were removed from storage weekly. The monomeric content of rAAT was assayed by SE-HPLC. The enzymatic activity of rAAT was measured by its elastase inhibition capacity.

A zero-order rate constant for the loss of monomeric rAAT was obtained from the slope of the linear regression line of the percent remaining versus time data for each formulation at each temperature. The zero-order kinetics was chosen for simplicity.

Table V shows the rate constants for various strengths of the rAAT formulation. A rank order was given to each rate constant to indicate the relative stabilization power of each adjuvant. The rank orders are summarized in Table VI. Tables V and VI reveal a few interesting phenomena. Among various categories of stabilizers, sugar seems to be most effective, although it is only effective at a relatively high concentration. The 10% sucrose formulation clearly stabilized while the 5% sucrose formulation was not much better than the control formulation. Chelating agents appeared to be effective stabilizers. Both citrate formulations (with and without EDTA) showed a stabilization effect for the 50 mg/ml rAAT solution; however, the effect was not clear for the 10 mg/ml solution. No synergistic effect was observed between EDTA and citrate. Among the three antioxidants studied, only thioglycerol appeared to have some stabilization effect. The other two, namely ascorbic acid and sodium formaldehyde sulfoxylate, actually showed an adverse effect on the monomeric stability of rAAT. Polyethylene glycol stabilized the 50 mg/ml, but destabilized the 10 mg/ml rAAT solution. This phenomenon apparently resulted from the fact that polyethylene glycol indiscriminately produced rate constants of about 1.0–1.5%, which is between the

Table V
Estimated Shelf Life and Half-Life for Various Formulations: Loss of Monomeric rAAT Monitored by SE-HPLC

	Storage temperature			
	50°C	37°C	25°C	4°C
rAAT 10 mg/ml formulation				
Rate constant (concentration hr^{-1})	18.9010	1.0099	0.2809	0.0773
Half-life (t_{50}, hr)	0.529	4.951	17.800	64.683
Shelf life (t_{90}, hr)	0.055	10.317	3.738	13.583
rAAT 20 mg/ml formulation				
Rate constant (concentration hr^{-1})	15.7782	2.2648	0.5518	0.1201
Half-life (t_{50}, hr)	0.634	4.415	18.123	83.264
Shelf life (t_{90}, hr)	0.127	0.882	3.625	16.65
rAAT 25 mg/ml formulation				
Rate constant (concentration hr^{-1})	10.5999	2.4830	0.4779	0.2547
Half-life (t_{50}, hr)	1.179	5.034	26.156	49.077
Shelf life (t_{90}, hr)	0.236	1.007	5.231	9.815

rate constants of the 10 mg/ml control (0.5–0.9%) and the 50 mg/ml control (2.8–3.8%). Zinc chloride precipitated rAAT after 3 weeks' storage at 4°C. Tween showed an adverse effect on the monomeric stability of rAAT.

The stabilization effect of the adjuvant on enzymatic activity of rAAT appeared to be obscured by the variability of the assay results. In many cases, activity increased substantially upon storage. The increased activity over time rendered regression analysis meaningless. No conclusion was drawn from the activity results. Sugars offered the best protection against aggregation. The formulations with sucrose, citrate, and citrate/EDTA are the top three choices; however, the safety of sucrose for parenteral use is unknown. Thus, the citrate formulation was chosen as the formulation for lyophilization.

5.2. Lyophilization

Lyophilization of pharmaceuticals can yield high-quality products with an increased shelf life. Proteins or small drug molecules with stability problems are good candidates for lyophilization. Lyophilization is a process in which water is removed from formulations by sublimation at low temperatures under vacuum. The biological activity of proteins dried at low temperature is fairly well preserved and their shelf life prolonged in the dry state.

Table VI
Rank Order and Rate Constant (k in % Loss of Monomeric AAT per Week) of Various Formulations

rAAT (50 mg/ml) liquid formulation with individual adjuvant			
At RT	k	At 4°C	k
10% sucrose	0.79	10% sucrose	0.44
Citrate	1.12	4% PEG	1.03
Zinc chloride	1.15	Citrate with EDTA	1.17
Citrate with EDTA	1.18	2% PEG	1.51
4% PEG	1.48	Citrate	1.62
2% PEG	1.58	Zinc chloride	1.77
5% Sucrose	2.39	Ascorbic acid	2.29
Control (no adjuvant)	2.89	5% sucrose	2.47
Ascorbic acid	3.09	Tween	3.19
		Control	3.80

5.2.1. Formulation Consideration

Lyophilization, or freeze-drying, can be divided into three stages: freezing, primary drying, and secondary drying. The success of lyophilization is dependent on the formulation constituents as much as the freeze-drying cycle (Pikal, 1990; Vemuri, 1991). Freezing of a protein solution can cause denaturation of the protein because of pH shift or ionic strength changes during freezing (Orii and Morita, 1977). The lyophilized cake can slowly lose integrity under conditions of inferior formulation and/or improper lyophilization cycle. Pharmaceutical scientists play a critical role in selection of proper excipients to ensure the success of lyophilization. Buffers, bulking agents, tonicity modifiers, antioxidants, and chelating agents are commonly used to stabilize a lyophilized product. Bulking agents such as mannitol, glycine, povidone, sucrose, and lactose are used to enhance product elegance and the mechanical strength of the lyophilized cake. These agents also prevent a "blowout" which may occur when the total solid is less than 1%.

When a protein solution is subjected to freezing, the protein is exposed to an environment which is quite different from the solution state. The salt begins to concentrate in the ice lattice and, as a result, a downward shift of the pH may occur in the freezing phase of the lyophilization. This is due to the partial crystallization of the salt in the buffer. For example, Na_2HPO_4 crystallizes faster than NaH_2PO_4 in the phosphate buffer (Larsen, 1973). The phenomenon causes a dramatic decrease in the pH of a phosphate buffer when the temperature decreases in the freezing phase of the lyophilization. Other salts such as sodium citrate, potassium citrate, and potassium phosphate tend to remain amorphous during freezing and thereby produce smaller pH changes. The pH change in the microenvironment of the ice lattice may promote denaturation and subsequent aggregation of the protein.

5.2.2. Effect of Bulking Agent

Lyophilization is employed to improve the shelf life of rAAT. Effects of several commonly used bulking agents, namely sucrose, lactose, and povidone (PVP), on the stability of lyophilized rAAT were studied. A solution of rAAT in a phosphate/citrate buffer (pH 7.0) was made with or without the bulking agent and placed in small vials. The vials were frozen at −40°C in a lyophilizer. At the end of the freezing cycle, half of the batch was removed and thawed at ambient temperature and the other half completed the lyophilization. Sample vials taken before freezing, after thawing, and after lyophilization were analyzed for monomeric rAAT content and activity. In addition, the samples were also analyzed for total rAAT protein by ELISA and for total nitrogen by chemiluminescence. Two concentrations of rAAT (10 and 50 mg/ml) were studied. The bulking agent concentration was 50 mg/ml.

Results of monomeric rAAT content, elastase inhibitory activity, total rAAT proteins, and total nitrogen content before and after lyophilization are shown in Tables VII and VIII for the 50 and 10 mg/ml rAAT formulations, respectively. Each value represents the mean of three determinations. PVP reduces the monomeric content of rAAT and, possibly as a result, lowers the activity of rAAT. PVP also severely interferes with the assay of total nitrogen because of its high nitrogen content. Sucrose formulation produces a 10–15% loss of the total nitrogen, possibly due to physical adsorption. As a result, the activity is correspondingly lower than for the control. None of the bulking agents studied interferes with the ELISA. Results of monomeric rAAT content reveal the existence of a polymeric form of rAAT even before freezing. The monomeric purity of the formulation is on average, less than 90%. The PVP formulation was as low as 65% monomeric pure. A comparison of the 10 and 50 mg/ml rAAT data reveals that the above observations hold true for the 10 mg/ml rAAT formulations. Small differences observed are within experimental variation. This finding suggests that the rAAT concentration is not critical for the current lyophilization process; however, lactose was chosen as bulking agent to obtain an elegant cake in vials after lyophilization.

To simplify interpretation of the results, the activity data and ELISA data in Tables VII and VIII may be divided by the total nitrogen to give

Table VII
Results of Various Cryoprotectant-Containing rAAT (50 mg/ml) Formulations[a]

Formulation	Monomeric rAAT (%)	Active rAAT (mg/ml)	Total rAAT (mg/ml)	Total nitrogen as BSA (mg/ml)
Citrate (control)				
Before freezing	81.7 (1.0)	43.6 (1.7)	44.9 (2.6)	52.2 (0.4)
After thawing	81.5 (2.1)	40.9 (1.1)	45.2 (1.3)	51.2 (0.3)
After lyophilization	81.5 (2.1)	42.1 (1.1)	47.7 (1.2)	51.2 (0.3)
With PVP				
Before freezing	62.9 (2.3)	33.8 (3.8)	50.6 (3.2)	—[b]
After thawing	65.3 (1.1)	33.3 (2.7)	44.8 (2.3)	—[b]
After lyophilization	70.2 (1.2)	33.2 (0.1)	49.2 (1.0)	—[b]
With sucrose				
Before freezing	79.8 (0.3)	38.9 (1.7)	47.0 (0.7)	44.4 (0.3)
After thawing	79.5 (0.3)	36.3 (0.5)	45.3 (1.0)	44.0 (1.2)
After lyophilization	79.7 (1.6)	33.7 (0.6)	47.3 (2.1)	40.0 (0.4)
With lactose				
Before freezing	84.6 (0.2)	49.1 (0.7)	50.4 (1.7)	50.8 (0.9)
After thawing	84.3 (0.3)	46.7 (0.8)	46.7 (1.2)	51.4 (0.8)
After lyophilization	87.3 (1.2)	41.3 (0.6)	45.6 (1.9)	46.9 (1.0)

[a] Samples were analyzed before freezing, after freeze-thawing, and after lyophilization–reconstitution. Numbers in parentheses represent one standard deviation of the mean.
[b] Interference in the assay.

percent specific activity and percent specific ELISA values. Contrary to general belief, neither freeze–thaw nor lyophilization–reconstitution does any damage to the rAAT protein. The citrate formulation (control) performs as well as those containing the various bulking agents.

5.2.3. Effect of Freezing Rate

The effect of freezing rate on the stability of rAAT was studied by "fast-freezing" and "slow-freezing" the rAAT solution (20 mg/ml in a phosphate–citrate buffer, pH 7.0) in liquid nitrogen (−196°C) or on a −40°C freezer shelf, respectively. Frozen samples were kept at −20°C for a predetermined period until assay. Effects of the freezing rate on the monomeric rAAT content and the elastase inhibitory capacity of rAAT were examined over the stability study period of 8 weeks. Neither of the freezing conditions caused a decrease in the monomeric content or the elastase inhibitory activity of rAAT.

Table VIII
Results of Various Cryoprotectant-Containing rAAT (10 mg/ml) Formulations[a]

Formulation	Monomeric rAAT (%)	Active rAAT (mg/ml)	Total rAAT (mg/ml)	Total nitrogen as BSA (mg/ml)
Citrate (control)				
Before freezing	88.6 (1.2)	9.4 (0.7)	9.8 (0.3)	10.6 (0.2)
After thawing	86.9 (2.8)	9.3 (0.1)	9.5 (0.3)	10.8 (0.1)
After lyophilization	86.9 (2.8)	9.6 (0.2)	9.7 (0.1)	10.3 (0.1)
With PVP				
Before freezing	65.0 (2.5)	5.0 (0.2)	9.6 (0.4)	—[b]
After thawing	61.0 (2.0)	5.2 (0.1)	9.4 (0.1)	—[b]
After lyophilization	67.3 (1.4)	9.1 (0.1)	9.5 (0.1)	—[b]
With sucrose				
Before freezing	84.6 (0.3)	7.5 (0.2)	9.8 (0.1)	8.6 (0.1)
After thawing	83.8 (0.3)	7.8 (0.1)	9.6 (0.3)	8.8 (0.1)
After lyophilization	82.7 (2.8)	6.7 (0.5)	9.2 (0.3)	8.3 (0.2)
With lactose				
Before freezing	87.9 (0.3)	8.1 (0.3)	9.7 (0.1)	9.9 (0.2)
After thawing	88.7 (0.8)	8.9 (0.1)	9.4 (0.2)	10.0 (0.4)
After lyophilization	90.6 (2.7)	6.1 (0.4)	9.6 (0.2)	9.6 (0.2)

[a] Samples were analyzed before freezing, after freeze-thawing, and after lyophilization–reconstitution. Numbers in parentheses represent one standard deviation of the mean.
[b] Interference in the assay.

5.2.4. Long-term Stability of Lyophilized rAAT

Lactose was chosen as the bulking agent for the lyophilization of rAAT. A 12-month stability study was conducted to investigate the long-term stability of the lyophilized rAAT formulation. The phosphate–citrate formulation without a bulking agent served as the control.

A sterile solution of 40 mg/ml rAAT in phosphate–citrate buffer (pH 7.0) with or without lactose (40 mg/ml) was divided into 2.0-ml aliquots and filled into 5-ml glass vials. The vials were capped loosely with gray butyl rubber lyo-stoppers and lyophilized under 150 mtorr at −43 to −45°C shelf temperature in a VirTis lyophilizer. The vials and the rubber stoppers were sterilized before use. After lyophilization, the vials were securely stoppered and capped with aluminum seals. The vials were then kept at 4°C or room temperature (22°C) for up to 12 months before assay.

Figure 13 shows the results of the assay for monomeric rAAT content (top panel) and the elastase inhibitory activity of rAAT (lower panel) over the 12-month storage period. No loss of monomeric rAAT was observed for the lactose formulation or the control at either 4°C or room temperature; however, the elastase inhibitory activity of rAAT decreased to about 70–90% of the initial activity, depending on the storage temperature.

6. CONCLUSION

rAAT is unstable in phosphate-buffered solutions. Addition of citrate buffer to the phosphate-buffered rAAT solution improves its stability. The pH of the solution is critical for the stability of rAAT. A V-shaped pH stability profile is observed, the most stable pH for rAAT being 7.5. Ionic strength suppresses rAAT protein aggregation and potassium chloride is more effective than sodium chloride for the inhibition of rAAT aggregation. Sucrose stabilizes the liquid formulation better than the other stabilizers examined. Lyophilization dramatically improves the stability of rAAT. Bulking agents such as lactose and providone have little effect on the stability of freeze-dried rAAT. Freezing rate does not affect rAAT stability, either. A lyophilized rAAT containing lactose as the bulking agent shows acceptable stability for the monomeric content of rAAT over a storage period of 12 months at 4°C

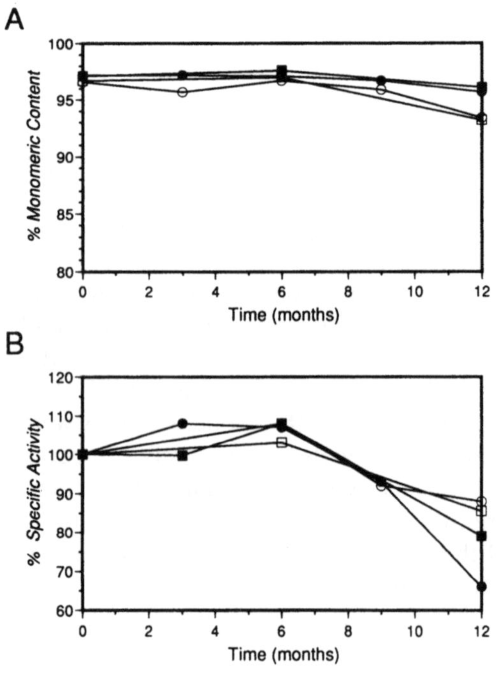

Figure 13. Monomeric rAAT content (top panel) and elastase inhibitory activity of rAAT (bottom panel) over a 12-month period for lyophilized formulations. □, rAAT in citrate buffer control at 4°C; ○, rAAT in citrate buffer control at room temperature; ■, citrate formulation with lactose as bulking agent at 4°C; ●, citrate formulation with lactose as bulking agent at room temperature.

and at room temperature; however, the elastase inhibitory activity of rAAT decreases to about 70–90% of the initial activity over the same period of time, depending on the storage temperature. It appears that rAAT is more stable at 4°C than at room temperature. The product stability of rAAT can be extended beyond 12 months if it is lyophilized and stored in a refrigerator at 2–8°C.

Acknowledgment

The data presented in this chapter were collected when the authors were at Cooper Labs, Inc., Mountain View, California 94039.

REFERENCES

Arakawa, T., Hsu, Y. R., and Yphantis, D. A., 1987, Acid unfolding and self-association of recombinant Escherichia coli derived human interferon gamma, *Biochemistry* **26**:5428–5432.

Barrett, A. J., 1975, Protein degradation in health and disease, *Ciba Found. Symp.* **75**:1–13.

Beatty, K., Bieth, J., and Travis, J., 1980, Kinetics of association of serine proteinases with native and oxidized alpha-1-proteinase inhibitor and alpha-1-antichymotrypsin, *J. Biol. Chem.* **225**:3931–3934.

Brems, D. N., Plaisted, S. M., Harvel, H. A., and Tomich, C. S., 1988, Stabilization of an associated folding intermediate of bovine growth hormone by site-directed mutagenesis, *Proc. Natl. Acad. Sci. USA* **85**:3367–3371.

Eriksson, S., 1965, Studies in alpha-1-antitrypsin deficiency, *Acta Med. Scand. Suppl. 432* **177**:1–85.

Fennema, O. (ed.), 1979, *Proteins at Low Temperatures,* American Chemical Society, Washington, D.C., Chapter 1.

Fishbein, W. N., and Winkert, J. W., 1977, Parameters of biological freezing damage in simple solution: Catalase I: The characterization pattern of intracellular freezing damage exhibited in a membraneless system, *Cryobiology* **14**:389–398.

Fishbein, W. N., and Winkert, J. W., 1979, Parameters of freezing damage to enzymes, in: *Proteins at Low Temperatures* (O. Fennema, ed.), American Chemical Society, Washington, D.C.

Gadek, J. E., and Crystal, R. G., 1983, Alpha-1-antitrypsin deficiency, in: *The Metabolic Basis of Inherited Diseases,* 5th ed. (J. B. Stanbury, J. B. Wyngaarden, and D. S. Frederickson, eds.), McGraw–Hill, New York, pp. 1450–1467.

Johnson, D., and Travis, J., 1979, The oxidative inactivation of human alpha-1-proteinase inhibitor. Further evidence for methionine at the reactive center, *J. Biol. Chem.* **254**:4022–4026.

Larsen, S., 1973, Stability of drugs in frozen-systems. VI. Effects of freezing upon pH for buffered aqueous solutions, *J. Biochem.* **81**:163–168.

Laurell, C. B., and Eriksson, S., 1963, The electrophoretic alpha-1-globulin pattern of serum in alpha-1-antitrypsin deficiency, *Scand. J. Clin. Lab. Invest.* **15:** 132–140.

Layne, E., 1957, in: *Methods in Enzymology,* Vol. 3 (S. P. Colowich and N. O. Kaplan, eds.), Academic Press, New York, pp. 447–454.

Lieberman, J., 1973, Heat lability of alpha-1-antitrypsin variants, *Chest* **64:**579.

Mittman, C. (ed.), 1971, *Pulmonary Emphysema and Proteolysis,* Academic Press, New York.

Orii, Y., and Morita, J., 1977, Measurement of the pH of frozen buffer solutions by using pH indicators, *J. Biochem.* **81:**163–168.

Pikal, M. J., 1990, Freeze-drying of proteins. Part I. Process design, *BioPharm* **3:**18–27.

Rosenberg, S., Barr, P. J., Nagarian, R. C., and Hallewell, R. A., 1984, Synthesis in yeast of a functional oxidation-resistant mutant of human alpha-1-antitrypsin, *Nature* **312:**77–80.

Simpson, R. B., and Kauzmann, W., 1953, The kinetics of protein denaturation. I. The behavior of the optical rotation of ovalbumin in urea solutions, *J. Am. Chem. Soc.* **75:**5139–5152.

Travis, J., and Salvesen, G. S., 1983, Human plasma proteinase inhibitors, *Annu. Rev. Biochem.* **52:**655–709.

Travis, J., Owen, M., George, P., Careel, R., Rosenberg, S., Hamewell, R., and Barr, P., 1985, Isolation and properties of recombinant DNA produced variants of human alpha-1-proteinase inhibitor, *J. Biol. Chem.* **260:**4384–4389.

Tsuchiya, T., Tsuchiya, Y., Nonomura, Y., and Matsumoto, J., 1975, Prevention of freeze denaturation of carpactomyosin by sodium glutamate, *J. Biochem.* **77:**853–862.

Vemuri, S., 1991, Lyophilization cycle development for interleukin-2, *Dev. Biol. Stand.* **74:**341–351.

Vemuri, S., Yu, C., Wangsatornthankun, V., and Roosdorp, N., 1989, Effect of concentration on aggregation of recombinant alpha-1-antitrypsin injection, *Pharm. Res.* **6:**5–31.

Vemuri, S., Yu, C., and Roosdorp, N., Effect of lyoprotectants on freezing, lyophilization and storage of recombinant alpha-1-antitrypsin formulations, *Drug Dev. Ind. Pharm.* submitted for publication.

Yu, C., Roosdorp, N., and Pushpala, S., 1988, Physical stability of a recombinant alpha-1-antitrypsin injection, *Pharm. Res.* **5:**800–802.

Chapter 10

Fibrolase

A Fibrinolytic Protein from Snake Venom

Denise Pretzer, Brenda S. Schulteis,
Christopher D. Smith,
David G. Vander Velde,
James W. Mitchell,
and Mark C. Manning

1. INTRODUCTION

Since the pioneering research of Mitchell and Reichert in the 1800s (Mitchell, 1860; Mitchell and Reichert, 1886), a number of laboratories (Eagle, 1937; Didisheim and Lewis, 1956; Bajwa *et al.*, 1982) have demonstrated the existence of a group of proteases in the venom of snakes from the pit viper family. Further investigation has shown that these proteases disrupt the nor-

Denise Pretzer • Merck Research Laboratories, West Point, Pennsylvania 19486. *Brenda S. Schulteis* • Marion Merrell Dow Research Institute, Kansas City, Missouri 64134. *Christopher D. Smith* • Department of Pharmaceutical Chemistry, University of Kansas, Lawrence, Kansas 66045. *David G. Vander Velde* • NMR Laboratory, University of Kansas, Lawrence, Kansas 66045. *James W. Mitchell* • Oread Laboratories, Lawrence, Kansas 66044. *Mark C. Manning* • School of Pharmacy, University of Colorado Health Sciences Center, Denver, Colorado 80262.

Stability and Characterization of Protein and Peptide Drugs: Case Histories, edited by Y. John Wang and Rodney Pearlman. Plenum Press, New York, 1993.

mal hemostatic mechanisms in warm-blooded animals to produce adverse physiological effects following bites from reptiles in this group. These proteases basically fall into two classes: those which are thrombinlike and have coagulating activity, and those which dissolve clots via action primarily attributed to fibrinolytic activity (Eagle, 1937; Didisheim and Lewis, 1956; Bajwa *et al.*, 1982). Enzymes of the latter class are under investigation for potential applications as thrombolytic agents. Fibrolase is one such enzyme. First isolated from *Agkistrodon contortrix contortrix* (southern copperhead), it was shown to fall into the latter class of venom proteases having fibrinolytic activity (Bajwa *et al.*, 1982).

Fibrolase is being considered for possible indications in a number of disease states, such as following myocardial infarction, where prompt administration of a thrombolytic agent can prevent irreversible loss of heart muscle and markedly improve the individual's prognosis. *In vitro* evaluation of the proteolytic activity of fibrolase showed that it degrades fibrin as well as fibrinogen via hydrolysis of their α and β chains. The γ chain appears to be less susceptible to this enzyme (Ahmed *et al.*, 1990a). Specific *in vitro* hydrolysis of fibrin, by fibrolase, is accompanied by the appearance of a number of large-molecular-mass peptides of 49, 46, 44, 20, and 13 kDa. Fibrolase has also been shown to act directly on the α and β chains of fibrinogen. Peptides obtained following fibrolase degradation of fibrinogen differ in molecular mass from those produced by fibrin degradation (with masses of 49, 36, 22, 21, and 20 kDa). The relative rates of fibrin and fibrinogen hydrolysis are unknown. Fibrolase does not act through the activation of plasminogen as do thrombolytic agents such as tissue plasminogen activator and streptokinase (Markland *et al.*, 1988). Ahmed *et al.* (1990a) have shown the lack of activity of fibrolase with and without plasminogen using a plasmin-specific substrate (S-2251). In addition, fibrolase incubated with plasminogen failed to produce plasmin as determined by gel electrophoresis. The direct proteolysis of fibrin thrombi by fibrolase is due to a different mechanism of action and may be of considerable importance relative to currently available thrombolytics.

Incubation of fibrolase *in vitro* in the presence of plasma or blood from different species results in differential inactivation that is species specific. The greatest extent of inactivation was seen with rat plasma followed by rabbit, dog, and human plasma. Inactivation was attributed to varying concentrations of α_2-macroglobulin, a known inhibitor of this class of proteases. Further evaluation has demonstrated that this inhibition can be overcome by increasing the amount of fibrolase administered via titration (Ahmed *et al.*, 1990b).

In vivo efficacy of fibrolase has been studied using an animal model of arterial thrombosis (Ahmed *et al.*, 1990b). It has been shown that following a single i.v. bolus administration of fibrolase, femoral arterial clots are dissolved and time to reperfusion is dose-dependent and consistent with that observed following administration of streptokinase. No adverse effects on blood pressure and heart rate were observed. Evaluation of the effect of fibrolase on circulating levels of fibrinogen, clotting times, and levels of specific coagulation factors (e.g., factors V, VIII, and XII) has also been completed.

2. PHYSICAL PROPERTIES

Fibrolase is composed of a single chain of 203 amino acids (Fig. 1) in known sequence (Randolph *et al.,* 1992). It has a mass of approximately 23 kDa. It contains six cysteines, each of which is involved in the formation of one of the three disulfide bonds within the molecule (F. S. Markland, unpublished data). One disulfide connects residues 118 and 193; the locations of the other two disulfides have not been determined because of the difficulty in digesting the portion of protein containing the remaining cysteines. Fibrinolytic activity of fibrolase was shown to be dependent on intact disulfides (Ahmed *et al.,* 1990a). The protease has a pI of approximately 6.7 (Pretzer *et al.,* 1992) and has been classified a metalloproteinase based on its zinc requirement (Markland *et al.,* 1988). In addition, it has appreciable α-helical content (Pretzer *et al.,* 1991, 1992) and does not appear to be glycosylated (F. S. Markland, unpublished data). Table I summarizes these physical properties.

Isoenzymes of fibrolase are thought to exist because of evidence of sequence heterogeneity from the natural source, and observation of multiple bands on IEF gels of relatively pure fibrolase preparations (Markland *et al.,* 1988; Pretzer *et al.,* 1992). Purification of fibrolase was accomplished by the protocol of Markland *et al.* (1988).

Gln	Gln	Arg	Phe	Pro	Gln	Arg	Tyr	Val	Gln	Leu	Val	Ile	Val	Ala	Asp	His	Arg
Met	Asn	Thr	Lys	Tyr	Asn	Gly	Asp	Ser	Asp	Lys	Ile	Arg	Gln	Trp	Val	His	Gln
Ile	Val	Asn	Thr	Ile	Asn	Glu	Ile	Tyr	Arg	Pro	Leu	Asn	Ile	Gln	Phe	Thr	Pro
Val	Gly	Leu	Glu	Ile	Trp	Ser	Asn	Gln	Asp	Leu	Ile	Thr	Val	Thr	Ser	Val	Ser
His	Asp	Thr	Leu	Ala	Ser	Phe	Gly	Asn	Trp	Arg	Glu	Thr	Asp	Leu	Leu	Arg	Arg
Gln	Arg	His	Asp	Asn	Ala	Gln	Leu	Leu	Thr	Ala	Ile	Asp	Phe	Asp	Gly	Asp	Thr
Val	Gly	Leu	Ala	Tyr	Val	Gly	Gly	Met	Cys	Gln	Leu	Lys	His	Ser	Thr	Gly	Val
Ile	Gln	Asp	His	Ser	Ala	Ile	Asn	Leu	Leu	Val	Ala	Leu	Thr	Met	Ala	His	Glu
Leu	Gly	His	Asn	Leu	Gly	Met	Asn	His	Asp	Gly	Asn	Gln	Cys	His	Cys	Gly	Ala
Asn	Ser	Cys	Val	Met	Ala	Ala	Met	Leu	Ser	Asp	Gln	Pro	Ser	Lys	Leu	Phe	Ser
Asp	Cys	Ser	Lys	Lys	Asp	Tyr	Gln	Thr	Phe	Leu	Thr	Val	Asn	Asn	Pro	Gln	Cys
Ile	Leu	Asn	Lys	Pro													

Figure 1. Primary sequence of fibrolase.

Table I
Physical Properties of Fibrolase

Molecular mass 22,904 Da
203 amino acids
Three disulfide bonds; no free sulfhydryls
pI ~ 6.7
Zinc required for biological activity
Approximately 25% α-helical content
No glycosylation present

Fibrolase may be obtained from the venom of the southern copperhead snake by a multistep fractionation procedure. These steps, all performed at 4°C, include cation-exchange chromatography on CM-cellulose, gel filtration on Sephadex G-100, *p*-aminobenzamidine-agarose affinity chromatography, and anion-exchange chromatography on DEAE-cellulose (Ahmed *et al.*, 1990a). The purity of fibrolase in all preparations was greater than 90% as determined by gel electrophoresis.

3. ANALYTICAL METHODS

3.1. Proteolytic Activity with Oxidized Insulin B Chain as a Substrate

The proteolytic activity assay using the B chain of insulin was adapted from Wilkinson (unpublished data). The polypeptide was oxidized at the sulfhydryls in order to prevent aggregation, and any reference to insulin B chain refers to the oxidized form. The activity of fibrolase versus insulin B chain was determined by incubating fibrolase with the oxidized form of insulin B chain at 37°C. The molar excess of substrate was approximately 250 to 1500. After incubation, the reaction was quenched by addition of EDTA. The remaining insulin B chain was quantitated by HPLC. The HPLC conditions were: the mobile phase consisted of (1) 65% aqueous 25 mM sodium perchlorate and 0.7% (w/w) perchloric acid and (2) 35% 25 mM sodium perchlorate and 0.7% (w/w) perchloric acid in acetonitrile. The flow rate was 1.5 ml/min. A polymeric reversed phase (PRP-1) column (25 × 0.41 cm, 10 μm particle size) was employed using UV detection at 205 nm. The HPLC conditions resulted in a retention time of approximately 5 min for undegraded insulin B chain. To determine the proteolytic activity in samples, assay results were compared with results obtained concurrently for fibrolase standards. A single preparation of fibrolase was fully characterized and designated as the standard. All other lots were compared with this standard.

3.2. Proteolytic Activity Assay with Azocasein as a Substrate

Azocasein was synthesized by diazotization of casein as described by Charney and Tomarelli (1947). Fibrolase was incubated at 37°C with azocasein. The molar excess of substrate ranged from about 6 to 60. The reaction was quenched by addition of perchloric acid. Following centrifugation, acid-soluble azopeptides, resulting from the action of fibrolase on azocasein, were quantitated by absorbance at 390 nm. Proteolytic activity in samples was determined by comparison with standards from a fibrolase lot of known activity assayed concurrently.

Proteolytic activity assays were used to quantitate the enzymatic activity of the protein. Since fibrin degradation (clot dissolution) occurs via proteolysis, proteolytic activity of fibrolase should be an important corollary to bioactivity. Bioactivity would be best demonstrated using *in vitro* clot lysis or fibrin plate assays. Fibrolase reproducibly demonstrated activity in these assays, but their use on a routine basis would be time-consuming and tedious. The proteolytic assays are a useful method of measuring the effect of variables (such as pH and temperature) on the activity of fibrolase.

3.3. Circular Dichroism (CD) Measurements

All CD spectra were obtained on an Aviv 60DS spectrophotometer equipped with a Lauda temperature bath, using jacketed quartz cells (Hellma) with path lengths of 0.5–5.0 mm. During studies at elevated temperatures, the sample was equilibrated at each temperature for about 10 min, and the temperature of the water bath was recorded as the sample temperature. Maximum deviation of the sample and bath temperatures was estimated to be 2°C. Protein concentrations ranged from 0.060 to 0.800 mg/ml in 10 mM sodium acetate (pH 3 and 5), 10 mM sodium phosphate (pH 2, 6–8), or 10 mM sodium borate (pH 9) buffers. All spectra were smoothed and baseline corrected prior to plotting. A mean residue weight of 113.2 was employed in calculation of molar ellipticity values.

3.4. Nuclear Magnetic Resonance (NMR) Measurements

Samples for NMR were prepared by dialysis of fibrolase against sodium phosphate buffer in water (10 mM, pH 8.0) followed by dialysis against sodium phosphate buffer in D_2O (10 mM, pD 8.0). The D_2O was 99.9 atom% D. The pD was estimated by adjustment of the pH meter reading (Glasoe and Long, 1960). Protein concentration was 2.5 mg/ml (0.1 mM).

Spectra were taken on a Bruker AM-500 spectrometer with a dedicated proton probe equipped with a coil designed for optimum water suppression; presaturation was necessary, even in D_2O, at the low protein concentrations used. The spectra shown resulted from 200 scans with a 55° pulse and a 2.2-sec recycle delay. The temperature was controlled and measured by the spectrometer's variable temperature unit.

3.5. Electrophoresis and Protein Quantitation

The purity of fibrolase was monitored using various electrophoretic techniques. Determination of the protein concentration was accomplished by standard chemical methods described below.

3.5.1. Gel Electrophoresis

All gels were run on the Pharmacia Phast system, according to the manufacturer's instructions. Prepared gels were obtained from Pharmacia. SDS-PAGE and native gels were run on 8–25% gradients. IEF gels were done covering the pH range 3 to 9. For all gels, bands were detected by staining with Coomassie blue. Samples analyzed by SDS-PAGE contained approximately 0.13 M (3.7%) SDS. Reduced samples were prepared by addition of 0.66 M (5%) β-mercaptoethanol and heating at 95°C for 5 to 10 min. Samples for native gels and IEF gels were applied without dilution, or diluted with water or 10 mM Tris buffer (pH 8). In all cases, the sample volume was typically 1 μl.

3.5.2. Protein Quantitation

Protein concentrations were determined using the Pierce microBCA kit. The assay is based on the reaction of protein with Cu^{2+} to form Cu^{+}, which complexes with bicinchoninic acid (BCA) forming a colored product whose intensity at 562 nm is related to the protein concentration using bovine serum albumin as a standard. In addition, the extinction coefficient of fibrolase at 280 nm was determined based on measurements of several lots in Hepes buffer (pH 7.5). The average was 1.05 for a 1 mg/ml solution.

4. EFFECT OF pH ON FIBROLASE STABILITY

Total protein concentration in solution was determined, following centrifugation, over the pH range 1 to 10 (Fig. 2). Protein concentration was

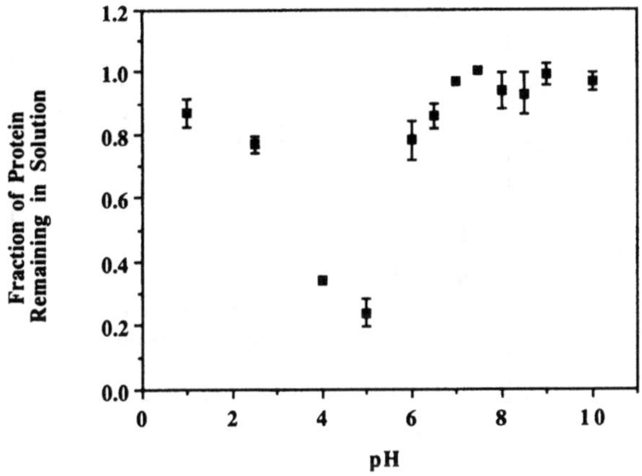

Figure 2. Protein concentration (as percentage of the initial concentration of 0.5 mg/ml) as a function of pH.

unchanged in the neutral to alkaline pH range (6.5 to 10). As the pH decreased, total protein in solution decreased from an initial concentration of 0.5 mg/ml, to an apparent minimum at pH 5. At lower pH values, substantial concentrations of protein (30 to 90% of initial protein at pH 1 to 4) remained in solution.

The effect of pH on specific proteolytic activity of fibrolase was evaluated over the pH range 1 to 10 using azocasein and insulin B-chain substrates (Fig. 3). At neutral to alkaline pH values (6.5 to 9), no significant change in proteolytic activity of fibrolase was observed. At pH 10, the data suggested a slight loss of activity. This decrease was greater for the azocasein substrate relative to the insulin B-chain substrate. As the pH was lowered, the activity of fibrolase decreased. At pH 5, approximately 85% of the initial activity remained. Furthermore, at pH 4, only 20% of the initial activity was detected. At pH 1 and 2.5, azocaseinolytic activity was less than 10%. Activity versus insulin B chain was also low at these pH values, but showed high variability in the assay results.

Gel electrophoresis results (data not shown) correlated well with activity and concentration results. SDS-PAGE gels of reduced and nonreduced samples showed no change in the location of the fibrolase band over the entire pH range 1 to 10. Band intensity corresponded qualitatively to protein concentration results, as expected. Native gels of samples at pH 5 to 10 showed no variation in migration of the fibrolase band. However, at more acidic pH values, no fibrolase band was seen. Instead, staining was observed as a streak or as multiple bands with increased mobility relative to fibrolase. These results were consistent with the lack of proteolytic activity at acidic pH.

Consideration of specific activity data together with the protein concen-

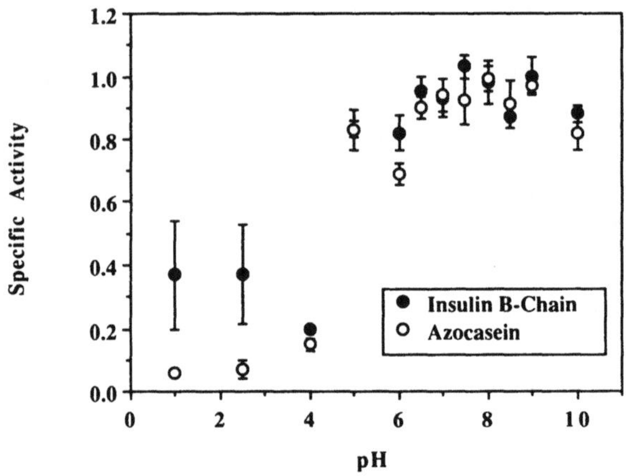

Figure 3. Specific proteolytic activity of fibrolase (expressed as fraction of initial activity) as a function of pH. Proteolytic activity versus insulin B chain and versus azocasein.

tration data over the pH range 1 to 10 indicates several interesting effects as a result of changing pH. In the neutral to alkaline pH range, both activity and total protein in solution were not significantly changed. As the pH was lowered, protein concentration decreased to an apparent minimum at pH 5, yet the protein remaining in solution had a specific activity of about 0.85. As the pH was lowered even further, most of the proteolytic activity of the protein was lost, yet substantial amounts of protein remained in solution.

The apparent minimum solubility of fibrolase at pH 5 does not correspond to the isoelectric pH of 6.7 previously determined for fibrolase. Minimum protein solubility would be predicted to occur near the isoelectric pH (Mulkerrin and Wetzel, 1989). Since the isoelectric point for fibrolase is close to neutral pH, the effect on solubility is a concern for formulation. However, the experimental data indicate that formulation of fibrolase at neutral pH, only slightly above the isoelectric pH, would avoid the apparent solubility minimum. Because our studies were not designed as solubility studies *per se* (i.e., an excess of protein was not present, thus equilibrium might not have been attained), the maximum solubility could not be deduced from the data.

Far-UV CD spectra of fibrolase displayed essentially no change in band shape or position over the pH range 5 to 9 (Fig. 4). However, some variations in intensity were observed. While some differences in intensity were observed in the far-UV CD spectra of fibrolase with change in pH (from 5 to 9), the band positions and shapes were invariant. Assuming a 10% error in determination of the protein concentrations, most of the variations in intensity fall within experimental error. These differences could also be accounted for by small increases (one or two turns) in the α-helix content of the protein. Together, it appears that small variation or error in protein concentration and/or CD measurements is sufficient to account for intensity variations of the far-UV CD spectra within the pH 5 to 9 range. At pH 2 and 3, significant differences were found relative to the pH range 5 to 9 (Fig. 5). The spectral changes at pH 2 and 3 were consistent with loss of α-helical structure and were similar to changes observed on thermal denaturation (Fig. 6). Varying pH also affected the thermal stability of fibrolase. At pH 5, the T_m dropped to 43°C, compared with 50°C at pH 8 (Fig. 7). In addition, at pH 5, the cooperativity of the transition was lower. Some destabilization of fibrolase with decreasing pH is indicated by the lower T_m observed at pH 5 relative to pH 8. Destabilization resulting in the lower T_m value would not necessarily be apparent in CD spectra taken at room temperature. The destabilization could be a result of unfavorable changes in the electrostatic network of the protein (Mulkerrin and Wetzel, 1989).

At acidic pH, the activity loss observed could be due to a pH-dependent chemical degradation mechanism. Oxidation or hydrolysis might be accelerated at acid pH values, while deamidation generally occurs more rapidly at neutral to alkaline pH values (Manning *et al.*, 1989). Autoproteolysis would probably not be accelerated at low pH given that the pH of maximal activity for fibrolase is in the neutral to alkaline range. It should be noted that structural changes at different pH values could result in accelerated chemical degradation because of exposure of previously protected residues. However, considering the rapidity of the loss of activity, chemical degradation is an unlikely mechanism.

Physical instability could also account for the observed loss of fibrolase activity at acidic pH. For metalloproteins, pH-dependent changes in the metal binding site leading to loss of the metal ion and subsequent changes in

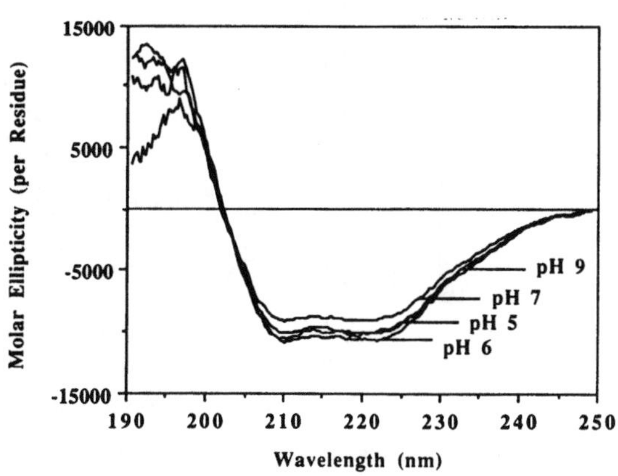

Figure 4. Far-UV CD spectra of fibrolase at pH 5, 6, 7, and 9. The solutions contained 10 mM phosphate buffer.

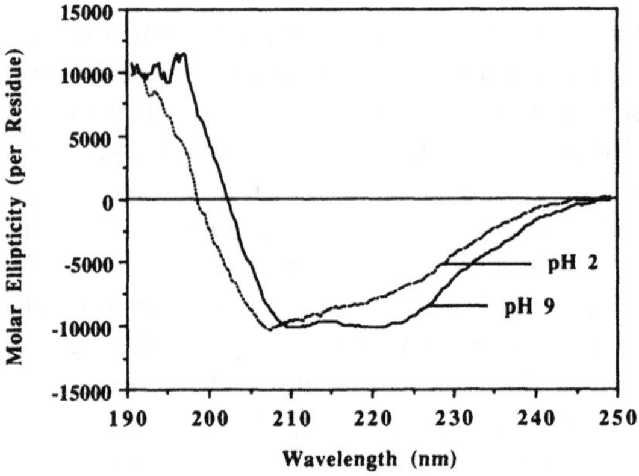

Figure 5. Far-UV CD spectra of fibrolase at pH 2 and 9.

structure or activity have been observed (Mock and Tsay, 1988; Parraga *et al.,* 1988). Although the metal binding site in fibrolase has not been characterized, the most common residues involved in zinc binding to proteins are cysteine and histidine (Vallee and Auld, 1990). The six cysteines in fibrolase are involved in covalent disulfide bonds. Of the ten histidine residues, a region containing two was identified as having significant homology to known metal binding sites of other zinc metalloproteases (see below). Chemical modification of histidine residues labeled only two of the ten, supporting that two are involved in zinc binding (F. S. Markland, unpublished results). Protonation of histidine as the pH is lowered would decrease protein affinity for zinc. The pK_a of histidine would be expected to be in the pH range 5 to 7, depending on the microenvironment of the residues (Kleanthous *et al.,*

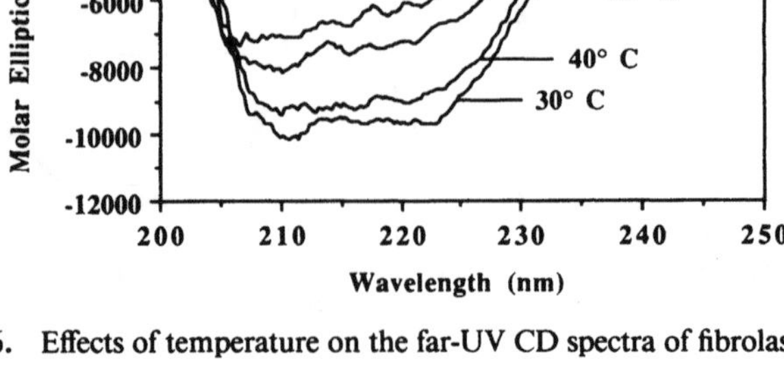

Figure 6. Effects of temperature on the far-UV CD spectra of fibrolase (pH 8).

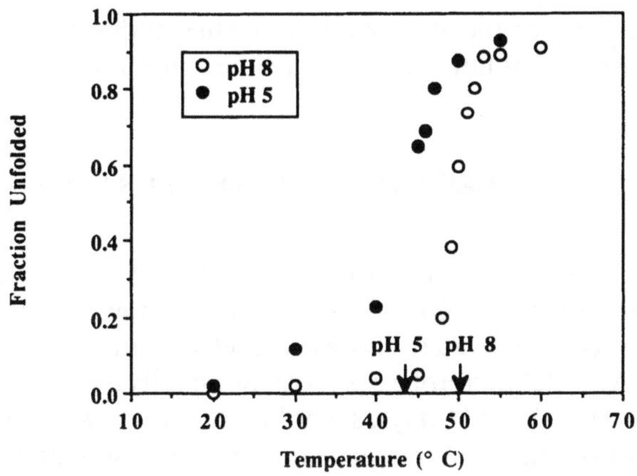

Figure 7. Thermal denaturation curves for fibrolase at pH 5 and 8, as determined by molar ellipticity at 222 nm.

1988; Marx, 1988; Mulkerrin and Wetzel, 1989). This pH range is where activity and solubility changes occurred for fibrolase. Alternatively, changes in ionization state of other ligands bound to the metal ion, such as a water molecule, would also alter metal binding characteristics and thus metalloprotease activity (Mock and Tsay, 1988). Decreased metal binding upon EDTA addition resulted in changes in the secondary structure of fibrolase (as determined by CD spectroscopy) and also produced a loss of activity (see below). The observed decrease in protein concentration at pH < 7 could result from precipitation induced by loss of secondary structure. Further study is needed to determine the exact mechanism responsible for loss of activity and protein concentration in the pH range 4 to 7.

A different mechanism is most likely operative at low pH (<pH 4–5) to account for the higher protein concentrations observed relative to the minimum at pH 5. At acidic pH, proteins typically unfold due to electrostatic repulsion from developing positive charges. In some cases, addition of counterions may mask the positive charges, decreasing electrostatic repulsion, and permit the protein to refold. The refolded state produced under such conditions is termed a molten globule state and is different from either the native or unfolded state. The molten globule state retains compact globular form and some secondary structure but has little or no well-defined tertiary structure (Baum *et al.*, 1989; Goto and Fink, 1989; Koseki *et al.*, 1988). The molten globule state would not be expected to have the activity of the native state since the active site would probably not be refolded correctly. Partial refolding of fibrolase, possibly to a molten globule state, may account for the

observed behavior at acidic pH. However, further studies are needed to characterize the structure of fibrolase under acidic conditions.

5. STABILITY OF FIBROLASE AND TEMPERATURE EFFECTS

The proteolytic activity of fibrolase at pH 8 was followed during storage at room temperature (~22°C) and 37°C. Figure 8 shows the specific activity of fibrolase for two substrates, azocasein and insulin B chain, during the 16-day study. At ~22°C, fibrolase proteolytic activity did not change significantly over 10 days. After 16 days, slight loss of activity was observed versus azocasein; however, there did not appear to be a corresponding loss of activity versus insulin B chain. SDS-PAGE gels run under reducing and nonreducing conditions showed no difference between initial samples and samples taken at day 16 (data not shown). Additionally, there was no loss of total protein in solution over 16 days. While these methods do not detect all possible changes which could occur for the protein, the results showed that fibrolase was stable at room temperature for 10 days. Slight loss of azocaseinolytic activity at 16 days indicated that fibrolase would not be stable over long periods of time (weeks to months) in solution at room temperature. These data suggested that the stability of fibrolase would probably not be adequate to formulate as a solution for long-term storage at room temperature.

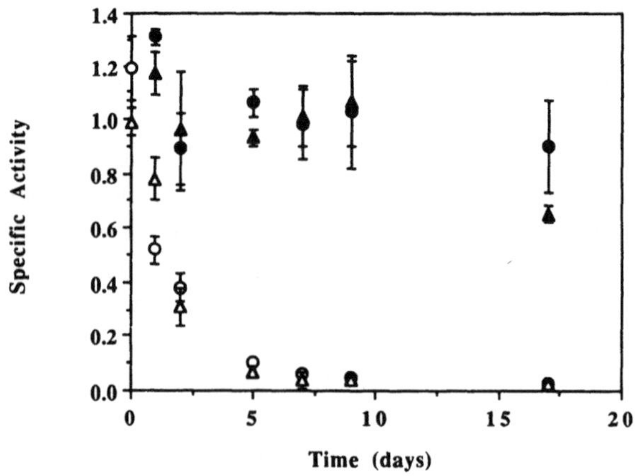

Figure 8. Specific proteolytic activity of fibrolase (expressed as fraction of initial activity) over time at room temperature (RT, ~22°C) and 37°C. Proteolytic activity versus insulin B chain at RT (●) and 37°C (○), and versus azocasein at RT (▲) and 37°C (△).

At 37°C, rapid loss of proteolytic activity was observed. Greater than 10% of the activity was lost in less than 1 day, and essentially no activity against either substrate remained after 10 days. SDS-PAGE gels of nonreduced samples (Fig. 9) showed loss of intensity of the fibrolase band and lower molecular weight bands. Similar results were observed for reduced samples. In contrast to the activity and gel results, total protein remained unchanged in solution over 16 days.

The observed changes could be due to chemical or physical degradation mechanisms (Manning *et al.*, 1989). Fibrolase contains potential sites for deamidation, oxidation, and hydrolysis, although in order to account for the observed loss, the rates of these reactions would have to be much greater than predicted for the experimental conditions (Manning *et al.*, 1989). One other mechanism specific to proteases which could account for the observed instability of fibrolase at 37°C would be autoproteolysis. Activity results were not inconsistent with an autoproteolytic mechanism assuming that degradation

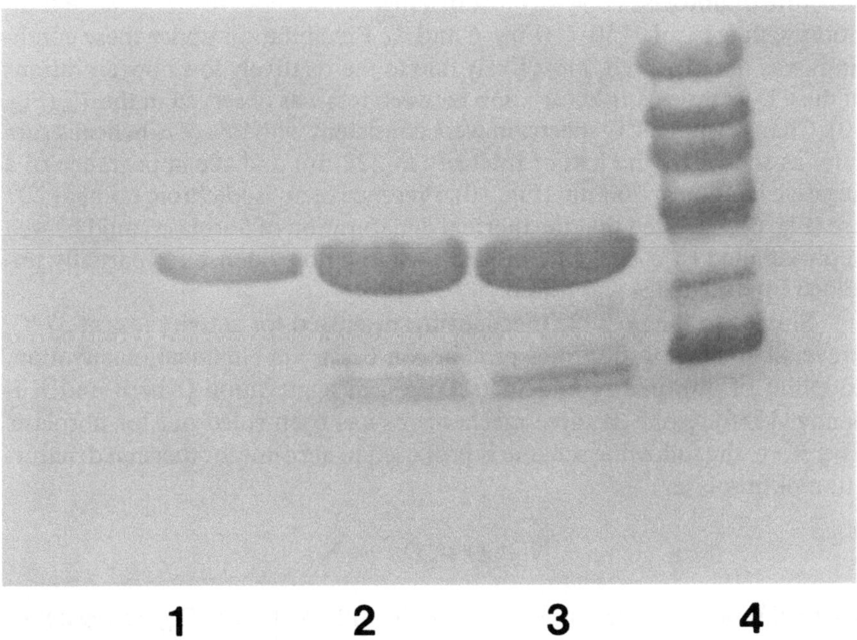

Figure 9. SDS-PAGE gels of nonreduced fibrolase samples stored at 37°C. Lane 1: sample after 7 days; lane 2: sample after 5 days; lane 3: sample after 0 days; lane 4: molecular mass standards, 14.4 to 94 kDa. Each lane was loaded with 4 μl of undiluted sample (initial fibrolase concentration was ~500 μg/ml).

products are inactive, although the large difference in stability between 22 and 37°C would seem to preclude chemical decomposition of any kind. Electrophoresis did not detect smaller protein species, yet autoproteolysis at multiple cleavage sites might produce fragments of sufficiently low molecular weights that would not be visible on the gels.

Physical instability such as adsorption, aggregation, or denaturation could account for the observed loss of activity at 37°C. Significant physical changes commonly result in precipitation of protein, which was not observed. In addition, CD and NMR did not detect significant conformational changes at 37°C (see below), although these samples were maintained at 37°C for a maximum time of only about 10 min. Visible absorbance did increase somewhat upon heating to 37°C. Heating above 37°C led to irreversible unfolding of fibrolase as shown by both CD and NMR (see below).

The far-UV CD spectrum of fibrolase at pH 8 indicates that it possesses a significant amount of α-helical structure (~25%; see Fig. 6). This observation was contrary to the predicted predominance of β-sheet structure (F. S. Markland, unpublished data). The globular structure appears to be stable over the temperature range 15–40°C. Above 40°C, some of the secondary structure of fibrolase is rapidly lost in a highly cooperative, irreversible transition with a T_m of ~50°C (Figs. 6 and 7). Precipitation under these conditions was not observed, most likely due to the relatively low concentrations of the CD samples. Little variation between lots was observed in the T_m (Fig. 10). Changes in the CD spectrum were consistent with loss of α-helical structure, as shown by the loss of intensity at 222 nm and the appearance of a negative band near 200 nm (Fig. 10). Presence of an isodichroic point at 207 nm (Fig. 6) suggested that the thermal denaturation of fibrolase could be well approximated by a two-state model, meaning no evidence for partially unfolded intermediates was found.

Similar to the possible mechanisms discussed for activity loss at 37°C, irreversible denaturation of a protein can occur via chemical inactivation, adoption of improperly folded structures, or aggregation (Ahern and Klibanov, 1986). None of these mechanisms has been ruled out for fibrolase. Therefore, the following scheme is proposed to account for thermal denaturation of fibrolase:

$$\mathrm{N} \rightleftarrows \mathrm{D} \rightleftarrows \mathrm{D}n \rightarrow \mathrm{X}$$

where N represents the native state, D, the unfolded state, D_n, aggregates of the denatured state, and X, an inactivated form of fibrolase arising from D. While D is described as an ensemble of unfolded structures, it is clear that thermally inactivated fibrolase maintains a significant amount of secondary structure, even at 70°C. Fibrolase heated past the T_m still displays significant

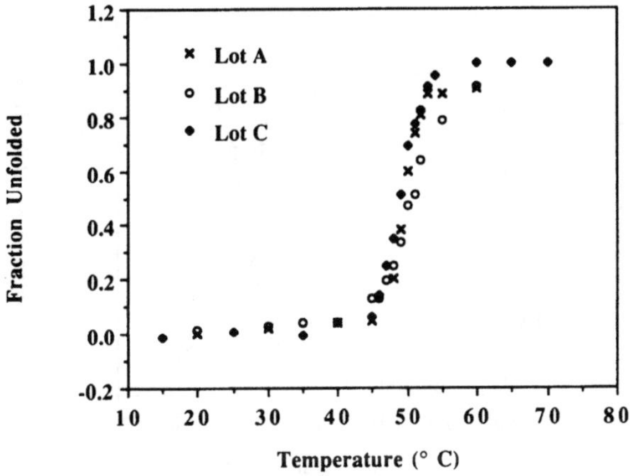

Figure 10. Lot-to-lot variation in T_m for fibrolase as determined by monitoring the mean residue ellipticity at 222 nm.

negative ellipticity (~ −3000) at 222 nm, while $[\theta]_{222}$ for fibrolase in 6 M guanidine hydrochloride is only ~ −500 deg cm^2/dmole (M. C. Manning, unpublished data). Deconvolution of the CD spectra (Chang *et al.*, 1978) during heating suggests that the changes observed are due to loss of α-helical structure (Fig. 11).

NMR spectra at elevated temperatures (Fig. 12) also showed evidence of irreversible denaturation in good qualitative agreement with the CD results. As the temperature was increased from 5°C (data not shown) to the onset of denaturation, the spectrum remained essentially the same, although many peaks narrowed. Above 35°C, linewidths remained approximately constant but the spectrum underwent gross changes, as numerous peaks disappeared. These observations are consistent with conversion of native protein to a random coil form. The transition was complete by 55°C. In particular, the unresolved amide peaks from 8.5 to 10.0 δ, which were protected from D_2O exchange in the native structure, were progressively lost. The complex pattern of native form methyl signals between 1.5 and −0.6 δ (the very high-field signals coming from methyl groups shielded by aromatic rings) collapsed to a few peaks at average random coil shift values (Fig. 13). The largest signal was at 0.9 δ, presumably due to methyl protons from leucine, isoleucine, and valine. From the loss of intensity of the peak at approximately 0 δ, which is due only to native protein and is readily measured, the T_m was calculated to be 46°C (Fig. 14). Smaller peaks in the same region showed identical behavior but were less intense and their plotted intensities contained more scatter. The spectrum was unchanged after cooling the sample back to room temperature (data not shown), as expected for an irreversible denaturation.

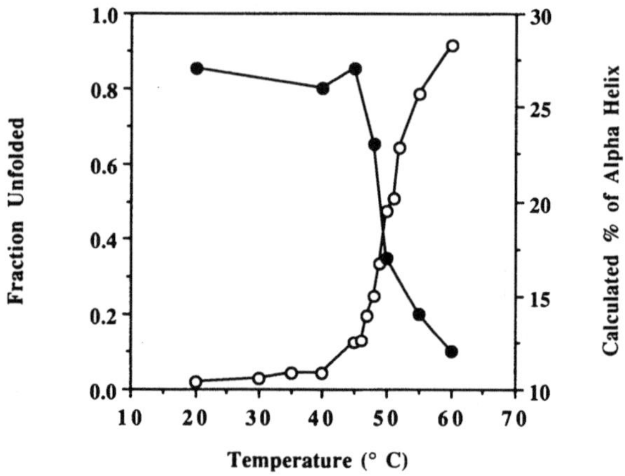

Figure 11. Thermal denaturation of fibrolase as determined by monitoring mean residue ellipticity at 222 nm. The degree of unfolding is given by both the fraction of unfolded protein (○) and the calculated α-helix content (●) using the method of Chang *et al.* (1978).

Although fibrolase is a larger protein than any whose solution structures have been determined by NMR, and assignment of any peaks in the spectrum to specific residues in the protein would require multidimensional experiments on cloned, overexpressed, and isotopically labeled protein, some useful information can be extracted from the variable temperature NMR spectra of the native protein. As mentioned previously, the signals at very high field in the spectrum of native fibrolase are not found in small peptides or in the random coil form of fibrolase. They result from highly shielded methyl groups most likely near an aromatic ring. Consequently, they are a useful marker of the amount of native material and can be used to prepare a melting curve (Figs. 13 and 14). The most upfield signal, at approximately $-0.6\ \delta$, which is also the smallest resolved signal in the spectrum, seems to result from a single methyl group. If the integral of this signal is assigned a value of three protons, the entire spectrum at 20°C has an integral of approximately 1550 protons. The presence of the water line makes the integral inexact, but this number is surprisingly close to the total number of protons expected from the consensus sequence. There are approximately 1190 nonexchangeable protons and 370 D_2O-exchangeable protons in the consensus sequence of 203 residues (there is some uncertainty in this assessment because of the unknown number of asparagine versus aspartic acid, and glutamine versus glutamic acid residues). Integration of the region from 10 to 8 δ suggests there are about 45 exchangeable amide protons protected from solvent in the native structure (not exchanged after several days in D_2O). Previous studies on large proteins whose x-ray crystal structures

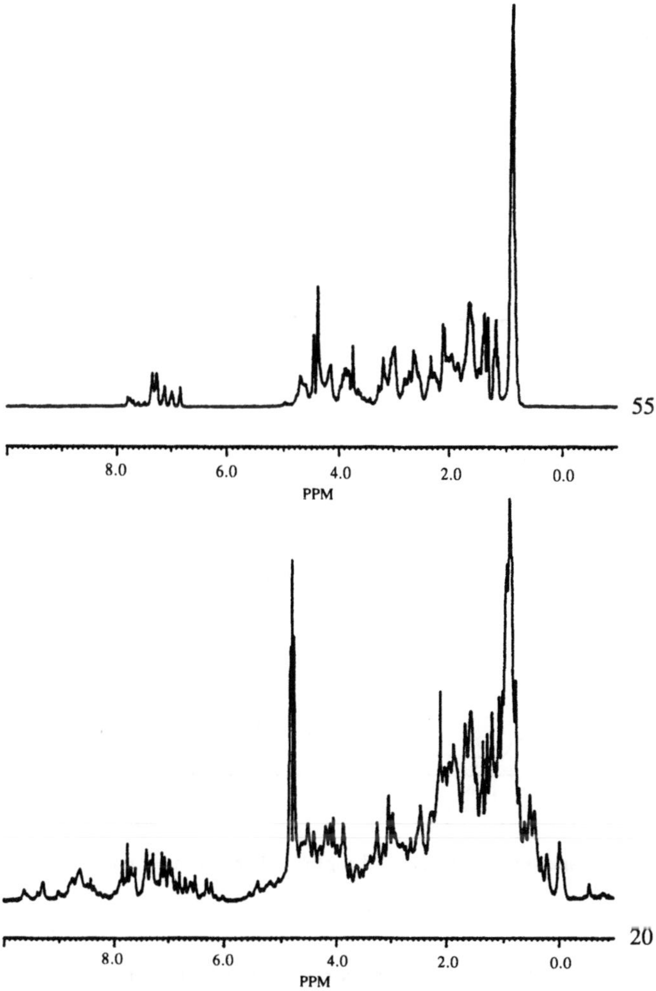

Figure 12. NMR spectra of fibrolase (2.5 mg/ml, pD 8) at 20 and 55°C.

have been solved have shown that the nonexchanged amide protons are involved in strong hydrogen bonds in α-helices or β-sheets (Redfield and Dobson, 1988).

Using the assumption that the peak at $-0.6\ \delta$ is a single methyl group (see Fig. 13), the peak at $0\ \delta$, which was used to prepare the melting curve, has an integral of 21 protons (seven methyl groups). It is therefore not surprising that the T_m values calculated from the NMR and CD data are slightly different, because the CD data reflect global structural changes, especially loss of

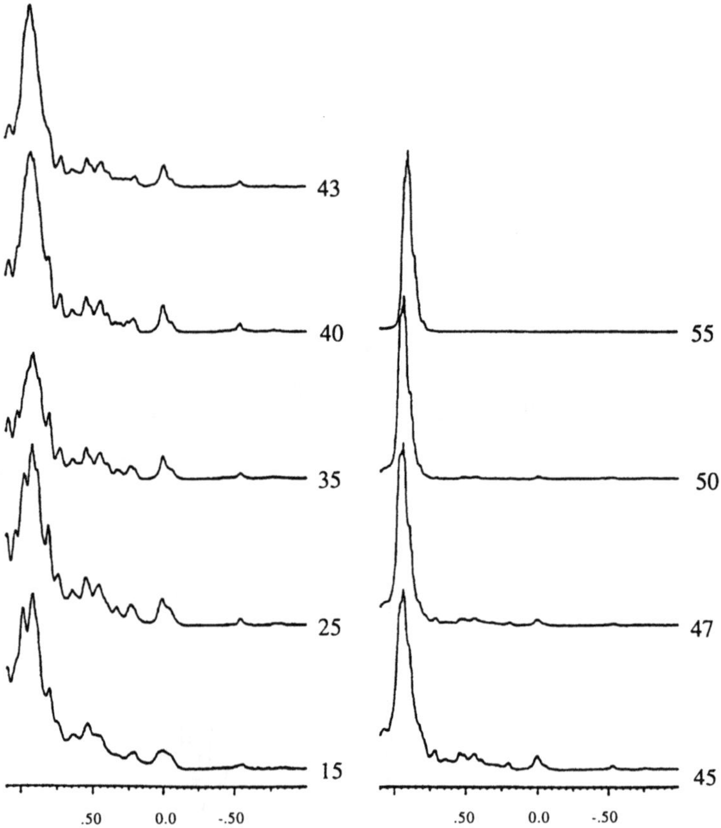

Figure 13. NMR spectra of fibrolase (2.5 mg/ml, pD 8) from $-1.0\ \delta$ to $+1.0\ \delta$ at increasing temperatures.

helical regions, while the NMR data give a composite picture of local structural changes at seven discrete sites in the protein.

Denatured fibrolase also displayed an increased tendency to aggregate and precipitate. After heating fibrolase past 50°C, NMR samples were very cloudy and a significant precipitate could be isolated. On the other hand, CD samples appeared to be only slightly turbid, although the concentrations were more than an order of magnitude less for the CD samples.

Aggregation and precipitation were also observed during attempts to quantitate protein aggregation as a function of temperature using UV-visible absorbance spectroscopy. After 10 min at 37°C, absorbance increased by approximately twofold at 400 nm. No visible turbidity or precipitate was observed. This apparent increase in absorbance is due to scattered light from

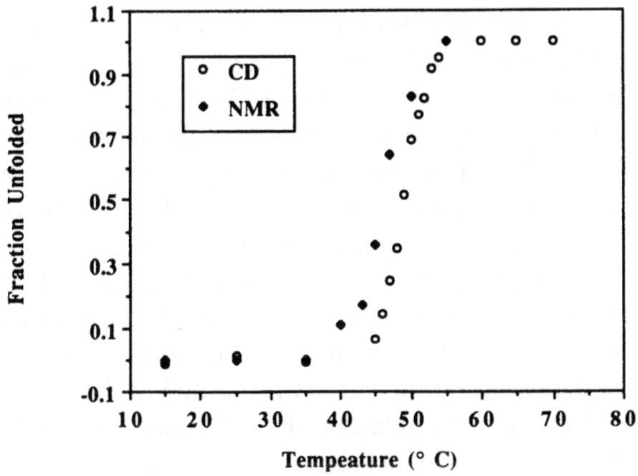

Figure 14. Thermal denaturation curves for fibrolase as determined by CD (○, 478 μg/ml, pH 8) and by NMR (♦, 2.5 mg/ml, pD 8) spectroscopy.

soluble protein aggregates (Glatz, 1992). Further heating to 45°C for less than 5 min resulted in extensive precipitation.

6. EFFECTS OF ZINC BINDING ON THE STABILITY AND STRUCTURE OF FIBROLASE

The presence of an essential zinc ion (Markland *et al.*, 1988; F. S. Markland, unpublished results) is a crucial consideration in evaluating the stability of fibrolase. Loss of structure and activity upon exposure to extremes of pH and temperature can be rationalized on the basis of removal of the zinc atom. Denaturation of fibrolase under these conditions leads to a marked tendency to aggregate and precipitate (see below). Various studies were performed in order to examine the strength of zinc binding, the effect of zinc on the structure and activity and secondary structure, and to determine the metal binding site.

6.1. Effects of EDTA Addition

The effect of EDTA on the activity and conformation of fibrolase was of interest because of the classification of fibrolase as a zinc metalloprotein (Markland *et al.*, 1988; Pretzer *et al.*, 1992). Other investigators had previously observed inhibition of azocaseinolytic activity of fibrolase by EDTA, and, in fact, these studies contributed to the characterization of fibrolase as a metalloprotein (Markland *et al.*, 1988; Ahmed *et al.*, 1990a). It was of interest to investigate what concentrations of EDTA affected fibrolase activity, and what other changes (e.g., structural) were caused by EDTA.

The effect of EDTA on proteolytic activity of fibrolase was dependent on the EDTA concentration added (see Table II). Partial inhibition was observed for EDTA concentrations as low as 0.1 mM. Complete inhibition occurred at 0.9 mM EDTA for activity versus the insulin B chain. Further increases in EDTA concentration had no effect. Activity toward insulin B chain was inhibited to a greater extent than was the azocaseinolytic activity. About half of the azocaseinolytic activity remained in the presence of 1 mM EDTA. A higher EDTA concentration of 5 mM was required for essentially complete (98%) inhibition of azocaseinolytic activity.

The reason for the difference in effective inhibitory concentrations of EDTA might be related to the difference in molecular size of the two substrates. The molecular weight of insulin B chain is about 3800, while that of azocasein is about 25,000. Because of the size difference, the substrates might be subject to different constraints around the active site of fibrolase. The data indicated that the active site constraints, or importance of metal binding, might be greater for the insulin B chain, since this activity was

Table II
Proteolytic Activity of Fibrolase in the Presence of EDTA

[EDTA] (mM)	% activity remaining relative to control[a]	
	Versus insulin B chain	Versus azocasein
0.1	51.4 (9.3)	55.2 (12.0)
0.2	51.8 (15.8)	50.1 (12.5)
0.5	32.7 (6.9)	49.6 (14.5)
0.9	0	nd
1.0	nd[b]	42.1 (12.5)
2.0	nd	22.6 (3.8)
5.0	nd	2.3 (0.1)
6.0	0	nd
20.0	0	nd

[a] No EDTA was added to the control samples. Initial fibrolase concentration was ~100 μg/ml. Numbers reported are the mean of at least three determinations, with standard deviations given in parentheses.

[b] nd, not determined.

inhibited at lower EDTA concentrations. It is interesting to note that these observations are in contrast to activity results for earlier mentioned studies (Fig. 3) which suggested that azocaseinolytic activity was lost more readily than was activity for the insulin B chain at pH extremes and after extended storage at room temperature. These results suggested that different effects might occur in the protein upon EDTA addition. The time course of EDTA inhibition was rapid (see Table III). Inhibition was complete at the first measurable time point for 6 and 20 mM EDTA, and remained complete over 1 hr. The time course of inhibition was not followed for other EDTA concentrations.

While submillimolar concentrations of EDTA (<1 mM) resulted in partial loss of activity versus insulin B chain and azocasein, changes in CD spectra were also noted at these EDTA concentrations (Fig. 15). These amounts of EDTA represent only a small molar excess of EDTA relative to fibrolase. At 0.1 mM EDTA, the molar excess of EDTA over fibrolase was only about 25 for studies examining the effect on activity and 10–15 in the case of the CD work. The low concentration and molar excess of EDTA effective in inducing activity and structural changes in fibrolase indicated that the zinc ion might be bound quite loosely to the protein.

CD results (ellipticity at 222 nm) were correlated with the observed activity changes. At submillimolar EDTA concentrations, structural changes were evident by CD as indicated by decreasing elliptical intensity at 222 nm

Table III
Time Course of EDTA Inhibition of Fibrolase Proteolytic Activity[a]

	Peak area ($\times 10^{-2}$)[b]		
Time (min)	0.0 mM EDTA	6 mM EDTA	20 mM EDTA
0	8.16	8.48	8.69
5	4.78	—	8.84
10	4.07	—	8.73
15	—	8.13	—
20	2.89	—	8.52
30	2.48	9.16	8.55
45	2.06	—	7.82
60	1.49	8.00	8.63

[a] Proteolytic activity measured versus insulin B chain.
[b] Peak area of insulin B-chain substrate. Decreasing peak area indicates proteolysis by fibrolase. Peak areas are reported for one determination. Typical assay variability was ~5%.

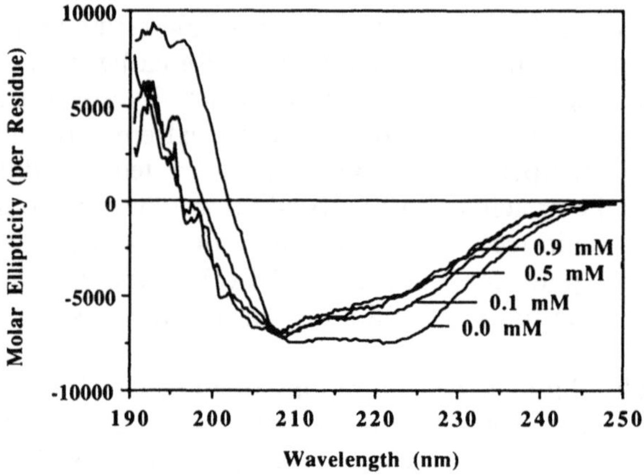

Figure 15. Effects of EDTA addition on the far-UV CD spectra of fibrolase.

(Fig. 15), and partial inhibition of activity was also observed (Table II). At EDTA concentrations ≥ 0.9 mM, structural changes observed by CD appeared to be complete, and correlated well with inhibition of activity for the insulin B chain. Changes similar to those observed upon addition of EDTA were noted upon addition of DTT (Fig. 16). While it is a well-known reducing agent, it is also a good chelating agent (Miller *et al.*, 1985), and it removed the zinc from fibrolase rapidly and easily. The resulting structure is similar to that found upon EDTA addition.

The structural changes observed by CD upon EDTA addition to fibrolase were corroborated by native gel electrophoresis. Fibrolase samples con-

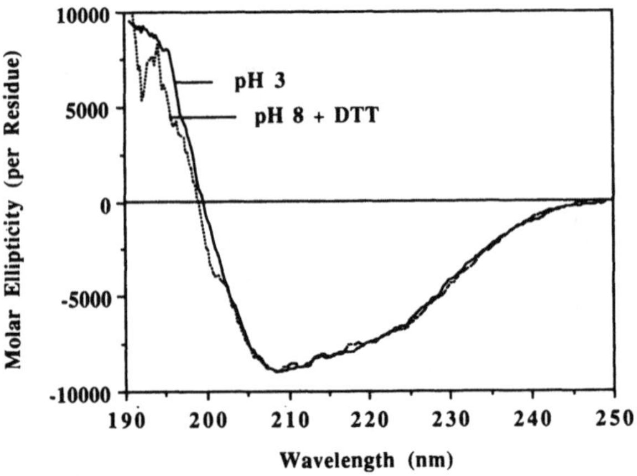

Figure 16. Far-UV CD spectra of fibrolase at pH 3 and at pH 8 in the presence of DTT.

taining EDTA (>1 mM) consistently migrated farther on native gels relative to fibrolase without EDTA present. These observed changes in the activity and conformation of fibrolase with EDTA supported involvement of zinc in the active site of the protein. The data showed not only that the presence of the metal ion is critical for proteolytic activity, but that it is also important to the structure of the protein. The metal ion may bind to several remote sites of the protein, and when the ion is removed by EDTA treatment, these residues might not retain the driving force for association, resulting in extensive structural perturbation.

6.2. Proposed Zinc Binding Site

While the exact location of the zinc binding site is not known for fibrolase, such sites have been identified for many zinc-containing enzymes, primarily through x-ray crystallography (Vallee and Auld, 1990). Considering that fibrolase possesses proteolytic activity, its sequence was compared with other metalloproteases. Comparison of the proposed site for fibrolase with the sequence from thermolysin and subtilisin showed a high degree of homology (Table IV). However, further analysis showed that the final glutamate ligand, located at position 166 of these two enzymes, appears to be missing from fibrolase. This may account for a relatively low binding affinity for the zinc in fibrolase. However, there is another class of zinc proteases which has an even higher degree of homology with fibrolase. These are the collagenases. The AHELGH sequence is identical to the central metal binding site in collagenases which have been sequenced to date. In addition, there is a conserved sequence in the COOH-terminal portion of collagenases which is reproduced in fibrolase as well (Table IV). Overall, this appears to substantiate that there is a well-defined zinc binding site in fibrolase and that it is similar to that found in other zinc-containing proteases.

Upon addition of EDTA, the zinc atom is irreversibly removed from fibrolase, with complete loss of enzymatic activity and an increased propensity toward aggregation and precipitation. Concomitant with the loss of activity is a substantial change in the far-UV CD spectrum (Fig. 15). Secondary structure analysis suggests that more than half of the α-helix content is lost in this transition. The α-helix content drops from ~24% to ~10%, based on deconvolutions using the data base of Chang *et al.* (1978). These results are consistent with the sequence analysis described below.

Based on secondary structure analyses (see below), a tentative binding site was proposed on an amphiphilic helix near the COOH-terminus (see Fig. 17). Note that previous work, using chemical modification, suggested two of

Table IV
Sequence Comparison for Proposed Fibrolase Zinc Binding Site and Other Zinc Proteases (see Vallee and Auld, 1990)

Fibrolase	T	M	A	**H**	**E**	L	G	**H**	N	L	G	(140–150)
Thermolysin	V	V	A	**H**	**E**	L	T	**H**	A	V	T	(139–149)
Aminopeptidase N, *E. coli*	V	I	G	**H**	**E**	Y	F	**H**	N	W	T	(139–149)
Subtilisin	V	T	A	**H**	**E**	M	T	**H**	G	V	T	(140–150)
Collagenase, human	V	A	A	**H**	**E**	L	G	**H**	S	L	G	(215–225)
Collagenase, rabbit	V	A	A	**H**	**E**	L	G	**H**	S	L	G	(215–225)
Collagenase IV	V	A	A	**H**	**E**	F	G	**H**	A	L	G	(214–225)
Stromelysin-2	V	A	A	**H**	**E**	L	G	**H**	S	L	G	(214–224)
Metalloproteinase-1	A	A	T	**H**	**E**	L	G	**H**	S	L	G	(211–221)
Metalloproteinase-3	V	A	A	**H**	**E**	I	G	**H**	S	L	G	(215–225)
Immunity protein region 2	V	C	S	**H**	**E**	L	G	**H**	S	L	G	(75–85)

Fibrolase	A	N	S	C	V	M	A	A	M	(162–170)
Thermolysin	G	A	I	N	E	A	I	S	D	(162–170)
Aminopeptidase N, *E. coli*	L	S	L	K	E	G	L	T	V	(162–170)
Subtilisin	G	A	L	N	E	S	F	S	D	(162–170)

Glu^{166} is the third ligand (after the two histidines) in thermolysin and subtilisin. There is no such ligand in fibrolase. There seems to be a better homology with collagenases, not only in the metal binding site but elsewhere in the C-terminus. This portion is given as significant in the Vallee and Auld (1990) review, but they do not appear to discuss it.

Fibrolase	**P**	S	K	L	**F**	S	D	C	S	(175–185)
Collagenase, human	**P**	S	Y	T	**F**	S	G	D	V	(238–246)
Collagenase, rabbit	**P**	N	Y	M	**F**	S	G	D	V	(238–246)
Collagenase IV	**P**	M	Y	R	**F**	T	E	G	P	(238–246)
Stromelysin, human	**P**	L	Y	N	S	**F**	T	E	L	(238–246)
Metalloproteinase-3	**P**	L	Y	H	S	L	T	D	L	(239–247)
Immunity protein region 2	E	R	T	L	**F**	S	V	D	K	(96–104)

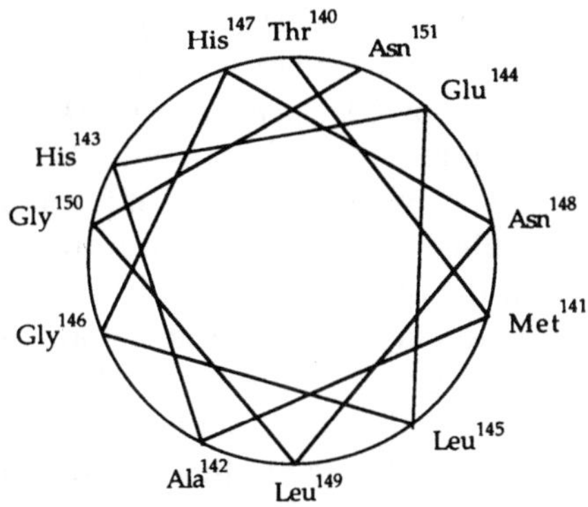

Figure 17. Helix wheel diagram for putative zinc-binding site in native fibrolase. It includes residues 142–153 (AHELGHNLGMNH).

the zinc ligands were histidines (F. S. Markland, unpublished results). Positioning of the ligands on an α-helix would be consistent with the loss of significant helical structure upon addition of EDTA. Beginning with Ala^{142}, the helix wheel diagram (Fig. 17) shows that His^{143} and His^{147} are on the same side of the helix, and are properly positioned to coordinate to a zinc ion. It is possible that Glu^{144}, which is conserved in many other sequences of zinc-containing enzymes, may also participate in coordination to the zinc atom.

Identification of α-helical segments in fibrolase was accomplished by secondary structure prediction methods. The primary scheme used was that of Cid *et al.* (1982), which employs hydrophobicity and solvent accessibility as guidelines to secondary structure types. A significant α-helix, which includes the histidines involved in metal binding, is predicted to extend from residues 140 to 152. The method of Garnier *et al.* (1978) also predicts this α-helix, although with a slightly different length. Each of these methods indicates an overall α-helix content of ~25%, in agreement with the CD analysis. Chou–Fasman parameters (Chou and Fasman, 1974) also gave a decided preference to the α-helix ($p = 1.08$ versus 0.95) over β-sheet for this region.

7. SUMMARY

Fibrolase is an active fibrinolytic agent and possesses potential for use in thrombolytic therapy. Its mode of action had been characterized, both *in vitro* and *in vivo.* Possessing three disulfide bonds, native fibrolase is nonglycosylated and binds an intrinsic zinc atom. The zinc is essential for retention of activity and structural integrity.

In solution, fibrolase is sensitive to changes in pH and temperature

(Pretzer *et al.,* 1991). At neutral to basic pH (pH 5–9), the solubility and stability of fibrolase is nearly constant. Little structural variation can be detected by CD spectroscopy. However, decrease in pH below 5 leads to a pronounced reduction in both the solubility and activity of fibrolase. At pH 3 and below, the solubility of fibrolase returns but the activity does not. This solubility profile is unusual in that the minimal solubility is well removed from the pI (which is 6.7). It is proposed that the behavior of fibrolase with variation in pH can be understood in terms of capacity to bind zinc. At pH 5 to 9, the protein binds zinc and the structure and activity are preserved. Near pH 5, the histidine residues which serve as ligands for the zinc become protonated and zinc binding is lost. Loss of zinc leads to local unfolding of a helical segment of fibrolase, exposing hydrophobic groups which allow the protein to rapidly aggregate. At lower pH values (1–3), the protein again adopts a more globular structure, similar to molten globule states, and the solubility increases. However, without the zinc, fibrolase remains inactive. Changes in pH also affect thermal stability. The T_m for fibrolase moves from 50°C at pH 8 to 43°C at pH 5.

Increases in temperature also lead to removal of the zinc ion, again producing a partially denatured protein with a marked tendency to aggregate. In both cases (decrease in pH and increase in temperature), analysis of the CD spectra indicates that the protein has primarily lost α-helical secondary structure. A major change in structure can also be observed using NMR spectroscopy. At temperatures below 35°C, the globular structure of fibrolase remains intact, although some increase in chain mobility can be noted with increased temperature. Upon melting, numerous signals collapse as the protein unfolds. Transition temperatures (T_m) as measured by CD and NMR are in good agreement.

Similar structural changes can be induced by adding zinc chelators such as EDTA and DTT. This leads to complete loss of activity at EDTA concentrations above 1.0 mM. Changes in the CD spectra correlate well with loss of proteolytic activity. However, there are subtle differences in the NMR and CD spectra obtained upon addition of EDTA, increase in temperature, and decrease in pH, suggesting that there may be some dissimilarity in the structures of the denatured forms.

REFERENCES

Ahern, T. J., and Klibanov, A. M., 1986, Why do enzymes irreversibly inactivate at high temperatures? in: *Protein Structure, Folding, and Design* (D. L. Oxender, ed.), Liss, New York, pp. 283–289.

Ahmed, N. K., Tennant, K. D., Markland, F. S., and Lacz, J. P., 1990a, Biochemical characteristics of fibrolase, a fibrinolytic protease from snake venom, *Haemostasis* **20**:147.

Ahmed, N. K., Gaddis, R. R., Tennant, K. D., and Lacz, J. P., 1990b, Biological and thrombolytic properties of fibrolase—A new fibrinolytic protease from snake venom, *Haemostasis* **20**:334.

Bajwa, S. S., Kirakossian, H., Reddy, K. N. N., and Markland, F. S., 1982, Thrombin-like and fibrinolytic enzymes in the venom from the gaboon viper (*Bitis gabonica*), eastern cottonmouth moccasin (*Agkistrodon p. piscivorus*) and southern copperhead (*Agkistrodon c. contortrix*) snakes, *Toxicon* **20**:427.

Baum, J., Dobson, C. M., Evans, P. A., and Hanley, C., 1989, Characterization of a partly unfolded protein by NMR methods: Studies on the molten globule state of guinea pig alpha-lactalbumin, *Biochemistry* **28**:7.

Chang, C. T., Wu, C.-S. C., and Yang, J. T., 1978, Circular dichroic analysis of protein conformation: Inclusion of beta turns, *Anal. Biochem.* **91**:13.

Charney, J., and Tomarelli, R. M., 1947, A colorimetric method for the determination of the proteolytic activity of duodenal juice, *J. Biol. Chem.* **171**:501.

Chou, P. Y., and Fasman, G. D., 1974, Prediction of protein conformation, *Biochemistry* **13**:222.

Cid, H., Bunster, M., Arrigada, E., and Campos, M., 1982, Prediction of secondary structure of proteins by means of hydrophobicity profiles, *FEBS Lett.* **150**:247.

Didisheim, P., and Lewis, J. H., 1956, Fibrinolytic and coagulant activities of certain snake venoms and proteases, *Proc. Soc. Exp. Biol. Med.* **93**:10.

Eagle, H., 1937, The coagulation of blood by snake venoms and its physiological significance, *J. Exp. Med.* **65**:613.

Garnier, J., Osguthorpe, D. J., and Robson, B., 1978, Analysis of the accuracy and implication of simple methods for predicting the secondary structure of globular proteins, *J. Mol. Biol.* **120**:97.

Glasoe, P. K., and Long, F. A., 1960, Use of glass electrodes to measure acidities in deuterium oxide, *J. Phys. Chem.* **64**:188.

Glatz, C., 1992, Modeling of aggregation—Precipitation phenomena, in: *Pharmaceutical Biotechnology,* Vol. 2 (T. J. Ahern and M. C. Manning, eds.), Plenum, New York, pp. 135–166.

Goto, Y., and Fink, A. L., 1989, Conformational states of β-lactamase: Molten-globule states at acidic and alkaline pH with high salt, *Biochemistry* **28**:945.

Kleanthous, C., Wemmer, D. E., and Schachman, H. K., 1988, The role of an active site histidine in the catalytic mechanism of aspartate transcarbamoylase, *J. Biol. Chem.* **263**:13062.

Koseki, T., Kitabatake, N., and Doi, E., 1988, Conformational changes in ovalbumin at acidic pH, *J. Biochem.* **103**:425.

Manning, M. C., Patel, K., and Borchardt, R. T., 1989, Stability of protein pharmaceuticals, *Pharm. Res.* **6**:903.

Markland, F. S., Reddy, K. N. N., and Guan, L., 1988, Purification and characterization of a direct-acting fibrinolytic enzyme from southern copperhead venom, in: *Hemostasis and Animal Venoms* (H. Pirkle and F. S. Markland, eds.), Dekker, New York, pp. 173–189.

Marx, G., 1988, Zinc binding to fibrinogen and fibrin, *Arch. Biochem. Biophys.* **266**:285.

Miller, J., McLachlan, A. D., and Klug, A., 1985, Repetitive zinc-binding domains in the protein transcription factor IIIA from *Xenopus* oocytes, *EMBO J.* **4**:1609.

Mitchell, S. W., 1860, Upon the venom of the rattlesnake, *Smithson. Contrib. Knowl.* **12**:145.

Mitchell, S. W., and Reichert, E. T., 1886, Researches upon the venoms of poisonous serpents, *Smithson. Contrib. Knowl.* **26**:186.

Mock, W. L., and Tsay, J.-T., 1988, pK values for the active site residues of carboxypeptidase A, *J. Biol. Chem.* **263**:8635.

Mulkerrin, M. C., and Wetzel, R., 1989, pH dependence of the reversible and irreversible thermal denaturation of γ interferons, *Biochemistry* **28**:6556.

Parraga, G., Horvath, S. J., Eisen, A., Taylor, W. E., Hood, L., Young, E. T., and Klevit, R. E., 1988, Zinc-dependent structure of a single-finger domain of yeast ADR1, *Science* **241**:1489.

Pretzer, D., Schulteis, B. S., Vander Velde, D. G., Smith, C. D., Mitchell, J. W., and Manning, M. C., 1991, Stability of the thrombolytic protein fibrolase. Effect of temperature and pH on activity and conformation, *Pharm. Res.* **8**:1103.

Pretzer, D., Schulteis, B. S., Vander Velde, D. G., Smith, C. D., Mitchell, J. W., and Manning, M. C., 1992, Effect of zinc binding on the structure and stability of fibrolase, a fibrinolytic protein from snake venom, *Pharm. Res.* **9**:870.

Randolph, A., Chamberlain, S. H., Chu, H.-L. C., Retzios, A. D., Markland, F. S., Jr., and Masiarz, F. R., 1992, Amino acid sequence of fibrolase, a direct-acting fibrinolytic enzyme from *Agkistrodon c. contortrix* venom, *Protein Sci.* **1**:590.

Redfield, C., and Dobson, C. M., 1988, Sequential ^{1}H NMR assignments and secondary structure of hen egg white lysozyme in solution, *Biochemistry* **27**:122.

Vallee, B. L., and Auld, D. S., 1990, Zinc coordination, function, and structure of zinc enzymes and other proteins, *Biochemistry* **29**:5647.

Chapter 11

Insulin Structure and Stability

Jens Brange and Lotte Langkjær

1. INTRODUCTION

Since its isolation by Banting and Best (1922), insulin has been one of the most extensively studied molecules in biochemistry. With the size of a polypeptide, but having all of the structural features of a large protein, insulin has been an ideal model compound for investigating the structure, properties, and characteristics of proteins. Thus, its primary structure (Sanger *et al.*, 1955) and chemical synthesis (Meienhofer *et al.*, 1963; Katsoyannis, 1964) were the first established for a protein, and its crystallization in a variety of forms (Schlichtkrull, 1958), elucidation of the biosynthetic pathway in the pancreas (Steiner, 1967) and of the hormone's three-dimensional structure (Adams *et al.*, 1969; Baker *et al.*, 1988) all represented major scientific achievements and important milestones in our understanding of protein structure and function. More recently, human insulin has been the first protein to be biosynthesized on a large scale in microorganisms encoded by DNA recombinant methods, and insulin analogues made through recombinant DNA technology represent the first successful attempts to tailor protein hormones for therapy (Brange *et al.*, 1988; Markussen *et al.*, 1988).

Insulin has been in therapeutic use for nearly 70 years, but its stability in pharmaceutical formulation has so far only been systematically studied in terms of biological potency (Storvick and Henry, 1968; Pingel and Vølund, 1972). Whereas the physical stability of insulin under a variety of conditions has been extensively studied throughout the years, especially in connection

---

Jens Brange and Lotte Langkjær • Novo Research Institute, Novo Nordisk A/S, DK-2880 Bagsvaerd, Denmark.

Stability and Characterization of Protein and Peptide Drugs: Case Histories, edited by Y. John Wang and Rodney Pearlman. Plenum Press, New York, 1993.

with its more recent use in infusion pumps (for review see Brange *et al.*, 1987a), very little information has appeared in the literature on the chemical transformation of insulin during storage of insulin preparations. The few investigations in the past have all dealt with its hydrolysis in acid medium into insulin desamido products (Sundby, 1962; Slobin and Carpenter, 1963; Berson and Yalow, 1966).

The purpose of the present chapter is to briefly summarize the structural elements relevant for the physicochemical and chemical characteristics of insulin, especially when formulated for therapy, and to review the present knowledge as to the physical and chemical stability of insulin pharmaceutical preparations.

2. INSULIN STRUCTURE

2.1. Primary Structure

The human insulin molecule contains 51 amino acid residues in two polypeptide chains (A and B) linked by two disulfide bonds as illustrated in Fig. 1. The A chain consists of 21 residues with an additional disulfide loop between A6 and A11 whereas the longer B chain holds 30 residues. Primary structures are now known for insulins from more than 50 animal species. Insulins from the hystricomorph rodent family show the highest number of amino acid substitutions; guinea pig insulin, for example, contains 18 residues different from the respective positions in human insulin. Of the other insulins used for pharmaceutical preparations, porcine and bovine insulins contain alanine in position B30 and bovine insulin contains, in addition, alanine in A8 and valine in A10.

The invariant residues (ten in each chain; see Fig. 1) are typically the residues responsible for the structural integrity of the molecule and which help to define the folding of the molecule, e.g., the cysteine residues, the nonpolar residues constituting the hydrophobic core, and other residues interacting in the folding of the molecule into its three-dimensional structure.

2.2. Secondary and Tertiary Structure

Despite variations in primary structure, the folding and packing of the two chains into their three-dimensional conformation are essentially the same in all insulins. The A chain forms two nearly antiparallel α-helices, A2 to A8 and A13 to A20. The B chain forms a single α-helix from B9 to B19 followed by a turn and a β-strand from B21 to B30. The arrangement of the

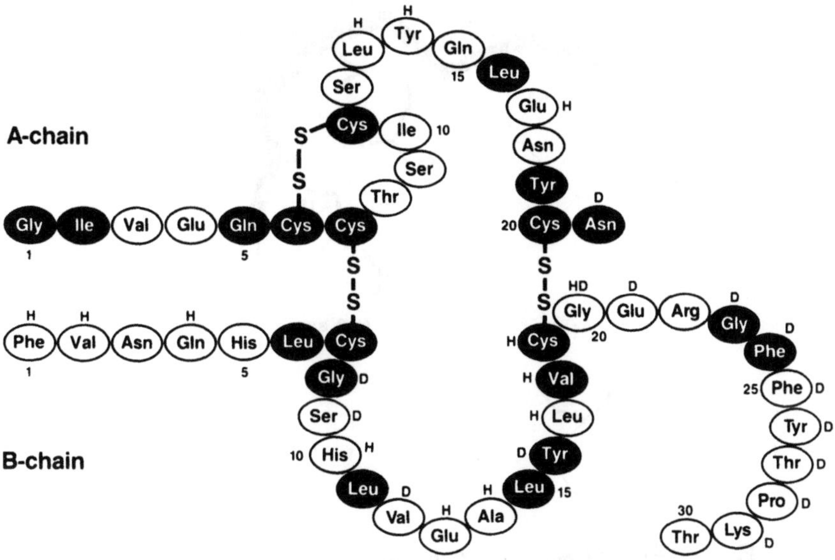

Figure 1. The primary structure of human insulin. The black residues designate the amino acids which are invariant among species of insulin. The letters indicate the residues involved in association of the molecule: D, dimer formation; H, hexamer formation. Porcine insulin only differs in position B30 (Ala instead of Thr). Bovine insulin has the same substitution and contains, in addition, Ala in position A8 and valine in position A10.

chains buries the cysteine A6–A11 and the aliphatic side chains of residues A2, A16, B11, and B15 in the nonpolar core. The surface of the insulin monomer is covered by both polar and nonpolar residues. Whereas the A chain and the B9–B19 helix form a stable structural unit, the B25–B30 and the B1–B8 segments are variable in conformation. Recently it has been discovered that phenol promotes formation of an additional helical segment from B1 to B8 (Wollmer *et al.*, 1987; Derewenda *et al.*, 1989) which involves movement of more than 25 Å of residue B1. Figure 2 shows the structure of the insulin monomer with and without this extra α-helix.

2.3. Quaternary Structure (Self-Association)

Insulin exists as a monomer only at low concentration (<0.1 μM, ~0.6 μg/ml). It dimerizes at higher concentrations relevant for pharmaceutical formulation, and in the pH range 4–8, in the presence of zinc ions, three dimers assemble at concentrations >0.01 mM further into a hexamer (Fig. 3). At concentrations ≥ 2 mM the hexamer is formed at neutral pH without the assistance of zinc ions (Hansen, 1991). The two insulin molecules in the

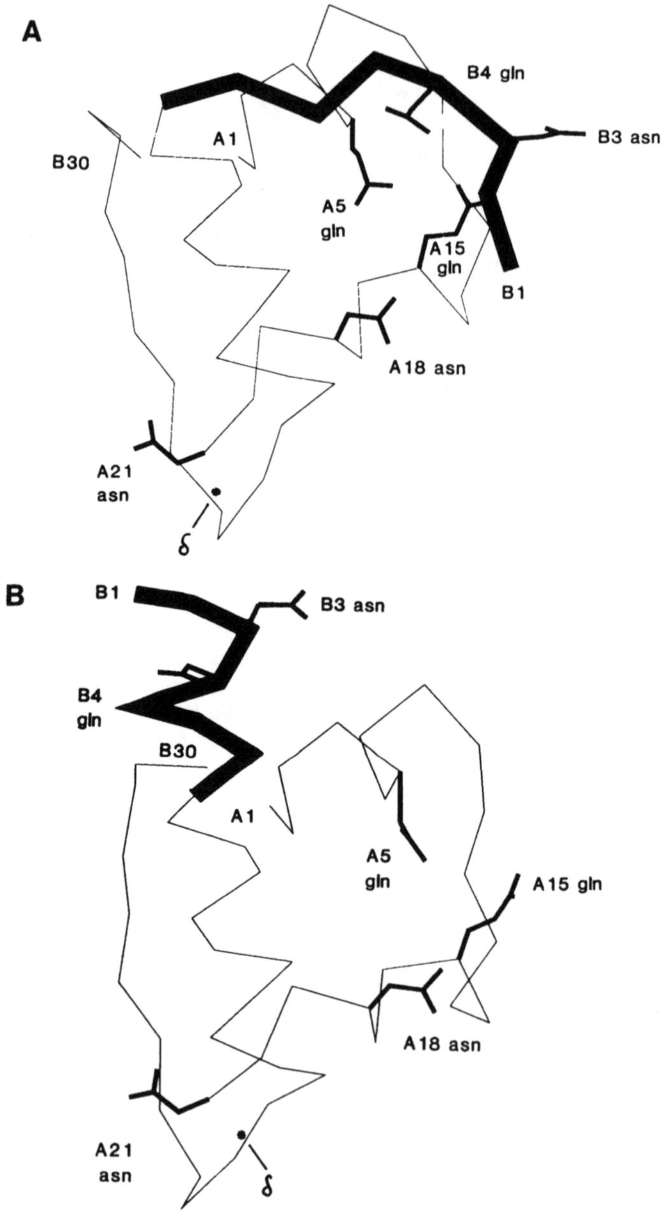

Figure 2. Tertiary structures of the insulin molecule. The panels show the peptide backbone atoms of the A and B chains and their arrangement in the three-dimensional packing of the molecule. The A chain and most of the B chain are drawn with thin bonds. The B chain from B1 to B8 is drawn with very broad bonds. Only the side chains of the six potentially labile amide residues are shown (thicker bonds). (A) The "normal," most common structure with extended B-chain NH_2-terminal. (B) The structure with a switch into α-helix of residues B1 to B8 promoted by phenol or phenol derivatives.

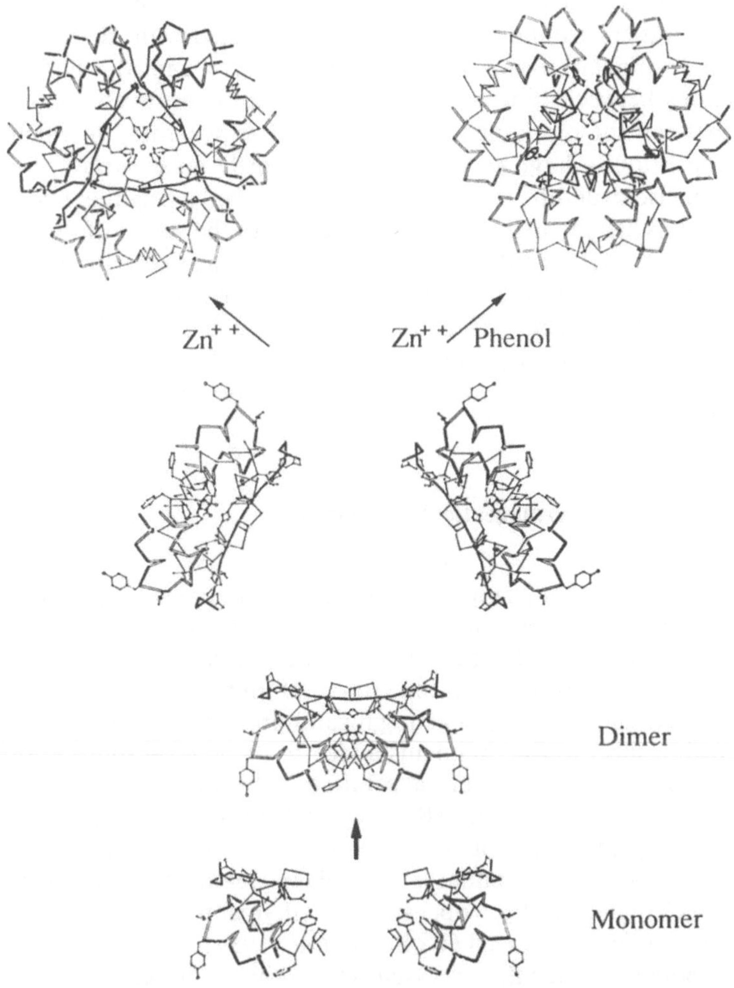

Figure 3. The self-assembly of insulin from the monomer to the dimer, and of three dimers into a hexamer in the presence of zinc ions. When, in addition, phenol (or phenol derivatives like *m*-cresol) is present, the gathering of the three dimers is accompanied by a conformational transformation of B1 to B8 from an extended chain into an α-helix. Only the α-carbon atoms of the main chains are indicated. The A chains are drawn with open bonds, the B chains with thinner full bonds, and B1–B8 with thick bonds. The atoms in the side chains interacting in dimer and hexamer formation (see Fig. 1) are drawn with thin bonds.

dimer are held together by nonpolar forces and four hydrogen bonds (between B24 and B26 main chain atoms) arranged as an antiparallel β-sheet structure between the two COOH-terminal strands of the B chain. The packing of the dimers around two zinc ions is associated with the burial of the remaining nonpolar surface (A13–A14, B1–B2, B14, and B17–B18), but the interactions between dimers in the hexamer are considerably looser than those between the monomers in the dimer. The hexamer, in which the A chain constitutes a large proportion of the mostly polar surface, forms an almost spheric structure with a diameter of 50 Å and a height of about 35 Å. The zinc ions are situated in the polar channel in the center of the hexamer coordinated to B10 histidines. The above-mentioned structural transition induced by phenol requires the presence of zinc ions and thus is a property of the hexamer. The phenol molecules bind in nonpolar cavities in the interface between dimers created by packing of A-chain residues from one dimer and the B1–B8 helical segment from the adjacent dimer (Derewenda *et al.*, 1989).

2.4. Structure in Pharmaceutical Formulations

Today, all insulin formulations are either neutral solutions or suspensions of zinc insulin containing 40, 80, or 100 IU/ml (100 IU/ml ~0.6 mM ≈4 mg/ml). Consequently, these pharmaceutical preparations all share the hexamer of insulin as the common structural unit.

The formulation of rapid-acting insulin preparations (Regular insulin) varies with respect to the excipients added to the neutral solution of hexameric insulin. The original neutral solution (Schlichtkrull *et al.*, 1965) contains NaCl and methylparaben (Regular I) whereas a later version is formulated with glycerol and phenol or *m*-cresol (Regular II). In the latter formulation the phenol content is most likely creating the previously mentioned α-helix whereas the propensity of methylparaben to induce the same change is very much reduced (Wollmer *et al.*, 1987).

The potential of insulin to crystallize in a variety of forms, a reflection of the molecule's diverse association patterns and ligand binding properties, has been utilized in the formulations with retarded onset of action (Brange *et al.*, 1987a). Although crystals of the insulin monomer, dimer, and hexamer have been obtained (Derewenda *et al.*, 1989), the common unit in the pharmaceutically relevant crystals is the hexamer.

When a neutral solution of hexameric zinc insulin, in the presence of phenol (or *m*-cresol), is mixed with protamine (an arginine-rich and therefore strongly basic polypeptide (~4.3 kDa) isolated from salmon sperm), a precipitate is formed which converts to orthorhombic crystals with a strong tetragonal appearance (Krayenbühl and Rosenberg, 1946; Baker and Dodson, 1970). These crystals contain approximately 0.9 molecule of protamine and 2 atoms of zinc per insulin hexamer. The protamine is rather loosely bound in the crystal but mediates interactions between dimers in the hexamer where it penetrates the polar central channel. In the interstices between hexamers it is also involved in hexamer–hexamer contacts (Simkin *et al.*, 1970; Balschmidt *et al.*, 1991). Six *m*-cresol molecules occupy cavities in the hexamer created by A-chain residues from one dimer and the B1–B8 helical structure from the adjacent dimer (Balschmidt *et al.*, 1991). The pharmaceutical preparation based on this principle is known as NPH (neutral protamine Hagedorn) or Isophane insulin and has an intermediate release profile after subcutaneous injection.

A possibility of gradation of the prolongation of the release of insulin was accomplished with the Lente insulin principle (Hallas-Møller, 1956), also named insulin zinc suspensions (IZS). It is based on the discovery that a protracted effect, depending on the physical state of the insulin, could be obtained by addition of small amounts of extra zinc ions to hexameric 2 zinc insulin in neutral medium. Amorphously precipitated zinc insulin complex (Semilente or IZS, amorphous) shows a slight retardation whereas rhombohedral insulin crystals show a very prolonged action profile when additional zinc ions are added (Ultralente or IZS, crystalline). An intermediate action profile was obtained with a 3 : 7 mixture of amorphous and crystalline insulin particles (Lente or IZS, mixed).

Biphasic insulin (Schlichtkrull *et al.*, 1965), a stable mixture of 1 part porcine insulin in solution and 3 parts bovine insulin in crystalline suspension, is an example of elegant utilization of the difference in solubility of porcine and bovine insulins at neutral pH. Mixtures of NPH crystals and a neutral solution of insulin also show a biphasic action profile, although part of the dissolved insulin after mixing becomes loosely bound to the surface of the crystals (Brange *et al.*, 1987a).

2.5. Stability Overview

Maintenance of the structural integrity of a therapeutic protein is essential for its efficacy in relation to physiological and pharmacological activity. The structural changes taking place during its storage and use can be categorized into those involving either physical or chemical alterations in the molecule. Physical instability refers to changes in conformation (secondary to quaternary structure), also called unfolding or "denaturation." Chemical instability involves covalent modification of the sequence (primary structure) leading to bond formation or cleavage. In practice, however, these two distinct routes of transformation of the structure have a mutual influence, as changes in conformation affect the susceptibility of the molecule for chemical attack and vice versa.

3. PHYSICAL STABILITY

Like other globular proteins, insulin tends to adopt and maintain a three-dimensional structure in which its hydrophobic surfaces are buried by folding or assembly of individual molecules. However, changes of the native conformation or of its "normal" pattern of assembly can be provoked by a variety of factors. The most frequent (macroscopically visible) events encountered in handling or use of insulin are increase in solution viscosity, precipitation from solutions, changes in physical appearance of suspensions, and loss of insulin due to adsorption.

3.1. Isoelectric and Metal-Ion-Induced Precipitation

Formation of amorphous or crystalline precipitates from solutions of insulin is usually caused by reduction in insulin solubility which can be due to changes in pH (isoelectric precipitation) or introduction of Zn^{2+} or certain other divalent metal ions. Thus, in unbuffered, neutral solutions of insulin, even small amounts of acidic substances, e.g., leachables from polymer materials used in devices for insulin delivery (Melberg *et al.*, 1988), can result in sufficient decrease in pH to bring insulin into its isoelectric precipitation zone (pH 4.5–6.5). The source of metal ions is often rubber, e.g., vial stoppers, in contact with the insulin solution.

3.2. Fibrillation

The inherent tendency of insulin to undergo conformational changes resulting in aggregation and formation of a viscous gel or insoluble precipitates was observed early in the insulin era (du Vigneaud *et al.*, 1928; Krogh and Hemmingsen, 1928). This so-called fibrillation of insulin undoubtedly caused the early period's low yields in extraction of insulin from the pancreas, and later on, hampered the development of chromatographically purified insulin (Schlichtkrull *et al.*, 1974). More recently, insulin aggregation has complicated the use of insulin in delivery systems (Lougheed *et al.*, 1980; Brange *et al.*, 1987a), especially in implantable insulin pumps.

The mechanism of insulin fibril formation has been studied in detail by Waugh (1946, 1957; Waugh *et al.*, 1953), who found that the interactions between subunits are mainly hydrophobic and that an initial lag time apparently was necessary in order to obtain stable fibril nuclei able to start and seed subsequent aggregation. The subunits of these interactions have been suggested to be either the dimer or the monomer of insulin (Waugh *et al.*, 1953; Burge and Rougvie, 1972). Neutral solutions of hexameric insulin exhibit a quite anomalous aggregation behavior in being more prone to fibrillation at reduced concentration of insulin (Brange and Havelund, 1983; Dathe *et al.*, 1990; Sluzky *et al.*, 1991). In sharp contrast, if zinc-free insulin in neutral solution is monomerized by introduction of 60% ethanol, the extent and rate of aggregation are proportional to insulin concentration (Brange *et al.*, 1987b). Therefore, formation of fibril nuclei probably proceeds via monomerization of oligomeric insulin. This is supported by the fact that hydrophobic surfaces like air–water or plastic–water interfaces increase the tendency to fibrillation (Brange *et al.*, 1982; Lougheed *et al.*, 1983; Thurow and Geisen, 1984; Feingold *et al.*, 1984; Dathe *et al.*, 1990; Sluzky *et al.*, 1991, 1992) probably via interaction with and subsequent exposure of the insulin monomer's hydrophobic interfaces normally buried in the dimer or hexamer (Thurow and Geisen, 1984). The promoting effect of organic solvents, which weaken hydrophobic interactions and dissociate insulin assemblies, on the formation of insulin fibrils can probably be ascribed to the same phenomenon. Investigations on different insulin derivatives have revealed that exposure of the hydrophobic amino acid residues A2 Ile, B11 Leu, and B15 Leu to the surface of the insulin monomer (by displacement of the B-chain COOH-terminal from its normal position) is a prerequisite for formation of insulin fibrils. Thus, if up to five or eight residues are removed from this terminal, an increasing tendency to fibrillation is observed. In contrast, when movement of the B-chain COOH-terminal is blocked partly (i.e., proinsulin) or wholly (i.e., miniproinsulin) by linking the terminal to the nearby A-chain NH_2-terminal, the fibrillation process proceeds more slowly or not at all, respectively (Brange *et al.*, 1987b).

In the electron microscope, insulin fibrils are seen as long fibers with a diameter of 10–50 nm (Brange *et al.*, 1987a). By dynamic light-scattering technique it has recently been found that insulin fibrils are of narrow size distribution with a mean diameter of 200 nm (Dathe *et al.*, 1990). Whereas the initial steps in association of insulin monomers into fibril nuclei probably are common for any insulin fibril formation, the further growth (aggregation) into macroscopic fibrils depends on the experimental conditions (Brange *et al.*, 1987a). Based on computer simulations of insulin aggregation, a mechanism for fibrillation, in which partially unfolded insulin monomers associate and form intermediate assemblies, has recently been proposed (Sluzky *et al.*, 1991, 1992). When such intermediate fibril nuclei have reached a critical size (170 nm in diameter corresponding to ~100 insulin molecules) they have sufficient surface area for stability and for interactions with native insulin. The lag phase, often observed in insulin fibrillation experiments, was explained by the slow formation of stable intermediates as no significant aggregation was noticed until a population of nucleating species was established.

Insulin fibrils are insoluble in most aqueous media, including mineral acids, but can be dissolved and regenerated (renatured) into fully biologically active insulin in alkaline medium (pH > 11, followed by neutralization) or by using strong organic acids, e.g., formic acid.

Numerous ways of stabilizing insulin solutions against insulin fibrillation for use in delivery systems (insulin pumps) have been reported (for a review see Brange *et al.,* 1987a). Most of these methods are either impractical, inconvenient, or unsafe in the sense that the additives used are unphysiological or have a deteriorating effect on the chemical and biological stability of the insulin. A marked fibrillation-inhibitory effect is obtained by low concentrations of lecithins (Hansen *et al.,* 1986) or of a synthetic detergent (Grau and Saudek, 1987). Great improvements of the physical stability of neutral solutions of insulin are also achieved, without compromising the quality of the insulin, by addition of extra zinc ions which apparently further stabilize the hexameric structure (Brange *et al.,* 1986). The most effective stabilization is seen when these methods are combined (Table I).

3.3. Adsorption Phenomena

Loss of insulin due to adhesion to surfaces is a well-known problem to all who work with dilute solutions of insulin. The normally small fraction of insulin adsorbed decreases with increasing insulin concentration and varies with the material as more insulin is adsorbed to hydrophobic than to hydrophilic surfaces (Sefton and Antonacci, 1984). Addition of 0.1–1% of albumin is normally used to counteract insulin adsorption. The adsorption phenomena are less of a concern when concentrations higher than 5 IU/ml ($\approx$ 0.2 mg/ml $\approx$ 0.03 mM) are used as the relative loss is negligible. Adsorption is probably preceded by dissociation of insulin hexamers and conformational changes of the monomer which, as mentioned above, may be the initial stages in the insulin fibrillation process.

3.4. Influence of Insulin Species and Purity

Human, porcine, and bovine insulins vary in their tendency to form insulin fibrils, bovine insulin being more prone to fibrillation than the other two species of insulin. Apparently the A8 residue (Ala in bovine and Thr in

Table I
Physical Stabilization of Neutral Insulin Solutions[a]

Zinc ions/hexamer	DOPC concentration (μg/ml)	Days to precipitation in shaking test (37°C)
2.2	0	<1
2.2	30	18
2.9	30	32
3.9	30	40
4.6	30	>100

[a] The table illustrates the combined effect of stabilization against insulin fibrillation by additional zinc ions and by the semisynthetic lecithin, dioctanoylphosphatidylcholine (DOPC). Normal vials of insulin (Regular II, porcine, 100 IU) were tested by shaking in a water bath placed in vertical position (amplitude 5 cm, frequency 100/min) as described by Brange *et al.* (1986) until an increase in turbidity as measured by nephelometry was observed (Brange, Hansen, and Havelund, unpublished).

porcine and human insulin), which is on the surface of the hexamer, plays a role in directing the interactions of insulin with hydrophobic surfaces, the first steps in the fibrillation process.

With the introduction of chromatographic purification methods in the 1970s (Schlichtkrull *et al.*, 1972) it became clear that there is a reverse relationship between the tendency to fibrillation and the content of protein impurities in the insulin. It was later confirmed that this effect was not a result of the purification processes as such, but was due to removal of the contaminants. If the impurities were reintroduced, the former lesser propensity to fibrillation could be reestablished. It was also shown that higher molecular contaminants such as covalent dimer, proinsulin, etc., were most effective in counteracting fibrillation (Brange *et al.*, 1982).

3.5. Physical Stability of Pharmaceutical Formulations

In addition to the above-mentioned stability problems the storage and use of the different pharmaceutical formulations give rise to the following stability concerns.

3.5.1. pH Changes

In connection with the use of neutral solutions of insulin in delivery systems (insulin infusion pumps), lowering of the pH, due to carbon dioxide diffusion or to leaching of acid substances from reservoir or tubing materials, may lead to precipitation of insulin (Melberg *et al.*, 1988). Introduction of phosphate buffer counteracts this instability (Ratner and Steiner, 1987; Eichner *et al.,* 1988). The slight decrease of pH during storage of insulin preparations containing methylparaben as preservative is the result of hydrolysis of a small fraction of this substance into *p*-hydroxybenzoic acid (Brange *et al.,* 1987a). However, only after prolonged storage at temperatures greater than 25°C may this lead to precipitation of the insulin.

3.5.2. Phase Stability

Despite the capacity of insulin to undergo structural changes even when present in a crystalline state (Bentley *et al.,* 1978), there is a remarkable integrity of the different phases of insulin during proper storage of pharmaceutical preparations comprising mixed physical phases of insulin. Thus, no crystallization of the amorphous insulin fraction takes place during storage of the Lente preparation as long as the amount of free zinc ions exceeds 6 atoms per hexamer of insulin. And no precipitation of dissolved porcine insulin or dissolution of bovine insulin crystals in the biphasic preparation occurs as long as the pH of the medium is kept within narrow limits. However, if the suspension medium is changed, a slow transformation of one crystal type into another may occur to a limited extent in NPH preparations as observed during dissolution experiments (Schade *et al.,* 1987).

The attachment of clumps of insulin to the sides of the vials (termed flocculation or frosting), often observed during use of human NPH insulin (Benson *et al.,* 1988), is most likely to be a macroscopic manifestation of insulin fibrillation. Due to the fact that human insulin is slightly more soluble and its hexamers have a greater tendency to dissociate than animal insulins (Brange *et al.,* 1990), the supernatant of the NPH human insulin suspension will presumably contain sufficient insulin in monomeric form to account for the denaturation (fibrillation) on the walls of the vial. This phenomenon is much less frequent with NPH of animal insulins. The same problem has not been reported with the Lente-type preparation in which the solubility of insulin is effectively suppressed by addition of surplus zinc ions.

3.5.3. Sensitivity to High and Low Temperatures

The physical state of insulin in pharmaceutical formulation is highly influenced in a rather unpredictable way by changes in temperature. Although neutral solutions withstand temperatures as high as 60°C for several hours without physical change, the concomitant influence of shear stress results in formation of precipitates (fibrils) within minutes. The effect of heat on insulin suspensions depends on the physical state of the insulin. Whereas wholly crystalline suspensions are relatively unaffected by an increase in temperature, multiphase insulin preparations are significantly affected by even moderate temperature increases. Especially when amorphously precipitated insulin is part of the suspension, temperature rise may lead to lump formation and deposits on the walls of the vial (Brange *et al.,* 1987a). If insulin solutions are frozen, the insulin will appear precipitated after thawing but will dissolve again when brought to ambient temperatures. After such treatment neutral insulin solutions are chemically unchanged whereas acid insulin solutions might be extensively deamidated (Berson and Yalow, 1966). Insulin suspensions exposed to freezing usually change in appearance and become lumpy or granular due to aggregation of the insulin particles (Graham and Pomeroy, 1978; Brange *et al.,* 1987a) which may influence the timing of action.

4. CHEMICAL STABILITY

4.1. Introduction

In contrast to the physical changes in structure, the chemical, covalent changes in the structure are irreversible and may lead to formation of molecules which are less active and potentially immunogenic. Proteins are known to be degraded nonenzymatically by various chemical reactions (Manning *et al.,* 1989). The most prevalent of these is deamidation, a reaction in which the side chain amide group in glutamine or asparagine residues is hydrolyzed to form a free carboxylic acid (Robinson and Rudd, 1974; Geiger and Clarke, 1987). Insulin contains six such residues: Gln^{A5}, Gln^{A15}, Asn^{A18}, Asn^{A21}, Asn^{B3}, and Gln^{B4} (Fig. 1). In general, asparagine residues are more labile than glutamine residues (Robinson and Rudd, 1974).

Chemical deterioration of insulin during storage of pharmaceutical preparations becomes apparent when aged samples are analyzed by disc electrophoresis (Fig. 4) or by size exclusion chromatography (SEC) (Fig. 5). More recently a two-step HPLC method (SEC-HPLC followed by reverse-phase HPLC of the individual peaks from the first run) has been used to demonstrate the formation of hydrolysis products as well as minor amounts of higher-molecular-weight transformation (HMWT) products (Brange *et al.,* 1986).

The sum of hydrolysis products (as revealed by disc electrophoresis, Fig. 4) and HMWT products accounts, depending on temperature and formulation, for 80–90% of the total chemical degradation of insulin observable by using reverse-phase HPLC. However, the remaining transformation products are distributed over more than ten distinct peaks in the chromatogram (Brange *et al.,* 1992c).

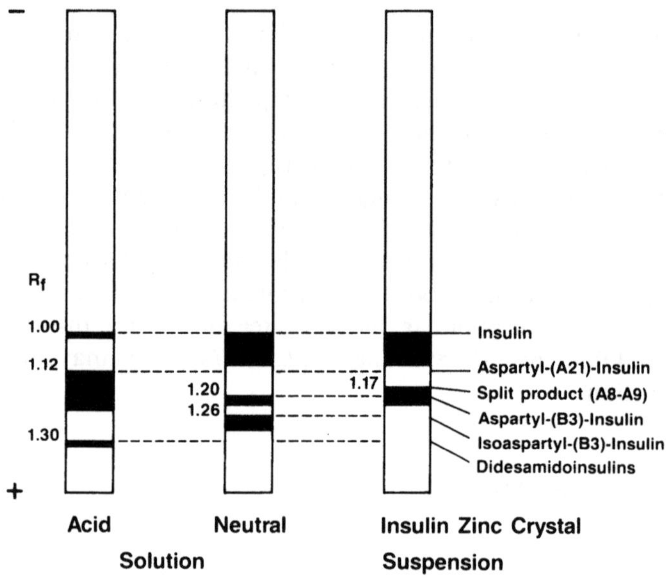

Figure 4. Disc electrophoresis (pH 8.4) pattern of different insulin preparations after storage for 6 months at 25°C.

4.2. Formation of Hydrolysis Products

Hydrolysis of insulin has been studied during storage of various insulin preparations at different temperatures. Insulin deteriorates rapidly in acid formulations due to extensive deamidation at residue A21 (Sundby, 1962). In neutral formulations hydrolysis takes place only at residue B3 and at a substantially reduced rate compared with the acid formulations. The B3 transformation results in formation of a mixture of isoAsp and Asp derivatives (Brange *et al.*, 1983, 1985, 1992a). In the IZS, crystalline (Ultralente) type of preparation an additional hydrolytic reaction leading to main chain cleavage of the A chain between A8 and A9 is observed (Brange *et al.*, 1985, 1992a).

4.2.1. Deamidation Reactions

An acid solution of insulin (pH 2–3) was for 40 years the only rapid-acting preparation available. In this formulation monodesamido-(A21)-insulin is formed by 1–2% per month at 4°C (Brange *et al.*, 1987a). Stored at 25°C this derivative constitutes more than 90% of the total protein after 6 months and only small amounts of didesamido derivatives can be observed

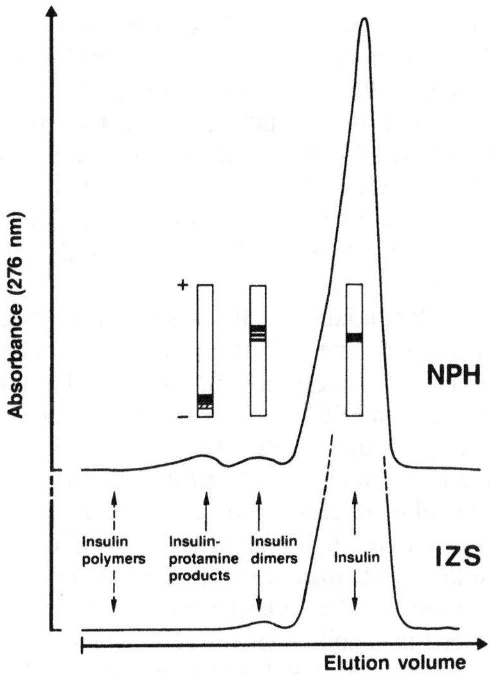

Figure 5. Size exclusion chromatography (SEC) on Bio-Gel P-30 columns of protracted insulin preparations after storage for 12 months at 25°C. (Upper) Isophane insulin or neutral protamine Hagedorn (NPH). (Lower) insulin zinc suspension, mixed (IZS). The inserted disc electrophoresis (pH 4) patterns show that the covalent insulin–protamine products formed in NPH preparation, and eluting in front of the covalent insulin dimers (CID) on the SEC column, are much more positively charged than the CID.

(Fig. 4). The particular lability of Asn^{A21} against hydrolysis is due to its position as a COOH-terminal residue where deamidation is catalyzed by the terminal carboxyl group (Leach and Lindley, 1953; Aldersley *et al.*, 1974).

In neutral medium, insulin deamidates by an entirely different mechanism involving an intermediate formation of a cyclic succinimide at residue B3 which hydrolyzes either under retention of a normal peptide bond B3–B4 (corresponding to simple deamidation) or an isopeptide bond (β-carboxyl linkage). This leads to a free β- (Asp derivative) or α-carboxyl group (isoAsp derivative), respectively. This transformation at B3 is independent of concentration (strength) and, except for IZS with crystals, of species of insulin (human, porcine, or bovine), but it is strongly influenced by the composition of the neutral solution (Table II) (Brange *et al.*, 1992a).

The B3 deamidation also takes place when insulin is formulated as a suspension with virtually no insulin in solution (<0.1%), but the rate varies with the composition of the medium and the physical state of the insulin (Table II, Fig. 6). In neutral suspensions of amorphously precipitated insulin (i.e., Semilente), deamidation proceeds with the same rate as in neutral solution but the hydrolysis of the succinimide intermediate now favors formation of the normal Asp derivative which accounts for 65% of the total hydrolysis. The same dominance of the Asp derivative over the isoAsp derivative is observed when neutral suspensions of insulin crystals (i.e., Ultralente or NPH preparations) undergo deamidation, but total deamidation is much slower. Thus, deamidation is considerably reduced when the suspended insulin is in the crystalline rather than the amorphous state which indicates that formation of the rate-limiting cyclic imide decreases when the flexibility of the tertiary structure is reduced (Brange, 1992). These results support tertiary structure as a major determinant of protein deamidation (Stephenson and Clarke, 1989; Lura and Schirch, 1988; Kossiakoff, 1988; Wright, 1991).

Deamidation at B3 was virtually undetectable in the bovine version of IZS, crystalline, reflecting the very compact and stable packing of the hexamer in these crystals. In contrast, the porcine and, especially, the human versions of IZS are much more prone to deamidation at B3 which might be explained by weaker intermolecular contacts between hexamers in the human insulin crystals (Chawdhury *et al.*, 1983). Neutral solutions containing phenol or *m*-cresol showed lower B3 transformation than the neutral solution with methylparaben as preservative (Brange *et al.*, 1992a), probably due to the stabilizing effect of phenol on the tertiary structure (α-helix formation) around the deamidating residue (Derewenda *et al.*, 1989) resulting in reduced probability for imide formation. The phenol-containing crystal suspensions, i.e., NPH and mixtures based on these tetragonal crystals, also exhibited relatively low rates of B3 transformation (Fig. 6).

The ratio in which the two B3 derivatives (isoAsp/Asp) are formed varies with the formulation but is independent of time and temperature, suggesting a pathway only involving intermediate imide formation without

Table II
Hydrolysis Reactions[a]

Preparation	Insulin species	Deamidation (%/month) 4°C	15°C	25°C	37°C	Ratio isoAsp/Asp	Chain cleavage A8–A9 (%/month) 4°C	25°C
Regular I	P or H	0.2	0.7	2.1	15	2.0	N.D.	
Regular II	P or H	0.1	0.3	1.1	5.1	1.4	N.D.	
IZS, amorphous	P	0.2	0.8	2.3		0.6	N.D.	
IZS, crystalline	H	0.03		1.0		0.6	<0.01	0.4
	P	0.02		0.2		0.5	0.01	1.0
	B			<0.01			0.1	5.9
Isophane (NPH)	H			0.6		0.6	N.D.	
	P			0.3		0.6	N.D.	

[a] Values are the initial, mean increase of hydrolysis products during storage of insulin preparations (data from Brange *et al.*, 1992a). Abbreviations: NPH, neutral protamine Hagedorn; P, porcine; H, human; B, bovine; N.D., not detectable.

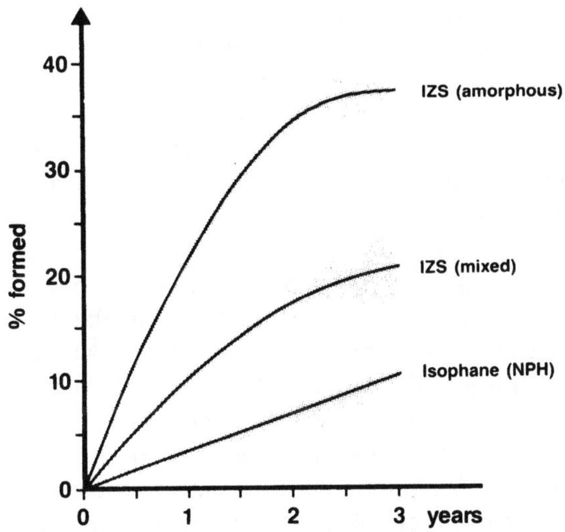

Figure 6. Time courses of formation of the total amounts of hydrolysis products during storage at 25°C of short- and intermediate-acting porcine insulin preparations: Isophane (NPH), and insulin zinc suspensions (IZS) in two different formulations (mean with 95% confidence interval).

any direct side chain hydrolysis. In oligopeptides the hydrolytic opening of the intermediate succinimide derivative generates a mixture of deamidated products with an approximately 3 : 1 predominance of the isoAsp residue (i.e., original Asn transformed to isoAsp with a free α-carboxylate group) relative to the normal peptide (Meinwald *et al.*, 1986; McFadden and Clarke, 1987; Lura and Schirch, 1988; Capasso *et al.*, 1989; Patel and Borchardt, 1990; Tyler-Cross and Schirch, 1991).

The observed reduction in the isoAsp/Asp ratio when the deamidation takes place in a rigid protein rather than in a conformationally much more flexible peptide (Table II; Brange *et al.*, 1992a) is in agreement with findings on deamidation of α-crystallin (Voorter *et al.*, 1988) and of serine hydroxymethyltransferase (Artigues *et al.*, 1990) (ratios of 1.6 and 1, respectively).

4.2.2. Chain Cleavage

In crystalline suspensions containing rhombohedral crystals in combination with a surplus of zinc ions in addition to the 2–4 structurally bound Zn^{2+} (IZS, crystalline and mixed, both with >20 zinc ions/hexamer), cleavage of the peptide bond A8–A9 can be observed during normal storage (Brange *et al.*, 1992a). The rate of formation of this A8–A9 split product is species dependent: bovine > porcine > human insulin. The hydrolytic cleavage of the peptide linkage A8–A9 requires the presence of additional free zinc ions. Thus, no cleavage takes place in biphasic insulin containing the same type of crystal with only 4 zinc ions per hexamer. The main chain hydrolysis is most likely due to an autoproteolytic effect of an adjacent insulin hexamer in the rhombohedral crystal resulting in a metalloproteinase-like cleavage of the peptide chain (Brange, 1992). The key element for enzyme activity, a Glu carboxylate group, is available from the adjacent hexamer, and a zinc ion, supposed to be coordinated to one Glu and two His in the Michaelis complex (Toma *et al.*, 1989), is available from the medium. In addition, two B5 His residues are in close contact (4 Å) across the hexamer boundary in proximity to the A8 and A9 residues (Baker *et al.*, 1988), and they actually form a zinc binding site within the crystals (Emdin *et al.*, 1980).

Rate data at 4 and 25°C for these reactions are shown in Table II. The insulin species variation supports the autocatalytic theory as bovine insulin with the fastest formation actually has the closest interactions in the stacking of hexamers in the crystal. The different rate of formation of the split product in bovine insulin relative to human and porcine insulin can also be accounted for by primary structure differences (A8 Ala and Thr, respectively). The slightly faster formation in porcine as compared with human insulin rhombohedral crystals can only be ascribed to differences in the three-dimensional packing of the hexamers in the crystal.

4.3. Di- and Polymerization

Formation of HMWT products during storage of insulin preparations was first reported by Schlichtkrull and co-workers (1975; Brange *et al.*, 1982, 1983, 1984). The main products formed are covalent insulin dimers (CID), but in protamine-containing preparations a concurrent formation of covalent insulin–protamine products (CIPP) takes place at all storage temperatures (Fig. 5). At temperatures ≥ 25°C parallel or consecutive formation of covalent oligomers and polymers can also be observed in certain preparations. CID and CIPP are mainly formed through reactions of amine groups in one molecule with side chain carboxyamide groups in another molecule (transamidation), whereas the polymer formation is mainly due to chain reactions involving disulfide interactions. Compared with the rate of hydrolysis (see above), the rate of HMWT formation is one order of magnitude slower except for NPH preparations in which the rates (4–25°C) are comparable mainly because of the concomitant formation of CID and CIPP (Table III; Brange *et al.*, 1992b).

4.3.1. Amine Reactions

Rate of CID formation varies with composition and formulation, and for Isophane (NPH) preparations (including CIPP formation) also with strength of the preparation. In neutral solution the rate of CID formation is independent of concentration in the range from U 40 (0.24 mM) to U 500 which strongly indicates that the intermolecular chemical reaction occurs mainly within the hexameric units, and not between the hexamers in the solution. As CID formation is of the same order of magnitude in neutral solutions and in suspensions, when these preparations contain the same excipients (Regular I versus IZS types), the covalent insulin dimers in the insulin suspensions most likely also arise from intermolecular reactions within the hexameric unit.

In most types of preparations, formation of CID products is only slightly influenced by species of insulin; thus, human and porcine insulin exhibit the same stability. During storage at 4°C the most unstable type of preparation is Isophane (NPH) with a 15-fold faster rate of HMWT formation than the IZS types of preparation. However, at 37°C the transformation in these types of preparation is of the same order of magnitude.

The stability of the neutral solutions is intermediate in this respect, with the glycerol containing Regular II as the most labile. This is in agreement with earlier observations that glycerol in high concentration causes substantially increased formation of covalent insulin dimers and polymers (Brange *et al.*, 1982; Blackshear *et al.*, 1983).

In the rhombohedral crystals in IZS, crystalline, the trimers and tetramers formed during prolonged storage at temperatures above 25°C are clearly products of consecutive reactions of insulin with already formed insulin dimers and trimers presumably in the form of successive transamidation (Brange *et al.*, 1992b).

4.3.2. Disulfide Interchange Reactions

During storage at ambient and higher temperatures, formation of covalent oligomers and polymers can be observed in the neutral solutions and in the preparations containing amorphously precipitated insulin (IZS, amorphous and mixed). These reactions occur in parallel to the formation of dimers, as deduced from the time courses, and are due to disulfide interchange reactions (Brange, 1992). Such chain reactions may be a result of initial thiol formation through hydrolysis or β-elimination (Florence, 1980; Manning *et al.*, 1989). Intermolecular disulfide exchange requires juxtaposition of disulfide bridges from different insulin molecules. Analyses of the crystal structures by molecular graphics reveal that this is not the case within the hexamer or when hexamers pack in the rhombohedral (IZS crystalline preparations) or monoclinic (NPH) crystals. It is therefore not surprising that covalent polymerization due to disulfide reshuffling is undetectable in these preparation types even above ambient temperatures. In contrast, disulfide interaction becomes possible when the hexamers approach one another in a random way as in neutral solution or when the insulin is amorphously precipitated (Brange *et al.*, 1992b).

4.4. Influence of pH and Auxiliary Substances

The chemical transformation of insulin formulated as Ultralente (IZS, crystalline) as a function of pH is illustrated in Fig. 7. There is an optimum for the formation of covalent dimers and oligomers around pH 4 whereas the deterioration due to different hydrolysis reactions is at a minimum around pH 6.5. Similar profiles have been obtained with other types of insulin formulations although in neutral and slightly alkaline media increasing formation of the B3 deamidation products with increasing pH can be observed instead of the formation of the A-chain split product seen in the Ultralente crystals (Brange and Langkjær, 1992).

In alkaline media (pH > 9) an accelerated formation of covalent oligomers and polymers is seen as a function of pH, probably arising as a result of disulfide interactions. Concomitantly lower-molecular-weight degradation products are formed most likely in the form of insulin A and B chains originating from extensive disulfide cleavage by hydrolysis or β-elimination (Florence, 1980).

The types of isotonicity substance and preservative have a profound effect on the extent of deamidation as well as of CID formation. There is a clear gradation in the stabilizing effect, for the preservatives: phenol > *m*-cresol > methylparaben, and for the isotonic agents: NaCl > glycerol > glucose (Brange and Langkjær, 1992). When used separately the effect of phenol was greater than that of NaCl, but they apparently exert their influence through independent mechanisms as an additive effect by using the two compounds together was observed.

4.5. Effect of Temperature

In all pharmaceutical insulin preparations temperature has a profound influence on the rate of deamidation and formation of HMWT products (Brange *et al.*, 1992a,b; Brange, 1992).

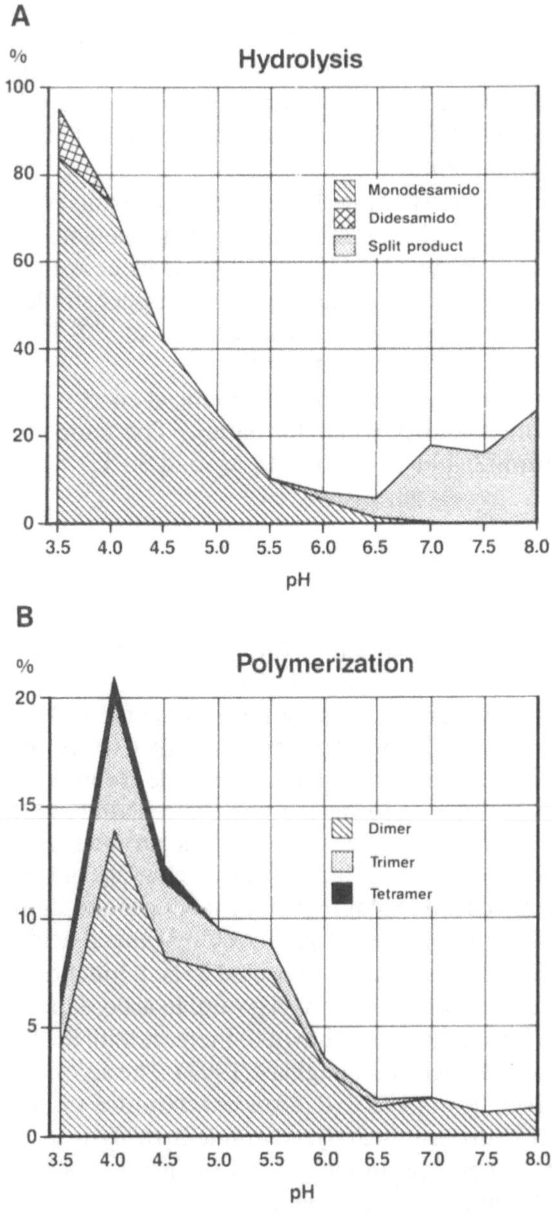

Figure 7. Chemical transformation of insulin during storage of rhombohedral insulin crystals (bovine insulin crystals, 0.7% NaCl, 0.2% phenol) as a function of pH during storage at 25°C for 12 months. (A) Formation of the hydrolysis products mono- and didesamido insulins, and the insulin split product (A8–A9). (B) Formation of covalent dimers and oligomers.

The effect of increasing the temperature by 10°C, coefficient Q_{10}, defined by the expression: $\log Q_{10} = 10 \times (\log k_2 - \log k_1)/(t_2 - t_1)$ with k = rate constant and t = temperature, on the deamidation at B3 in the Regular II formulation (pH 7.4) increases with increasing temperature from 2.6 around 10°C to 3.7 around 40°C, reflecting the increase in conformational freedom with increasing temperature.

From the linear parts of the slopes in the Arrhenius plots the activation energy (E_a) for the B3 deamidation reaction is calculated to 106 kJ/mole (25.3 kcal/mole) in Regular II (25–45°C). This value is in good agreement with the values 21.7 and 22.5 kcal/mole found under similar conditions for the deamidation of asparaginyl hexapeptides at 30–37°C (Geiger and Clarke, 1987; Patel and Borchardt, 1990).

Temperature has an even more pronounced effect on CID, CIPP, and, especially, oligomer and polymer formation. The temperature coefficient Q_{10} increases from 2–4 at 10°C to 4–16 around 30°C.

4.6. Kinetics and Mechanisms

The extreme susceptibility to hydrolysis of the COOH-terminal residue Asn^{A21} in acid medium, as compared with the other amide groups in insulin, is due to intramolecular catalysis by the protonated COOH-terminal carboxyl group involving a proton transfer from the COOH group (Leach and Lindley, 1953). After proton transfer to the amide group, nucleophilic attack by the carboxylate function forms the cyclic tetrahedral intermediate (Kirby and Lancaster, 1972) which after transfer of the proton from the hydroxyl to the nitrogen breaks down to the anhydride by cleavage of the C–N bond, the rate-determining step (Aldersley *et al.*, 1974; Kluger and Lam, 1978). Subsequently the anhydride is readily hydrolyzed by general acid catalysis.

The rate of hydrolysis can be expected to be a function of the degree of protonation of the terminal α-carboxyl group. In accordance, the rate of A21 deamidation of insulin has been shown to increase with decreasing pH in the pH interval 5 to 2 (Brange and Langkjær, 1992). The additional deamidation of insulin observed from pH 2 to 0.3 can be accounted for by further deamidation by direct hydronium-catalyzed amide hydrolysis, not only of Asn^{A21} but also of the other five amide groups in insulin, as actually observed (Sundby, 1962; Brange and Langkjær, 1992).

The rate-determining step in the deamidation reaction at Asn^{B3} in neutral medium is the formation of the intermediate cyclic imide, a much slower process than the subsequent hydrolysis into Asp and isoAsp derivatives (Meinwald *et al.*, 1986; Geiger and Clarke, 1987). The intramolecular imide is hydrolyzed by nucleophilic attack by hydroxide ions, resulting in a mixture of two different desamido products in which the polypeptide backbone is attached via an α-carboxyl linkage (Asp derivative) or via a β-carboxyl linkage (isoAsp derivative).

Formation of the succinimide intermediate requires a nucleophilic attack of the main chain peptide nitrogen on the carbonyl carbon of the side chain and this is only possible if these atoms are able to align properly for this reaction. This depends on the side chain bulkiness of the residue on the COOH-terminal side of the asparaginyl residue (Robinson and Rudd, 1974; Geiger and Clarke, 1987; Kossiakoff, 1988; Capasso *et al.*, 1989; Patel and Borchardt, 1990; Tyler-Cross and Schirch, 1991) as well as of the three-dimensional conformation and flexibility around the Asn residue (Clarke, 1987; Stephenson and Clarke, 1989; Lura and Schirch, 1988; Kossiakoff, 1988; Wright, 1991). Optimal conformation for imide formation occurs when the dihedral torsion angle ψ (defining the rotation around the α-carbon/peptide carbonyl carbon bond) is $-120°$ (Clarke, 1987). Molecular modeling has shown that succinimide formation at residue B3 requires a more than 100° rotation around the α-carbon/peptide carbonyl carbon which causes a movement of ~10 Å of the NH_2-terminal B1 residue (Brange, 1992). This is only possible because the NH_2-terminal residues are on the surface of the insulin hexamer and not very much restricted for such large conformational change.

It is striking that although Asn^{A18} actually has at least as favorable dihedral angles for imide formation as Asn^{B3} (Baker *et al.*, 1988), no deamidation can be detected in the A chain at neutral pH (Brange *et al.*, 1992c), probably due to the fact that the A18 residue is part of a helix and therefore with much more reduced conformational flexibility.

The decreasing isoAsp/Asp ratios observed when the deamidation takes place in a more or less rigid protein rather than in a conformationally much more flexible peptide without higher-order structural elements, probably do not originate from any direct amide hydrolysis. Instead the reversal in predominance of the individual hydrolytic products may be attributed to variation in local structure around the intermediate imide. Such differences could result in either reduced accessibility to hydrolysis of the succinimide peptide linkage of the imide (isoAsp formation) because of steric hindrance, or increased susceptibility of the imide β-peptide linkage to hydrolysis (Asp formation) due to catalytic effect by juxtaposed functional groups in the three-dimensional structure (Brange, 1992).

CIDs are formed between molecules in the dimeric or hexameric units common for all types of preparations mainly through transamidation reactions (aminolysis) by the B-chain NH_2-terminal amine group on Asn side chains of the A chain (Helbig, 1976; Brange *et al.*, 1984; Brange, 1992). A similar intramolecular nucleophilic displacement of an asparaginyl side chain amide by the NH_2-terminus involving formation of a seven-membered cyclic amide has been described for a tetrapeptide (Lura and Schirch, 1988).

The independence of CID formation on insulin concentration in the neutral solutions strongly indicates that the intermolecular chemical reaction occurs mainly within the hexameric units and not to any significant extent between the hexamers in the solution. As the crystalline and amorphous suspensions share with the solutions the hexamer as the common unit, it is conceivable that the covalent dimers also mainly form within the hexamer in these preparations. This is supported by the fact that CID formation is of the same order of magnitude in neutral solution and in suspensions when these preparations contain similar auxiliary substances.

In aminolysis of amides an increasing rate of reaction with the acidity of the medium, reaching an essentially constant value below pH 3, has been observed (Kirk and Cohen, 1972). This is in agreement with our finding that dimer formation (with consecutive transamidation resulting in formation of trimers and tetramers) increases substantially with acidity of the medium from pH 7 to 4 (Fig. 7B; Brange and Langkjær, 1992). The steep fall in CID formation from pH 4 to 3.5 most likely reflects the dissociation of the hexamer into mainly dimeric units (Blundell *et al.*, 1972) which gives further evidence for covalent dimer formation primarily being a reaction occurring within the hexamer.

The rate of CID formation is generally faster in glycerol-containing preparations, probably due to parallel amine reactions mediated by aldehyde impurities in the glycerol (Brange and Langkjær, 1992). Aldehydes are able to react chemically with the insulin amino groups through their carbonyl group under formation of Schiff base adducts. Such reaction products are able to undergo Amadori rearrangement (Acharya and Manning, 1983) which generates new carbonyl functions capable of forming Schiff base linkages with other amino groups. Eventually such reactions result in covalent cross-linking of the protein (Acharya and Manning, 1983; Bello and Bello, 1976). Thus, in addition to insulin's potential to form CID by transamidation reactions (Helbig, 1976; Brange 1992), the CID can also be generated via an initial reaction with aldehydes.

It is normally assumed that chemical decomposition of drug suspensions solely takes place in the part of the drug in solution (Carstensen, 1990). However, in the IZS types of formulation the amount of insulin in solution is extremely small (<0.1 U/ml). Nevertheless, CID forms at the same rate as in the neutral solutions containing 100 IU/ml (Table III), indicating that the transformation reactions in the insulin suspensions mainly take place within the solid crystalline phase. This is plausible because the hydrated molecules in an insulin crystal show plasticity and have conformational flexibility of their side chains and backbone for movements within the crystal lattice (Baker *et al.*, 1988; Caspar and Badger, 1991). Thus, the switch from extended to the α-helical fold of the NH_2-terminal B-chain segment is accomplished within the crystal lattice although some crystal cracking occurs (Bentley *et al.*, 1978).

CIPP formation probably proceeds via aminolysis by the single amino group (NH_2-terminal) in protamine. Although the content of protamine in the NPH crystals only corresponds to ~0.1 protamine molecule per insulin monomer the chemical reaction between protamine and insulin proceeds in the NPH preparation at all temperatures at a rate slightly higher than the rate of insulin dimer formation (Table III). In NPH (protamine insulin crystals) protamine is situated in the interstices between the hexamers and partly penetrates the central channel of the hexamer. Here it mediates interactions at the dimer–dimer interface but with large conformational flexibility (Dodson, personal communication). Because of this flexibility of the protamine in the crystals, its NH_2-terminal amino group has the capacity and possibilities to react with insulin by the same mechanisms as in CID formation.

Formation of covalent polymers in certain formulations at higher storage temperatures can be explained by a chain reaction involving disulfide interactions. The disulfide in insulin bridging the A7 and B7 residues is on the surface of the hexamer and therefore has the potential to participate in such intermolecular disulfide exchange between hexamers. Such reaction

Table III
Formation of Higher-Molecular-Weight Transformation Products[a]

Type of insulin preparation	Storage temperature 4°C	15°C % per year	25°C	37°C % per month
Regular I (human, porcine)	0.08	0.23	1.0	0.5
Regular II (human, porcine)	0.20	0.42	1.3	1.4
IZS, amorphous (porcine)	0.03	0.14	1.0	1.5
IZS, crystalline (bovine)	0.03	0.24	0.82	1.1
NPH (human, 100 IU/ml)	0.5	1.2	3.3	1.6

[a] Values are the initial, mean increase in the sum of covalent transformation products (covalent insulin dimer, covalent insulin–protamine complex, and covalent insulin oligomers and polymers) during storage of insulin preparations (data from Brange *et al.*, 1992b).

becomes possible when the individual hexamers are capable of approaching one another in a random way as in solution or when the insulin is amorphously precipitated. An increasing oligomer and polymer formation with increasing insulin concentration in the Regular formulation clearly indicates a reaction between the population of hexamers in the neutral solution (Brange *et al.*, 1992b). Whereas the initial disulfide rupture is a slow process, and therefore rate-determining, the subsequent interchange reactions are fast, and as every single interchange leaves a new highly reactive thiolate ion, the initial hydrolysis or β-elimination starts a chain reaction resulting in fast polymer formation. Therefore, accumulation of oligomeric disulfide exchange products is small.

4.7. Properties of Transformation Products

The different insulin derivatives and HMWT products formed during accelerated storage of insulin preparations have been isolated and characterized with respect to their biological potency and immunogenicity in rabbits.

Biological potencies varied from essentially unchanged for the desamido products down to ≤2% for the A8–A9 split product and the covalent polymerization products (Table IV). The low potency of the split product explains the relatively low biological stability observed during storage of bovine IZS (crystalline) (Pingel and Vølund, 1972).

The immunogenicity (in rabbits) of monodesamido-(A21)-insulin, CID, the covalent insulin polymers (Schlichtkrull *et al.*, 1975), monodesamido-(B3)-insulin, CIPP (Brange *et al.*, 1987a), and the A8–A9 split product (Brange *et al.*, 1984; Brange *et al.*, 1992c) has been found not to be significantly different from the parent insulin.

4.8. Influence on the Quality of Pharmaceutical Preparations

Among the chemical reactions affecting insulin during storage of the preparations, the hydrolytic reactions dominate and cause transformation of larger amounts of insulin than the di- and polymerization reactions. With respect to the influence on the quality of the insulin preparations, the molecular changes induced by deamidation are relatively small although the isoAsp formation, in addition to changing the uncharged Asn residue into a charged Asp group, also introduces an extra carbon atom into the peptide backbone which may cause more extensive structural changes. However, the deamidation products have essentially the same *in vivo* biological potency as the

Table IV
Biological Potency of Insulin Derivatives Formed during Storage of Insulin Preparations

Derivative	Preparation isolated from	Insulin species	Bioassay method[a]	Potency relative to insulin[b]
Monodesamido-(A21)-insulin	Acid solution	Porcine	MBG and MCA	92%
		Bovine		85%
Monodesamido-(B3)-insulin	Neutral regular	Porcine	MCA	97%
Covalent insulin dimer	Insulin zinc suspension	Porcine	MBG	15%
Covalent insulin–protamine complex	Isophane	Bovine	MBG	4%
Covalent insulin polymerization product	Insulin zinc suspension (amorphous)	Porcine	MBG	<2%
Split product (A8–A9)	Insulin zinc suspension (crystalline)	Bovine	MBG	2%

[a] MBG, mouse blood glucose assay; MCA, mouse convulsion assay.
[b] Values are the *in vivo* biological activity relative to an insulin standard (data from Brange *et al.*, 1987a; Brange *et al.*, 1992c).

intact molecule, and, as previously mentioned, the immunogenicity in rabbits was unchanged.

The extent of deamidation during shelf life (2 years; storage at the recommended temperature of 2–8°C) is ≤7%, resulting in insignificant change of the biological potency of the preparations. The cleavage of the peptide backbone of the A chain between A8 and A9 represents a much more dramatic change than the deamidation reactions and may induce more serious alterations in the three-dimensional structure of the molecule. In accordance, the *in vivo* potency of the split product is only about 2% of that of the parent hormone (Table IV). However, when the preparations are stored as recommended (temperature interval 2–8°C), the fall in potency during 2 years' shelf life is less than 5% (Brange *et al.*, 1992a).

The formation of HMWT products is generally much slower than the chemical decomposition of the insulin due to hydrolytic reactions. The impact on the quality and therapeutic usefulness of the preparations might, however, be more serious as some of the immunological side effects associated with insulin therapy have been asserted to be due to the presence of covalent aggregates of insulin in the therapeutic preparations (Robbins *et al.*, 1987; Robbins and Mead, 1987; Ratner *et al.*, 1990). Stored as recommended (2–8 °C), the content of CID products at expiry of the preparations is ≤0.5%. In NPH (100 IU/ml) additional formation of CIPP products amounts to ≤0.8% (Brange *et al.*, 1992b).

5. SUMMARY AND CONCLUSIONS

Insulin is composed of 51 amino acids in two peptide chains (A and B) linked by two disulfide bonds. The three-dimensional structure of the insulin molecule (insulin monomer), essentially the same in solution and in solid phase, exists in two main conformations. These differ in the extent of helix in the B chain which is governed by the presence of phenol or its derivatives. In acid and neutral solutions, in concentrations relevant for pharmaceutical formulation, the insulin monomer assembles to dimers and at neutral pH, in the presence of zinc ions, further to hexamers. Many crystalline modifications of insulin have been identified but only those with the hexamer as the basic unit are utilized in preparations for therapy. The insulin hexamer forms a relatively stable unit but some flexibility remains within the individual molecules. The intrinsic flexibility at the ends of the B chain plays an important role in governing the physical and chemical stability of insulin.

A variety of chemical changes of the primary structure (yielding insulin derivatives), and physical modifications of the secondary to quaternary structures (resulting in "denaturation," aggregation, and precipitation) are known to affect insulin and insulin preparations during storage and use (Fig. 8).

The tendency of insulin to undergo structural transformation resulting in aggregation and formation of insoluble insulin fibrils has been one of the most intriguing and widely studied phenomena in relation to insulin stability. Although the exact mechanism of fibril formation is still obscure, it is now clear that the initial step is an exposure of certain hydrophobic residues, normally buried in the three-dimensional structure, to the surface of the insulin monomer. This requires displacement of the COOH-terminal B-chain residues from their normal position which can only be accomplished via monomerization of the insulin. Therefore, most methods stabilizing insulin against fibrillation share the property of being able to counteract associated insulin from being disassembled.

Chemical deterioration of insulin during storage of pharmaceutical preparations is mainly due to two categories of chemical reactions, hydrolysis and intermolecular transformation reactions leading to insulin HMWT products. The predominant hydrolysis reaction is deamidation of Asn residues which in acid solution takes place at residue A21, in neutral medium at

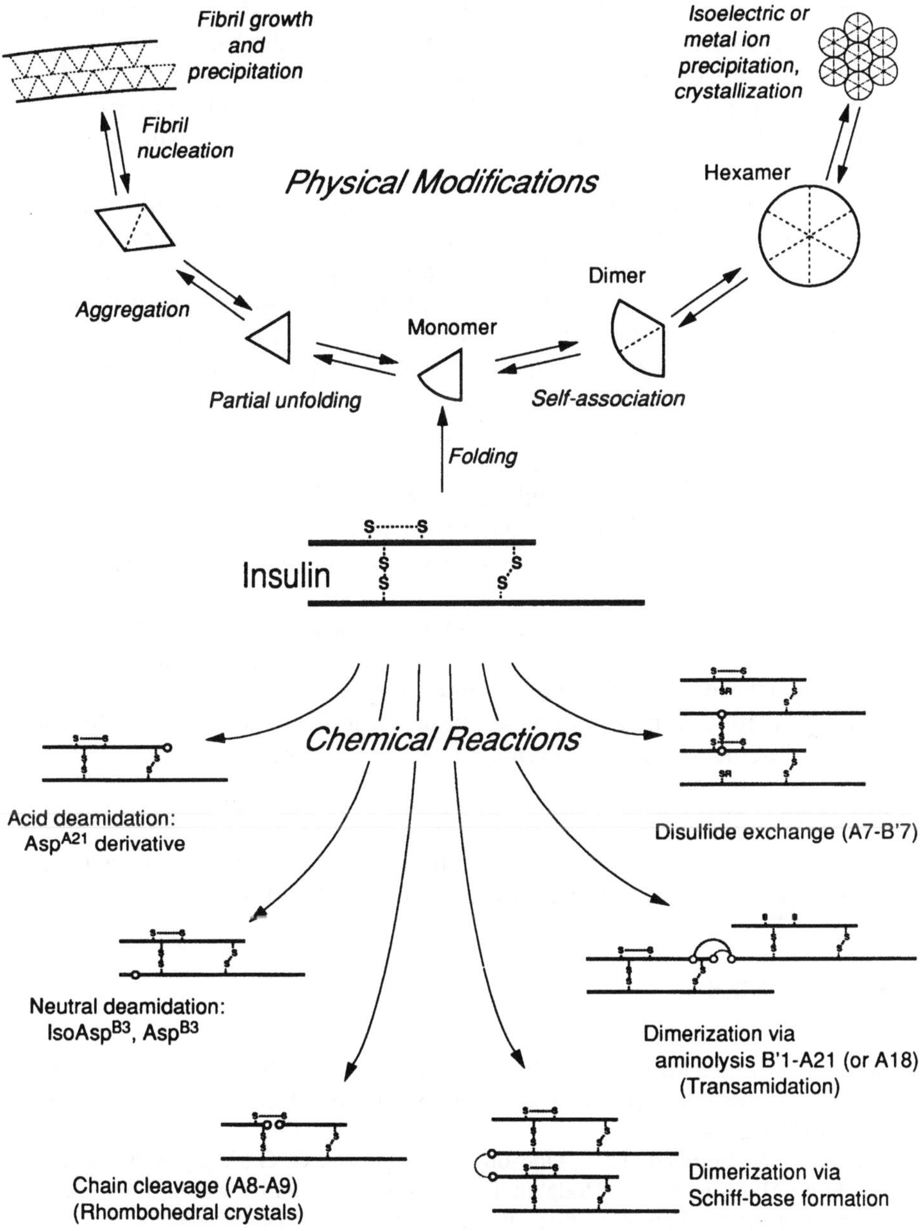

Figure 8. Schematic diagram of the main physical and chemical transformations affecting insulin during storage and use.

residue B3. An amazing hydrolytic cleavage of the backbone A chain, presumably autocatalyzed by an adjacent insulin molecule, has been identified in insulin preparations containing rhombohedral crystals in combination with free zinc ions. The HMWT products are mainly formed via intermolecular aminolysis between NH_2-terminal amine in insulin or auxiliary protamine and an amide side chain in the insulin A chain.

It is striking how physical and chemical instability of insulin can be assigned to the domains of the insulin molecule exhibiting the highest degree of conformational flexibility. Thus, the predominant physical process leading to precipitation and inactivation of insulin (insulin fibrillation) can be attributed to translocation of the B-chain COOH-terminal residues, and the predominant chemical degradation of insulin in pharmaceutical formulation, namely deamidation in neutral solution, takes place at the likewise flexible B-chain NH_2-terminal. Also, covalent dimerization is mainly mediated via the "elastic" B1 residue. When addition of phenol or zinc ions constrains the molecules and reduces their flexibility and tendency to disassemble, or if crystal formation renders the molecular assembly more rigid and stable, insulin becomes less prone to fibrillation and chemical degradation.

It can be concluded that conformational flexibility within the insulin molecule plays a crucial role in insulin stability. Thus, physical as well as chemical deterioration of insulin is almost entirely governed by the capacity of the B-chain terminals to undergo the structural changes necessary for the physical interactions or the chemical reactions to take place.

ACKNOWLEDGMENTS

The authors are grateful to Xiao Bing and Guy Dodson, University of York, U.K., for providing the drawings for Figs. 2 and 3.

REFERENCES

Acharya, A. S., and Manning, J. M., 1983, Reaction of glycolaldehyde with proteins: Latent crosslinking potential of α-hydroxyaldehydes, *Proc. Natl. Acad. Sci. USA* **80**:3590–3594.

Adams, M. J., Blundell, T. L., Dodson, E. J., Dodson, G. G., Vijayan, M., Baker, E. N., Harding, M. M., Hodgkin, D. C., Rimmer, B., and Sheat, S., 1969, Structure of rhombohedral 2 zinc insulin crystals, *Nature* **224**:491–495.

Aldersley, M. F., Kirby, A. J., Lancaster, P. W., McDonald, R. S., and Smith, C. R., 1974, Intramolecular catalysis of amide hydrolysis by the carboxy-group. Rate determining proton transfer from external general acids in the hydrolysis of substituted maleamic acids, *J. Chem. Soc. Perkin Trans. 2* **1974**:1487–1495.

Artigues, A., Birkett, A., and Schirch, V., 1990, Evidence for the in vivo deamidation and isomerization of an asparaginyl residue in cytosolic serine hydroxymethyltransferase, *J. Biol. Chem.* **265**:4853–4858.

Baker, E. N., and Dodson, G., 1970, X-ray diffraction data on some crystalline varieties of insulin, *J. Mol. Biol.* **54**:605–609.

Baker, E. N., Blundell, T. L., Cutfield, J. F., Cutfield, S. M., Dodson, E. J., Dodson, G. G., Hodgkin, D. M. C., Hubbard, R. E., Isaacs, N. W., Reynolds, C. D., Sakabe, K., Sakabe, N., and Vijayan, N. M., 1988, The structure of 2Zn pig insulin crystals at 1.5 Å resolution, *Philos. Trans. R. Soc. London* **319**:369–456.

Balschmidt, P., Hansen, F. B., Dodson, E. J., Dodson, G. G., and Korber, F., 1991, Structure of porcine insulin cocrystallized with clupeine Z, *Acta Crystallogr. Sect. B* **47**:975–986.

Banting, F. G., and Best, C. H., 1922, Pancreatic extracts, *J. Lab. Clin. Med.* **7**:464–472.

Bello, J., and Bello, H. R., 1976, Chemical modification and cross-linking of proteins by impurities in glycerol, *Arch. Biochem. Biophys.* **172**:608–610.

Benson, E. A., Benson, J. W., Jr., Fredlund, P. N., Mecklenburg, R. S., and Metz, R., 1988, Flocculation and loss of potency of human NPH insulin, *Diabetes Care* **11**:563–566.

Bentley, G., Dodson, G., and Lewitova, A., 1978, Rhombohedral insulin crystal transformation, *J. Mol. Biol.* **126**:871–875.

Berson, S. A., and Yalow, R. S., 1966, Deamidation of insulin during storage in frozen state, *Diabetes* **15**:875–879.

Blackshear, P. J., Rohde, T. D., Palmer, J. L., Wigness, B. D., Rupp, W. M., and Buchwald, H., 1983, Glycerol prevents insulin precipitation and interruption of flow in an implantable insulin infusion pump, *Diabetes Care* **6**:387–392.

Blundell, T., Dodson, G., Hodgkin, D., and Mercola, D., 1972, Insulin: The structure in the crystal and its reflection in chemistry and biology, *Adv. Protein Chem.* **26**:279–402.

Brange, J., 1992, Chemical stability of insulin: 4. Kinetics and mechanisms of chemical transformations in pharmaceutical formulation, *Acta Pharm. Nord.* **4**:209–222.

Brange, J., and Havelund, S., 1983, Properties of insulin in solution, in: *Artificial Systems for Insulin Delivery* (P. Brunetti, K. G. M. M. Alberti, A. M. Albisser, K. D. Hepp, and M. M. Benedetti, eds.), Raven Press, New York, pp. 83–88.

Brange, J., and Langkjær, L., 1992, Chemical stability of insulin: 3. Influence of excipients, formulation, and pH dependence, *Acta Pharm. Nord.* **4**:149–158.

Brange, J., Havelund, S., Hansen, P., Langkjaer, L., Sørensen, E., and Hildebrandt, P., 1982, Formulation of physically stable neutral solutions for continuous infusion by delivery systems, in: *Hormone Drugs* (J. L. Gueriguian, E. D. Bransome, and A. S. Outschoorn, eds.), US Pharmacopeial Convention, Rockville, Md., pp. 96–105.

Brange, J., Langkjaer, L., Havelund, S., and Sørensen, E., 1983, Chemical stability of insulin: Neutral insulin solutions, *Diabetologia* **25**:193.

Brange, J., Langkjaer, L., Havelund, S., and Sørensen, E., 1984, Chemical stability of insulin: Formation of covalent insulin dimers and other higher molecular weight transformation products in intermediate- and long-acting insulin preparations, *Diabetologia* **27**:259A–260A.

Brange, J., Langkjær, L., Havelund, S., and Sørensen, E., 1985, Chemical stability of insulin: Formation of desamidido insulins and other hydrolytic products in intermediate- and long-acting insulin preparations, *Diabetes Res. Clin. Pract.* (Suppl. 1):67.

Brange, J., Havelund, S., Hommel, E., Sørensen, E., and Kühl, C., 1986, Neutral insulin solutions physically stabilized by addition of Zn^{2+}, *Diabetic Med.* **3**:532–536.

Brange, J., Skelbaek-Pedersen, B., Langkjaer, L., Damgaard, U., Ege, H., Havelund, S., Heding, L. G., Jørgensen, K. H., Lykkeberg, J., Markussen, J., Pingel, M., and Rasmussen, E., 1987a, *Galenics of Insulin: The Physico-chemical and Pharmaceutical Aspects of Insulin and Insulin Preparations,* Springer-Verlag, Berlin.

Brange, J., Hansen, J. F., Havelund, S., and Melberg, S. G., 1987b, Studies of the insulin fibrillation process, in: *Advanced Models for the Therapy of Insulin-Dependent Diabetes* (P. Brunetti and W. K. Waldhäusl, eds.), Raven Press, New York, pp. 85–90.

Brange, J., Ribel, U., Hansen, J. F., Dodson, G., Hansen, M. T., Havelund, S., Melberg, S. G., Norris, F., Norris, K., Snel, L., Sørensen, A. R., and Voigt, H. O., 1988, Monomeric insulins obtained by protein engineering and their medical implications, *Nature* **333**:679–682.

Brange, J., Owens, D. R., Kang, S., and Vølund, A., 1990, Monomeric insulins, and their experimental and clinical implications, *Diabetes Care* **13**:923–954.

Brange, J., Langkjær, L., Havelund, S., and Vølund, A., 1992a, Chemical stability of insulin: 1. Hydrolytic degradation during storage of pharmaceutical preparations, *Pharm. Res.* **9**:715–726.

Brange, J., Havelund, S., and Hougaard, P., 1992b, Chemical stability of insulin: 2. Formation of higher molecular weight transformation products during storage of pharmaceutical insulin preparations, *Pharm. Res.* **9**:727–734.

Brange, J., Hallund, O., and Sørensen, E., 1992c, Chemical stability of insulin: 5. Isolation, characterization and identification of insulin transformation products, *Acta Pharm. Nord.* **4**:223–232.

Burge, M. J., and Rougvie, M. A., 1972, Cross-β protein structures. I. Insulin fibrils, *Biochemistry* **11**:2435–2439.

Capasso, S., Mazzrella, L., Sica, F., and Zagari, A., 1989, Deamidation via cyclic imide in asparaginyl peptides, *Pept. Res.* **2**:195–200.

Carstensen, J. T., 1990, *Drug Stability: Principles and Practices,* Dekker, New York.

Caspar, D. L. D., and Badger, J., 1991, Plasticity of crystalline proteins, *Curr. Opin. Struct. Biol.* **1**:877–882.

Chawdhury, S. A., Dodson, E. J., Dodson, G. G., Reynolds, C. D., Tolley, S. P., Blundell, T. L., Cleasby, A., Pitts, J. E., Tickle, I. J., and Wood, S. P., 1983, The crystal structure of three non-pancreatic human insulins, *Diabetologia* **25**:460–464.

Clarke, S., 1987, Propensity for spontaneous succinimide formation from aspartyl and asparaginyl residues in cellular proteins, *Int. J. Pept. Protein Res.* **30**:808–821.

Dathe, M., Gast, K., Zirwer, D., Welfle, H., and Mehlis, B., 1990, Insulin aggregation in solution, *Int. J. Pept. Protein Res.* **36**:344–349.

Derewenda, U., Derewenda, Z., Dodson, E., Dodson, G. G., Reynolds, C. D., Smith, G. D., Sparks, C., and Swenson, D., 1989, Phenol stabilizes more helix in a new symmetrical zinc insulin hexamer, *Nature* **338**:594–596.

du Vigneaud, V., Geiling, E. M. K., and Eddy, C. A., 1928, Studies on crystalline insulin. VI. Further contributions to the question whether or not crystalline insulin is an adsorption product, *J. Pharmacol. Exp. Ther.* **33**:479–509.

Eichner, H. L., Selam, J.-L., Woertz, L. L., Cornblath, M., and Charles, M. A., 1988, Improved metabolic control of diabetes with reduction of occlusions during continuous subcutaneous insulin infusion, *Diab. Nutr. Metab.* **1**:283–287.

Emdin, S. O., Dodson, G. G., Cutfield, J. M., and Cutfield, S. M., 1980, Role of zinc in insulin biosynthesis. Some possible zinc–insulin interactions in the pancreatic B-cell, *Diabetologia* **19**:174–182.

Feingold, V., Jenkins, A. B., and Kraegen, E. W., 1984, Effect of contact material on vibration-induced insulin aggregation, *Diabetologia* **27**:373–378.

Florence, T. M., 1980, Degradation of protein disulphide bonds in dilute alkali, *Biochem. J.* **189**:507–520.

Geiger, T., and Clarke, S., 1987, Deamidation, isomerization, and racemization at asparaginyl and aspartyl residues in peptides. Succinimide-linked reactions that contribute to protein degradation, *J. Biol. Chem.* **262**:785–794.

Graham, D. T., and Pomeroy, A. R., 1978, The effects of freezing on commercial insulin suspensions, *Int. J. Pharm.* **1**:315–322.

Grau, U., and Saudek, C. D., 1987, Stable insulin preparation for implanted insulin pumps. Laboratory and animal trials, *Diabetes* **36**:1453–1459.

Hallas-Møller, K., 1956, The 'Lente' insulins, *Diabetes* **5**:7–14.

Hansen, J. F., 1991, The self-association of zinc-free human insulin and insulin analogue B13-glutamine, *Biophys. Chem.* **39**:107–110.

Hansen, P. E., Brange, J., and Havelund, S., 1986, Stabilized insulin preparations and a process for preparation thereof, U.S. Patent 4,614,730.

Helbig, H.-J., 1976, Insulindimere aus der B-Komponente von Insulinpräparationen, Rheinish-Westfälishe Technische Hochschule, Aachen, Germany (dissertation).

Katsoyannis, P. G., 1964, The synthesis of the insulin chains and their combination to biologically active material, *Diabetes* **13**:339–348.

Kirby, A. J., and Lancaster, P. W., 1972, Structure and efficiency in intramolecular and enzymic catalysis. Catalysis of amide hydrolysis by the carboxy-group of substituted maleamic acids, *J. Chem. Soc. Perkin Trans. 2* **1972**:1206–1214.

Kirk, K. L., and Cohen, L. A., 1972, Intramolecular aminolysis of amides. Effects of electronic variations in the attacking and leaving groups, *J. Am. Chem. Soc.* **94**:8142–8147.

Kluger, R., and Lam, C.-H., 1978, Carboxylic acid participation in amide hydrolysis.

External general base and general acid catalysis in reactions of norbornenylanilic acids, *J. Am. Chem. Soc.* **100**:2191–2197.

Kossiakoff, A. A., 1988, Tertiary structure is a principal determinant to protein deamidation, *Science* **240**:191–194.

Krayenbühl, C., and Rosenberg, T., 1946, Crystalline protamine insulin, *Rep. Steno Mem. Hosp. Nord. Insulinlab.* **1**:60–73.

Krogh, A., and Hemmingsen, A. M., 1928, The destructive action of heat on insulin solutions, *Biochem. J.* **22**:1231–1238.

Leach, S. J., and Lindley, H., 1953, The kinetics of hydrolysis of the amide group in proteins and peptides. Part 2. Acid hydrolysis of glycyl- and 1-leucyl-1-asparagine, *Trans. Faraday Soc.* **49**:921–925.

Lougheed, W. D., Woulfe-Flanagan, H., Clement, J. R., and Albisser, A. M., 1980, Insulin aggregation in artificial delivery systems, *Diabetologia* **19**:1–9.

Lougheed, W. D., Albisser, A. M., Martindale, H. M., Chow, J. C., and Clement, J. R., 1983, Physical stability of insulin formulations, *Diabetes* **32**:424–432.

Lura, R., and Schirch, V., 1988, Role of peptide conformation in the rate and mechanism of deamidation of asparaginyl residues, *Biochemistry* **27**:7671–7677.

McFadden, P. N., and Clarke, S., 1987, Conversion of isoaspartyl peptides to normal peptides: Implications for the cellular repair of damaged proteins, *Proc. Natl. Acad. Sçi. USA* **84**:2595–2599.

Manning, M. C., Patel, K., and Borchardt, R. T., 1989, Stability of protein pharmaceuticals, *Pharm. Res.* **6**:903–918.

Markussen, J., Diers, I., Hougaard, P., Langkjaer, L., Norris, K., Snel, L., Sørensen, A. R., Sørensen, E., and Voigt, H. O., 1988, Soluble, prolonged-acting insulin derivatives. III. Degree of protraction, crystallizability and chemical stability of insulins substituted in position A21, B13, B23, B27 and B30, *Protein Eng.* **2**:157–166.

Meienhofer, J., Schnabel, E., Bremer, H., Brinkhoff, O., Zabel, R., Sroka, W., Klostermeyer, H., Brandenburg, D., Okuda, T., and Zahn, H., 1963, Synthese der Insulinketten und ihre Kombination zu insulinaktiven Präparaten, *Z. Naturforsch.* **18B**:1120–1121.

Meinwald, Y. C., Stimson, E. R., and Scheraga, H. A., 1986, Deamidation of the asparaginyl–glycyl sequence, *Int. J. Pept. Protein Res.* **28**:79–84.

Melberg, S. G., Havelund, S., Villumsen, J., and Brange, J., 1988, Insulin compatibility with polymer materials used in external pump infusion systems, *Diabetic Med.* **5**:243–247.

Patel, K., and Borchardt, R. T., 1990, Chemical pathways of peptide degradation. II. Kinetics of deamidation of an asparaginyl residue in a model hexapeptide, *Pharm. Res.* **7**:703–711.

Pingel, M., and Vølund, A., 1972, Stability of insulin preparations, *Diabetes* **21**:805–813.

Ratner, R. E., and Steiner, M. L., 1987, Insulin-pump therapy: Effect of phosphate-buffered human insulin, *Diabetes Care* **10**:787–788.

Ratner, R. E., Phillips, T. M., and Steiner, M., 1990, Persistent cutaneous insulin allergy resulting from high-molecular-weight insulin aggregates, *Diabetes* **39**:728–733.

Robbins, D. C., and Mead, P. M., 1987, Free covalent aggregates of therapeutic insulin in the blood of insulin-dependent diabetics, *Diabetes* **36**:147–151.

Robbins, D. C., Cooper, S. M., Fineberg, S. E., and Mead, P. M., 1987, Antibodies to covalent aggregates of insulin in blood of insulin-using diabetic patients, *Diabetes* **36**:838–841.

Robinson, A. B., and Rudd, C. J., 1974, Deamidation of glutaminyl and asparaginyl residues in peptides and proteins, in: *Current Topics in Cellular Regulation,* Volume 8 (B. L. Horecker and E. R. Stadtman, eds.), Academic Press, New York, pp. 247–295.

Sanger, F., Thompson, E. O. P., and Kitai, R., 1955, The amide groups of insulin, *Biochem. J.* **59**:509–518.

Schade, D. S., Morris, J., and DeLongo-Davis, J., 1987, Dissolution of human insulin, *J. Am. Med. Assoc.* **258**:2755.

Schlichtkrull, J., 1958, Insulin crystals. Chemical and biological studies on insulin crystals and insulin zinc suspensions, thesis, Munksgaard, Copenhagen.

Schlichtkrull, J., Munck, O., and Jersild, M., 1965, Insulin Rapitard and insulin Actrapid, *Acta Med. Scand.* **177**:103–113.

Schlichtkrull, J., Brange, J., Christiansen, A. H., Hallund, O., Heding, L. G., and Jørgensen, K. H., 1972, Clinical aspects of insulin—antigenicity, *Diabetes* **21**(Suppl. 2):649–656.

Schlichtkrull, J., Brange, J., Christiansen, A. H., Hallund, O., Heding, L. G., Jørgensen, K. H., Rasmussen, S. M., Sørensen, E., and Vølund, A., 1974, Monocomponent insulin and its clinical implications, *Horm. Metab. Res. (Suppl. Ser.)* **5**:134–143.

Schlichtkrull, J., Pingel, M., Heding, L. G., Brange, J., and Jørgensen, K. H., 1975, Insulin preparations with prolonged effect, in: *Handbook of Experimental Pharmacology, New Series,* Volume XXXII/2 (A. Hasselblatt and F. von Bruchhausen, eds.), Springer-Verlag, Berlin, pp. 729–777.

Sefton, M. V., and Antonacci, G. M., 1984, Adsorption isotherms of insulin onto various materials, *Diabetes* **33**:674–680.

Simkin, R. D., Cole, S. A., Ozawa, H., Magdoff-Fairchild, B., Eggena, P., Rudko, A., and Low, B. W., 1970, Precipitation and crystallization of insulin in the presence of lysozyme and salmine, *Biochim. Biophys. Acta* **200**:385–394.

Slobin, L. I., and Carpenter, F. H., 1963, The labile amide in insulin: Preparation of desalanine-desamido-insulin, *Biochemistry* **2**:22–28.

Sluzky, V., Tamada, J. A., Klibanov, A. M., and Langer, R., 1991, Kinetics of insulin aggregation in aqueous solutions upon agitation in the presence of hydrophobic surfaces, *Proc. Natl. Acad. Sci. USA* **88**:9377–9381.

Sluzky, V., Klibanov, A. M., and Langer, R., 1992, Mechanism of insulin aggregation and stabilization in agitated aqueous solutions, *Biotechnol. Bioengineer.* **40**:895–903.

Steiner, D. F., 1967, Evidence for a precursor in the biosynthesis of insulin, *Trans. N.Y. Acad. Sci. (Ser. II)* **30**:60–68.

Stephenson, R. C., and Clarke, S., 1989, Succinimide formation from aspartyl and asparaginyl peptides as a model for the spontaneous degradation of proteins, *J. Biol. Chem.* **264**:6164–6170.

Storvick, W. O., and Henry, H. J., 1968, Effect of storage temperature on stability of commercial insulin preparations, *Diabetes* **17**:499–502.

Sundby, F., 1962, Separation and characterization of acid-induced insulin transformation products by paper electrophoresis in 7 M urea, *J. Biol. Chem.* **237**:3406–3411.

Thurow, H., and Geisen, K., 1984, Stabilisation of dissolved proteins against denaturation at hydrophobic interfaces, *Diabetologia* **27**:212–218.

Toma, S., Campagnoli, S., Gregoriis, E. De, Gianna, R., Margarit, I., Zamai, M., and Grandi, G., 1989, Effect of Glu-143 and His-231 substitutions on the catalytic activity and secretion of Bacillus subtilis neutral protease, *Protein Eng.* **2**:359–364.

Tyler-Cross, R., and Schirch, V., 1991, Effects of amino acid sequence, buffers, and ionic strength on the rate and mechanism of deamidation of asparagine residues in small peptides, *J. Biol. Chem.* **266**:22549–22556.

Voorter, C. E., Haard-Hoekman, W. A. de, Oetelaar, P. J. M. van den, Bloemendal, H., and Jong, W. W. de, 1988, Spontaneous peptide bond cleavage in aging α-crystallin through a succinimide intermediate, *J. Biol. Chem.* **263**:19020–19023.

Waugh, D. F., 1946, A fibrous modification of insulin. I. The heat precipitate of insulin, *J. Am. Chem. Soc.* **68**:247–250.

Waugh, D. F., 1957, A mechanism for the formation of fibrils from protein molecules, *J. Cell. Comp. Physiol. (Suppl. 1)* **49**:145–164.

Waugh, D. F., Wilhelmson, D. F., Commerford, S. L., and Sackler, M. L., 1953, Studies on the nucleation and growth reactions of selected types of insulin fibrils, *J. Am. Chem. Soc.* **75**:2592–2600.

Wollmer, A., Rannefeld, B., Johansen, B. R., Hejnaes, K. R., Balschmidt, P., and Hansen, F. B., 1987, Phenol-promoted structural transformation of insulin in solution, *Biol. Chem. Hoppe-Seyler* **368**:903–911.

Wright, H. T., 1991, Sequence and structure determinants of the nonenzymatic deamidation of asparagine and glutamine residues in proteins, *Protein Eng.* **4**:283–294.

Index

MIX
Papier aus verantwortungsvollen Quellen
Paper from responsible sources
FSC® C105338

If you have any concerns about our products,
you can contact us on
ProductSafety@springernature.com

In case Publisher is established outside the EU,
the EU authorized representative is:
Springer Nature Customer Service Center GmbH
Europaplatz 3, 69115 Heidelberg, Germany

Printed by Libri Plureos GmbH
in Hamburg, Germany